AF790224

Plant Cell Monographs

Volume 20

Series Editor:

David G. Robinson
Heidelberg, Germany

Peter Nick
Karlsruhe, Germany

For further volumes:
http://www.springer.com/series/7089

# Plant Cell Monographs

Recently Published Titles

**The Plant Plasma Membrane**
Volume Editor: Murphy, A. S., Peer, W., Schulz, B.
Vol. 19, 2011

**Toxic Plant Proteins**
Volume Editor: Lord, J. M., Hartley, M. R.
Vol. 18, 2010

**Cell Biology of Metals and Nutrients**
Volume Editor: Hell, R., Mendel, R. R.
Vol. 17, 2010

**Lipid Signaling in Plants**
Volume Editor: Munnik, T.
Vol. 16, 2010

**Cell Biology of Plant Nematode Parasitism**
Volume Editor: Berg, R. H., Taylor, C. G.
Vol. 15, 2009

**Functional Organization of the Plant Nucleus**
Volume Editor: Meier, I.
Vol. 14, 2009

**The Chloroplast**
Interactions with the Environment
Volume Editors: Sandelius, A. S.
Aronsson, H.
Vol. 13, 2009

**Root Hairs**
Volume Editors: Emons, A. M. C. Ketelaar, T.
Vol. 12, 2009

**Plant Microtubules**
Development and Flexibility
2nd Edition
Volume Editor: Nick, P.
Vol. 11, 2008

**Plant Growth Signalling**
Volume Editors: Bögre, L., Beemster, G.
Vol. 10, 2008

**Cell Division Control in Plants**
Volume Editors: Verma, D. P. S., Hong, Z.
Vol. 9, 2008

**Endosperm**
Volume Editor: Olsen, O.-A.
Vol. 8, 2007

**Nitric Oxide in Plant Growth Development and Stress Physiology**
Volume Editors: Lamattina, L., Polacco, J.
Vol. 6, 2007

**The Expanding Cell**
Volume Editors: Verbelen, J.-P., Vissenberg, K.
Vol. 5, 2007

**The Plant Endoplasmic Reticulum**
Volume Editor: Robinson, D. G.
Vol. 4, 2006

**The Pollen Tube**
A Cellular and Molecular Perspective
Volume Editor: Malhó, R.
Vol. 3, 2006
Vol. 5/L, 2006

**Somatic Embryogenesis**
Volume Editors: Mujib, A., Samaj, J.
Vol. 2, 2006

**Plant Endocytosis**
Volume Editors: Šamaj, J., Baluška, F., Menzel, D.
Vol. 1, 2005

Jörg Fromm
Editor

# Cellular Aspects of Wood Formation

Springer

*Editor*
Jörg Fromm
University of Hamburg
Hamburg
Germany

*Series Editors*
David G. Robinson
Ruprecht-Karls-University of Heidelberg
Heidelberger Institute for Plant Sciences (HIP)
Department Cell Biology
Im Neuenheimer Feld 230
69120 Heidelberg
Germany

Peter Nick
Botanisches Institut
Molekulare Zellbiologie
Kaiserstr. 2
76131 Karlsruhe
Germany

ISSN 1861-1370 ISSN 1861-1362 (electronic)
ISBN 978-3-642-36490-7 ISBN 978-3-642-36491-4 (eBook)
DOI 10.1007/978-3-642-36491-4
Springer Heidelberg New York Dordrecht London

Library of Congress Control Number: 2013936094



Printed on acid-free paper

Springer is part of Springer Science+Business Media (www.springer.com)

# Preface

With today's ever growing economic and environmental problems, wood as a raw material takes on increasing significance as the most important renewable source of energy and as industrial feedstock of numerous products. Its chemical and anatomical structure and the resultant excellent properties allow wood to be processed into the most widely differing products: from logs to furniture and veneers and from wood chippings to pulp and paper. The aim of this book is to review research progress on the cellular aspects of cambial growth and wood formation in trees over the last decades. The book is divided into two major parts. The first part covers basic processes of wood biosynthesis, focusing mainly on five major steps that are involved in this process:

1. Cell division: The cambium is a secondary meristem responsible for wood increment. It is required for the regular renewing of xylem and phloem in trees and consists of dividing meristematic cells and their phloem and xylem derivatives. The latter give rise to various wood cells such as tracheids, vessels, fibres and parenchyma cells.
2. Cell expansion: During the formation of the primary wall, derivative cells expand in longitudinal, radial and tangential directions. This process is determined by various enzymes such as expansins, xyloglucan endotransglycosylases and pectinases as well as the availability of osmotically active ions (e.g. $K^+$). Hence, mineral nutrients play decisive physiological roles in xylogenesis. For example, during potassium starvation, cambial activity and wood increment and vessel size are significantly reduced. Molecular and electrophysiological studies indicate a strong involvement of specific $K^+$ channels in the regulation of wood formation.
3. Secondary cell wall formation: Formation of the secondary cell wall starts already before completion of cell expansion. In a coordinated manner cellulose, hemicellulose and lignin are synthesized by various enzymes to generate the different layers (S1, S2, S3) of the secondary wall which is mainly responsible for the outstanding mechanical properties of wood.

4. Programmed cell death: Water-conducting cells such as vessels and tracheids as well as most of the stabilizing cells like fibres die in a regulated manner before they take over their functions. After the collapse of the vacuole, specific enzymes that degrade the cytoplasm but not the cell wall are released.
5. Heartwood formation: Older annual rings no longer transport water or store nutrients but instead often store phenolic compounds as well as resins which provide long-term resistance to pathogens and thus increase natural durability.

The second part of the book deals with the regulation of wood formation by endogenous and exogenous factors. On the endogenous level, the emphasis is placed on two aspects:

- Control of wood formation by phytohormones: Auxins, gibberellins, cytokinins and ethylene have been shown to be involved in wood formation in a synergetic manner.
- Control of wood formation by molecular mechanisms: Wood formation is driven by the concerted expression of numerous genes. The use of molecular techniques in wood biology has enabled the identification of various transcription factors that are part of a network regulating secondary cell wall deposition.

Apart from endogenous factors, various exogenous effects are involved in wood formation:

- Climate factors: Environmental conditions such as temperature, precipitation, light intensity and duration have important effects on wood formation. While rising temperature is of significance for cambial reactivation in spring, a reduction in day length plays a major role in the early- to latewood transition as well as in cambial deactivation in trees growing in temperate zones. It is expected that with ongoing global warming, present-day tree species will not be optimally adapted hydraulically and mechanically to rising $CO_2$ levels and temperature. Apart from a shift of latitudinal and altitudinal ranges, these changing conditions also affect the structure of wood cells.
- Abiotic stress factors such as drought stress and salinity are becoming increasingly important and cause changes in wood structure and increment.
- Mechanical stress: Reaction wood is a mechanism by which trees respond to stem displacement. It is formed in response to prevailing winds, snow or slope and has the function to reorient a leaning stem or branch. Characteristic chemical and ultrastructural changes occur in tension wood of hardwood species as well as in compression wood of softwood species.

Due to modern microscopic as well as molecular techniques, the understanding of wood formation has progressed significantly in the last decade. By emphasizing the cellular aspects, this book gives first an overview of the basic processes of wood formation, before it focuses on factors involved in the regulation of this process.

Hamburg, Germany
December 2012

Jörg Fromm

# About the Editor

Jörg Fromm received his Ph.D. in 1986 at the University of Göttingen, Germany, where he also habilitated in Forest Botany after 5-year post-doctoral research on Tree Physiology. With a Heisenberg scholarship from the German Research Foundation, he shifted in 1992 to Cornell University, Ithaca, USA, to study signal transmission in trees. In 1996, he accepted a professor position at the University of Munich, and since 2007 he is a professor of Wood Biology at the University of Hamburg, Germany. His major research interests include the role of endogenous signals as well as environmental factors on tree growth, with a focus on wood formation.

# About the Editor



[illegible]

# Contents

# Part I
# Basic Processes of Wood Formation

# Xylem Development in Trees: From Cambial Divisions to Mature Wood Cells

Jörg Fromm

**Abstract** As one of the major parts of the biosphere, trees will play a significant role in the near future because of an increasing demand for wood as the most important natural raw material. Wood is generated by the vascular cambium and enables water transportation as well as providing mechanical support to the tree. Furthermore, it is the main renewable source for paper, buildings, furniture, boards and fuel. In recent decades intriguing developments in cell, molecular and structural biology have led to an integrated view of wood formation, from its start in the cambium by cell division, via cell expansion and cell wall thickening, to programmed cell death. These complex processes involve the interaction of both exogenous factors, such as photoperiod and temperature, and endogenous regulators, such as phytohormones. In addition, the coordinated expression of the numerous genes implicated in the biosynthesis of the major wood components—cellulose, hemicelluloses and lignin—drives the ordered development of wood. The huge amount of literature in the different fields of wood formation cannot be reviewed here in detail; rather, the aim of this chapter is to give a brief overview of the essential steps leading to mature wood cells, with an emphasis on current progress obtained by modern techniques which have increased our understanding of wood formation.

## 1 Introduction

Perennial woody plants dominate many natural land ecosystems. Their major difference to annual herbs is their long life cycle which, in trees, may span several centuries, encompassing germination, seedling, juvenility, maturity, senescence

J. Fromm (✉)
Institute for Wood Biology, University of Hamburg, Leuschnerstrasse 91, 21031 Hamburg, Germany
e-mail: joerg.fromm@uni-hamburg.de

J. Fromm (ed.), *Cellular Aspects of Wood Formation*, Plant Cell Monographs 20,
DOI 10.1007/978-3-642-36491-4_1, 

and finally death. To sustain their competitiveness through the course of this long life cycle, trees acquired multiple adaptation strategies which, as a whole, can only rarely be found in annual plant species. These strategies comprise fast and extensive metabolite translocation in response to sudden incidents in the environment, the storage of nutrients in vegetative organs, the responsiveness to seasonal changes in climate and water supply, a long distance communication system via various signals, as well as the competence of producing wood for durable stability and as a transport organ.

Biological production of wood depends on mineral and nutrient supply via the root system, on a balanced storage system and on a prompt and flexible metabolite translocation throughout the growing seasons, whether these are characterised by seasonal climates or by wet and dry periods. Today, wood as a raw material has taken an increasing significance as one of the most important renewable resources for meeting the growing demand for bioenergy, construction materials, wood-pulp for paper production and, of course, as the major onshore carbon sink. With respect to the worldwide on-going discussion on climate change and the need for alternative $CO_2$-neutral energy sources, typical plantation tree species have been the focus of renewed public interest. During the last decades, a few tree species have been established globally as model systems in tree research, such as *Populus spec.* in seasonal climates (Sterky et al. 2004; Tuskan et al. 2006) and *Eucalyptus* spec. in warmer and more clement climates (Foucart et al. 2006; Gion et al. 2011; Sexton et al. 2012). Poplar has become the model hardwood species in the northern hemisphere because of its relatively small genome, its ability to be easily propagated and to be genetically transformed (e.g. Tuominen et al. 1995; Hoenicka et al. 2012). Poplar and Eucalyptus are fast-growing trees, and hence, in their respective climate zones, they are also of growing economic importance for large-scale biomass production. In addition, other important tree systems are loblolly pine (e.g. Ralph et al. 1997) and black locust (Magel et al. 1995).

A lot of research on xylogenesis has also been conducted on primary growth systems such as the annual herbaceous species *Zinnia elegans* (Oda and Fukuda 2012). After isolating mesophyll cells from the leaves, they can be induced through appropriate hormonal treatments to re-differentiate into tracheary elements (Endo et al. 2008). This system can give valuable information on the role of the different substances required for wood development. Although this less-complex *Zinnia* cell culture system has been able to provide a lot of information on the signals that control plant vascular cell differentiation (Fukuda 2004), it has some shortcomings as a model for the process of wood formation in trees because the products remain as single cells and show no cycle of activity and dormancy (Chaffey 1999). *Arabidopsis*, however, which is the most advanced model system in plant molecular biology, under appropriate growth conditions, shows substantial secondary thickening in the hypocotyls (Chaffey et al. 2002) and can hence also provide valuable information with respect to wood formation (Zhang et al. 2011), despite its lack of perenniality.

In order to manage wood as an important resource in the future, it is necessary to understand the process of wood formation in trees more deeply. In recent decades considerable progress has been made concerning the cellular, physiological and molecular processes that underlie wood production. This research has revealed that the development of secondary xylem from the cambium is a very complex process which is under the control of numerous genes, as well as a combination of exogenous and endogenous factors involved in the different steps of cell differentiation. Most important are endogenous factors, such as hormones and various factors acting downstream of hormones, including transcription factors (TFs) and receptor kinases (RKs), together with their peptide ligands (Nieminen et al. 2012). This review, therefore, highlights current progress in research into these cellular processes of wood formation.

# 2 Structure and Function of the Cambium

## *2.1 Meristematic Features*

The cambium is a lateral secondary meristem that derives from the procambium, which in turn originates from the apical meristem (Larson 1994). The cambium develops out of the procambium when parenchyma cells between the vascular bundles start to divide in order to generate a vascular cylinder. As soon as such a cylinder is active, radial files of secondary xylem cells are delivered to the inside and radial phloem files to the outside by so-called periclinal cell divisions. Within each file of both phloem and xylem cells, one initial cell remains in the cambium and gives rise to daughter cells. These daughter cells either become phloem or xylem mother cells. Both types of cells, i.e. initials as well as mother cells, are cytologically almost identical and are called the cambial zone.

The cambial zone has two main functions: cell division and setting out patterns for differentiation. Initials retain the potential to differentiate into either xylem or phloem mother cells. They divide relatively infrequently, however, because of the importance of maintaining an undifferentiated state. Thus, the cambial zone consists of dividing initials, which maintain themselves, together with both xylem and phloem mother cells as their products. By intervening cell division, a mother cell differentiates until it becomes a mature cell. Since, in most species, xylem mother cells divide more than phloem mother cells, much more xylem is produced than phloem. Ratios vary between approximately 1:1 in the tropical hardwood *Mimusops elengi* (Ghouse and Hashmi 1983) and 15:1 in *Thuja occidentalis* (Bannan 1955). Most species have xylem to phloem ratios in the range between 4 and 10:1. Recently, progress has been made on the regulation of the xylem vs. phloem ratio in poplar. It has been found that when the transcription factor PtaLBD1 (which is a member of the LATERAL ORGAN BOUNDARIES DOMAIN (LBD) family) is overexpressed, secondary phloem formation increases, suggesting that this transcription factor is involved in the regulation of secondary phloem formation (Yordanov et al. 2010). In addition, the idea that the cambial zone consists of a meristematic region in the centre and two regions of differentiating phloem and xylem mother cells has been confirmed by gene expression data using microarrays from 20 μm thick sections of the cambial zone in aspen (Schrader et al. 2004a). Peak expression from the phloem mother cell region is most often associated with genes representing anticlinal divisions, whereas the xylem mother cell region is rich in cell cycle genes.

In contrast to apical meristems, cells of the cambial zone contain two morphologically distinct cell types: elongated fusiform cells that give rise to axially oriented cells within the xylem and phloem and almost isodiametric ray cells that generate horizontal cell systems such as parenchyma and tracheary elements within the rays. Fusiform initials show higher periclinal divisions than ray initials, which expand in a radial direction to keep step with incremental radial growth. The ratio between ray and fusiform initials depends mainly on the species. With increasing age and cambial dilatation, fusiform initials can differentiate to ray initials in order to keep the species-specific ratio between both constant. Microgenomic analysis has revealed cell type-specific gene expression patterns between ray and fusiform initials of poplar, indicating that photosynthesis genes are overrepresented in ray cambial cells in order to provide a photosynthetic system in rays (Goué et al. 2008). Regarding cell wall-related genes, those involved in pectin metabolism are overrepresented in ray cambial cells while those involved in xyloglucan metabolism are overrepresented in fusiform cambial cells (Goué et al. 2008).

Turning now to the cell position, the cambium can show a storeyed structure. This storeyed form occurs when fusiform cells are arranged in approximately horizontal layers with the cell ends on the same plane when viewed in the tangential section. Wood produced from a storeyed cambium shows a characteristic stratified structure, e.g. as in *Aesculus*, *Diospyros, Swietenia* and *Dalbergia* (Fig. 1a). A storeyed cambium is typical of tree species with short fusiform initials and is both phylogenetically more advanced and more frequent in tropical species than in species from temperate climate zones. In addition to fibres, both rays and vessel elements can show a stratified structure, e.g. as in *Dalbergia* (Fig. 1a). In contrast, in a non-storeyed cambium, the tips of fusiform cells end at different planes, leading to a wood cell structure with overlapping cell ends, e.g. as in *Populus, Fraxinus, Quercus* and conifers such as *Pinus* (Fig. 1b). Furthermore, species with non-storeyed cambia usually have long fusiform initials. The length of the fusiform cells depends on the species and on cambial age. The mean cell length of cambial initials and the final length of mature wood cells increase with tree age. In the case of fibres, their final length is determined by the length of fusiform initials and the degree of intrusive tip growth. In gymnosperm trees the length of fusiform cells ranges from 1,100 μm to 4,000 μm, while in angiosperm trees they vary between 170 μm and 940 μm (Larson 1994). As a consequence, therefore, of the complexity of the two differently oriented subsystems—axial and radial—many different phloem and xylem cell types are produced by the cambial zone within the tree. Tracheids, vessel elements, fibres and axial parenchyma cells emerge from fusiform

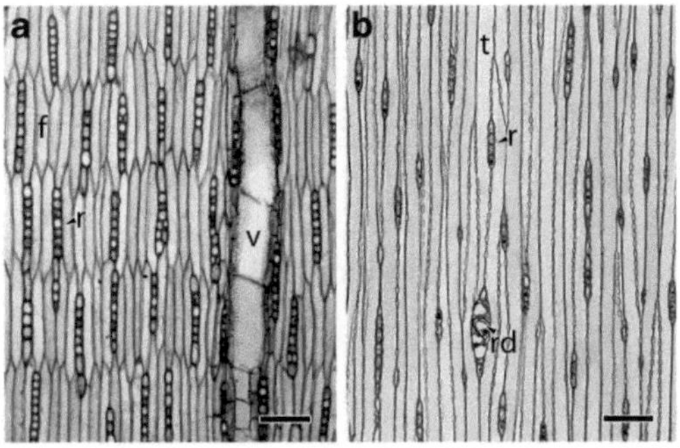


**Fig. 1** (**a**) Tangential section of developing wood from *Dalbergia riparia* showing the stratified structure of all cell types (Courtesy of Dr. A. Olbrich). (**b**) Tangential section of developing wood from *Pinus wallichiana* showing an interlocked non-stratified cell structure (Courtesy of V. Haag); *f* fibre, *r* ray, *rd* resin duct, *t* tracheid, *v* vessel element. Bar 100 μm

cambial cells whereas ray parenchyma cells and ray tracheids emerge from cambial ray cells (Fig. 2).

Regarding the regulation of the development of these specific cell types, the ectopic expression of some major transcription factors (VASCULAR-RELATED NAC-DOMAIN6 (VND6) and VND7) has been identified for vessels in *Arabidopsis* and *Populus* leaves (Yamaguchi et al. 2008). Microscopy of fusiform cells has shown all the characteristic features for high rates of protein biosynthesis and secretory activity (Catesson 1990; Arend and Fromm 2003). The new wall that needs to be formed during periclinal cell division is very large, and therefore, the rate of cell wall biosynthesis is extremely high in the active cambial zone. High rates of xylem cell formation also correlate with a high number of cambial cells (Gregory 1971; Uggla et al. 1998) because the latter differentiate to wood cells. Hence, both the number of xylem mother cells and the duration of the cell cycle in these cells are important for the rate of wood production. In various conifer species, for example, the shortest average duration of fusiform cambial cells across the cambial zone is in the range of 7–11 days (Mellerowicz et al. 2001). Both the number of mother cells and their division rate may be controlled by independent mechanisms. Since cell wall properties vary among cambial cells and their close derivatives, cell fate seems to be determined at an early stage (Catesson and Roland 1981; Catesson et al. 1994).

During periclinal division of fusiform cells, the phragmoplast and the newly formed cell plate must traverse the large central vacuole to reach the upper and lower ends of the cell. In fusiform cambial cells a cytoplasmic strand, derived from the Golgi vesicles, extends through the vacuole and the developing cell plate. In contrast, cambial ray cells do not possess large vacuoles; during division they show a complete, well-developed, mitotic apparatus which follows the usual mitosis pattern in meristematic plant cells (Fig. 3). The nascent periclinal walls of fusiform cambial cells are of a cellulosic nature with a high content of methylated pectin (Catesson 1989, 1990; Catesson et al. 1994). When the periclinal wall merges with

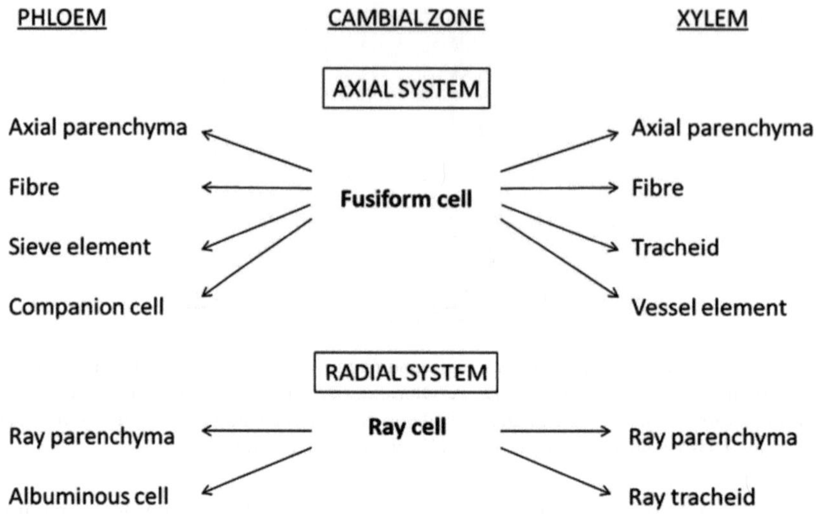


**Fig. 2** Phloem and xylem cell types differentiating from the cambial zone of angiosperm and/or gymnosperm trees

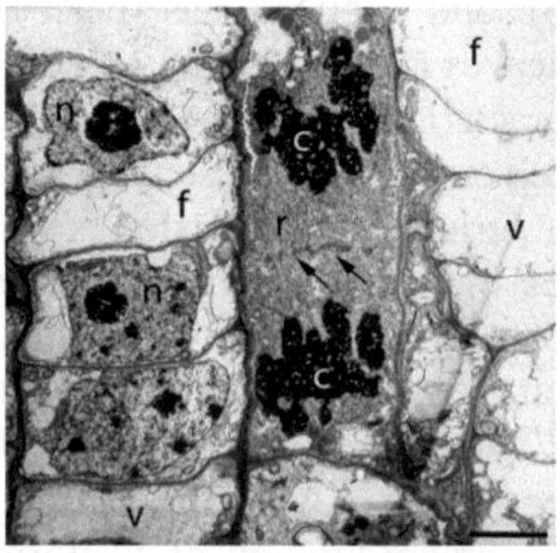


**Fig. 3** Transverse view of cells in the active cambial zone of spruce. Fusiform cells (f) show large vacuoles (v), thin cell walls and conspicuous large nuclei (n). A dividing ray cell (r) has no large vacuoles and exhibits a well-developed mitotic apparatus, as shown here in the state of a telophase with decondensing chromosomes (c) and the developing cell plate (*arrows*). Bar, 5 μm (After Arend and Fromm 2004)

the already existing radial wall, the latter is locally digested until the middle lamella is reached, and it then becomes continuous with the middle lamella of the newly formed tangential cell wall (Catesson and Roland 1981).

In addition to periclinal divisions, which lead to an increase in stem diameter, cambial initials show anticlinal divisions perpendicular to the stem surface, when wood production displaces the cambium outwards. In species with storeyed cambia, the newly formed cell wall is oriented radially, and new radial cell files are generated during anticlinal divisions. In species with non-storeyed cambia, anticlinal divisions occur pseudotransversely, i.e. the initials divide by generating a sloping anticlinal wall and by following intrusive tip growth. The orientation of this pseudotransverse division can be leftwards (S) or rightwards (Z) (Zagorska-Marek 1995). Tangential expansion of the cambium refers to both an increase in the number of cells and in their sizes. The majority of anticlinal divisions have been observed to occur within a single layer of cells within the cambial zone of aspen (Schrader et al. 2004a). The rate of anticlinal cell division is also much lower than the rate of periclinal cell division.

For vascular tissue development to occur normally, it has been shown that a peptide signal secreted from the phloem binds to a receptor-like kinase (PHLOEM INTERCALATED WITH XYLEM, PXY) in cambial cells (Fisher and Turner 2007; Hirakawa et al. 2008). Interestingly, xylem and phloem are no longer separated but are intermixed in the loss-of-function *pxy* mutant. Furthermore, it has been suggested that a *Populus* class III HD ZIP gene, popREVOLUTA (PRE) plays a fundamental role in regulating the patterning of secondary vascular tissues. In transgenic lines expressing a microRNA-resistant form of PRE abnormal formation of cambia occurs within the stem cortex with phloem developing inwards and xylem outwards (Robischon et al. 2011).

## 2.2 *Seasonal Activity*

In temperate latitudes, trees grow synchronously with the seasons and are able to endure periods unfavourable for growth by dormancy. According to Lang (1987), dormancy is the temporary absence of visible growth of any plant structure containing a meristem. Dormancy may also be regarded, however, as a developmental process parallel to active growth and not only a temporary inactive state. Especially in trees of boreal forests with very cold winters, the cycling between activity and dormancy is important for survival. Trees have to accurately synchronise the timing of their active and dormant states with the seasonal changes in order to be able to grow in tough climatic conditions. Cambial dormancy consists of two stages, rest and quiescence (Catesson 1994; Larson 1994, Fig. 4). Rest is controlled by endogenous signals and, in this stage, the cambium cannot divide. Rest can be overcome by giving a chilling treatment to the tree upon which the cambium makes the transition from rest to quiescence and regains responsiveness to auxin. After chilling occurs, therefore, warm temperatures can induce reactivation (Heide 1993). Recently, it has been shown in poplar that chilling of dormant buds hyperinduces FLOWERING LOCUS T and recruits GA-inducible 1,3-beta-glucanases to reopen signal conduits in order to trigger a release from dormancy (Rinne et al. 2011). In the ensuing quiescent stage, favourable growth conditions can induce cambial divisions. Generally, the quiescent dormant stage ends during cambial reactivation in spring. In trees from temperate climates, changes in photoperiod and temperature are the dominant environmental signals regulating this seasonal growth-dormancy cycling (Fig. 4).

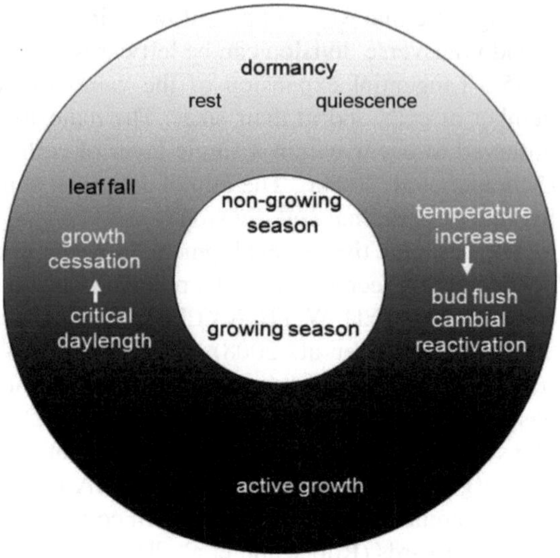


**Fig. 4** Transitions in seasonal growth-dormancy cycling in trees from temperate climates. Photoperiod is known to govern growth cessation in late summer/autumn, and in particular, short days are responsible for the cambial transition into the first resting phase (rest). Prolonged exposure to chilling temperatures in early winter initiates the second resting phase (quiescence) and will also release trees from dormancy. After the temperature passes a critical threshold in spring, growth resumes

Within the perennial life cycle of a tree, the cambium functions by renewal of xylem and phloem each year. Distinct differences occur in cambial cytology, however, between its active and dormant states. In several hardwood species, it has been shown that fusiform cambial cells are densely cytoplasmic with many small vacuoles during dormancy while during activity they are highly vacuolated (Robards and Kidway 1969; Sennerby-Forsse 1986; Farrar and Evert 1997; Arend and Fromm 2000). Cambial activity and the width of the cambial zone change during the season. The dormant cambial zone of poplar consists of only 3–4 layers of dense cytoplasmic cells with numerous small vacuoles, lipid droplets as storage material and thick cell walls. No cytoplasmic streaming occurs in these dormant cells. After reactivation, the cambial zone shows 6–7 cell layers per radial file, with thin cell walls and fewer larger vacuoles that appear electron transparent (Arend and Fromm 2000). These differences indicate that the thick walls during dormancy store material that can be metabolised in spring. Cambial cells also increase in radial width during reactivation, causing thinner radial walls. There is strong evidence that the wall thickening during dormancy can be attributed to the conspicuous development of cellulose microfibrils (Catesson 1994), while wall thinning is in part due to incomplete wall lysis (Funada and Catesson 1991). In *Aesculus hippocastanum*, the thicker wall of the dormant cambial cells is more highly structured than the amorphous cell wall of active cambia, as indicated by the presence of a 'herring-bone' lamellate structure (Chaffey et al. 1998). Also, cortical microtubules show a different orientation: they have been found to be axially oriented during dormancy but randomly orientated in active cambial cells, corresponding to the pattern of the microfibrils (Chaffey et al. 1998). In addition, the nature of pectins changes between activity and dormancy. Active cambium contains more hot-water-extractable pectin than dormant cambium (Baier et al.

1994; Ermel et al. 2000). Studies on poplar have indicated that during dormancy, pectin methyl esterase is upregulated while during cambial activity, pectin methylation is increased (Baier et al. 1994; Ermel et al. 2000; Follet-Gueye et al. 2000).

In temperate latitudes active and dormant states of the cambium cause a distinct annual ring structure with a characteristic earlywood/latewood pattern. Earlywood is produced during the first part of the growing season during peak radial growth and is characterised by high vessel or tracheid size and thin cell walls. It coincides with an increase in cambial zone width generated by the number of xylem mother cells. In contrast, latewood is formed during the last part of the growing season. With regard to xylem cell length, it has been found that it increases from a minimum in earlywood to a maximum in latewood (Bissett and Dadswell 1950) both because fusiform cambial cells elongate during the growing season and because during latewood formation intrusive fibre tip growth increases. Finally, xylem cell length decreases sharply at the annual ring boundary. Concerning the chemical composition, significant differences have been found in the distribution of sugar units in hemicelluloses between earlywood and latewood in Norway spruce. Latewood contains clearly more galactoglucomannan than earlywood and conversely less pectins. Lipophilic extractives are also less concentrated in latewood (Bertaud and Holmbom 2004).

The transition from early- to latewood is mainly controlled by environmental conditions. Willow twigs severed from an adult tree in August in mid-Europe are in the state of latewood production. When such cuttings are placed in a nutrient solution under increased temperature, the cambium begins to build earlywood (Fromm 1997). In the cambial region of *Pinus densiflora*, the transition from earlywood to latewood occurs concurrently with a decrease in the total amount of indole-3-acetic acid (IAA) after it has peaked, suggesting the involvement of IAA in the control of latewood formation (Funada et al. 2001). While the total amount of IAA does not change with latewood initiation in the cambial region of *Pinus sylvestris*, nonetheless, its radial distribution pattern is altered (Uggla et al. 2001), indicating that IAA probably has a role in defining the altered developmental pattern associated with latewood formation. In *Pinus radiata* and *P. sylvestris*, latewood formation seems to be correlated to an increase in the concentration of abscisic acid (ABA) (Jenkins and Shepherd 1974; Wodzicki and Wodzicki 1980).

In contrast to trees growing in temperate latitudes, trees of tropical regions with consistent day length and temperature generally do not exhibit a distinctive tree ring structure, and their cambium is more or less active over the whole year. Seasonality can occur in tropical zones, however, and wood structure also responds to changing weather conditions, i.e. during drought periods narrow wood cells are formed by the cambium, while during a rainy period cells with wide lumina are produced. For example, in Bolivian rainforests climate-growth analysis has indicated that rainfall plays a major role in tree growth (Brienen and Zuidema 2005). Amazonian trees growing in zones with dry periods drop their leaves during drought and build new leaves, as well as wood, shortly after the beginning of the rainy season (Vetter and Botosso 1989; Alves and Angyalossy-Alfonso 2000). Also flooding can cause cambial rest, and hence a ring structure, because root activity is reduced in oxygen-free conditions (Worbes 1985, 1995).

## 2.3 Cambial Reactivation

Well-coordinated changes in the cellular structure, physiology and metabolism of cambial cells define the transition between active and dormant states. The cytological and structural aspects of cambial reactivation in spring, as well as cambial activity and cessation, have been extensively studied (Catesson 1994; Larson 1994). Recently, it has been shown that tree social status also affects cambial activity. Activity starts earlier, stops later and lasts longer in dominant silver-fir trees than in intermediate and suppressed ones (Rathgeber et al. 2011). In general, in temperate climate zones, the formation of phloem starts before xylem differentiation in many diffuse-porous and coniferous species in spring. In contrast, in ring-porous species, phloem and xylem formation start simultaneously.

Cambial activity is a temporary event, characterised by a special physiological condition of the meristematic cells. A particularly future-oriented field of research in this area lies in the transduction of seasonally conditioned signals (e.g. day length, temperature) controlling cambial activity. Temperature plays a key role during cambial reactivation in spring. It has been shown that localised heating of tree stems can induce cambial reactivation in evergreen conifers (Oribe et al. 2001, 2003) and in hybrid poplar (Begum et al. 2007). In poplar, however, the buds of the trees had not yet burst, indicating that there is no close temporal relationship between bud burst and cambial reactivation. The heat-induced xylem differentiation in hybrid poplar stems was the same as that of xylem cells formed in the normal way, indicating that an increase in the temperature of the stem is one of the most important factors in cambial reactivation (Begum et al. 2007). Furthermore, in order to protect the sensitive cambial cells in spring from sudden reductions in temperature, several cold hardiness-related genes are superinduced during the early stage of reactivation (Druart et al. 2007).

Cambial reactivation in spring starts before any significant photosynthetic activity occurs in the tree. As a result, an alternative source of energy and carbon is needed for cell division in spring. Induction of sucrose synthase and various invertases during the early phase of reactivation in the cambium shows that sucrose is split into hexoses that can be metabolised via glycolysis. During cambial reactivation, high amounts of sucrose are required for cell growth. The uptake of sucrose into cambial cells might be under the control of a PM $H^+$-ATPase, which is demonstrated by immunolocalisation in the cambial and wood formation zone of poplar (Arend et al. 2002). Also, fats are metabolised via beta-oxidation, and the glyoxylate cycle and the level of amino acids are increased due to the degradation of storage proteins in spring (Druart et al. 2007). During cambial reactivation induced by artificial heating of *Cryptomeria japonica* stems, the levels of starch granules and lipid droplets decreased in the cambium, indicating that these might also be needed as sources of energy for the initiation of cambial cell division and xylem differentiation (Begum et al. 2010). In addition, immunolocalisation demonstrated plasmodesmatal trafficking of storage proteins during cambial reactivation in *Populus nigra*, indicating that lectin-like reserve proteins, or their degradation products, may be transferred through the plasmodesmata of phloem parenchyma and rays (Fuchs et al. 2010a). In cambial initials, the first cell division coincides with a massive increase in plasmodesmata numbers, in particular at the division wall (Fuchs et al. 2010b). In addition, the onset of rapid xylem production in spring correlates to a marked increase in stem respiration (Lavigne et al. 2004). Surprisingly, in wood rays of poplar, the metabolic pathways related to flower induction are already high in February (Larisch et al. 2012), indicating that reactivation from dormancy had already begun at this time of the year. In contrast, in July, the pathways related to active growth, such as lignin biosynthesis, nitrogen assimilation and defence, were enriched in rays (Larisch et al. 2012).

As early as the 1930s, studies based on bioassays supported the hypothesis that hormones from growing buds in tree stems can induce downward cambial reactivation in spring (Söding 1937; Avery et al. 1937). Subsequently, some of the key signalling molecules related to cambial reactivation, such as plant hormones, have been identified by biochemical and molecular approaches (Sundberg et al. 2000; Moyle et al. 2002; Tanino 2004). In recent years these hormone treatment studies have been complemented by studies using transgenic plants with modified hormonal signalling. An entire chapter of the present volume is dedicated on the role of hormones in regulating xylem development (see chapter "The Role of Hormones in Controlling Vascular Differentiation").

It is known that auxin plays a major role in wood formation. A radial concentration gradient of auxin has been found within the cambial zones of both *Populus* and *Pinus*, with the highest concentrations present in dividing cambial cells (Sundberg et al. 2000). This gradient correlated with the expression pattern of auxin signalling genes (Moyle et al. 2002) and is generated when auxin is synthesised at the shoot apex (Sundberg and Uggla 1997) and transported basipetally within the stem (Little and Savidge 1987; Schrader et al. 2003; Björklund et al. 2007). When auxin transport is inhibited, wood formation is suppressed in *Pinus* shoots (Sundberg et al. 1994), suggesting that auxin is required for secondary xylem development. Apart from the endogenous hormonal concentration, however, the sensitivity of cambial cells to hormones alters seasonally and plays a significant role in activation (Lachaud 1989). Transgenic poplar trees with decreased auxin responsiveness show fewer cambial cell divisions (Nilsson et al. 2008), and the level of cambial auxin signalling seems to be controlled separately by the rate of auxin transport and by the level of cambial responsiveness to auxin (Baba et al. 2011).

Apart from auxin, gibberellin acts as a mobile shoot-derived signal activating the onset of xylem production in *Arabidopsis* (Ragni et al. 2011). During cambial reactivation in spring, a transient induction of a gene encoding a gibberellin biosynthesis enzyme can be observed in poplar (Druart et al. 2007), suggesting a role for gibberellin in the activation of cambial growth. Furthermore, in transgenic plants with either enhanced gibberellin signalling (Mauriat and Moritz 2009) or biosynthesis (Eriksson et al. 2000), wood production is increased. Interestingly, increased gibberellin signalling also enhances polar auxin transport in poplar (Björklund et al. 2007) indicating an interrelationship between each of these hormones in the regulation of cambial activity.

Cytokinins are also involved in the regulation of auxin transport (Bishopp et al. 2011). Cytokinin signalling is necessary for cambial functioning in the roots of *Arabidopsis* (Matsumoto-Kitano et al. 2008) as well as during cambial development in poplar, where cytokinin primary response genes and cytokinin receptors are involved (Nieminen et al. 2008). Finally, it has been demonstrated that ethylene is involved in the formation of tension wood in poplar (Love et al. 2009, see also chapter "Biology and Chemistry of Tension Wood") as well as in tracheary element differentiation of *Zinnia* cell cultures (Pesquet and Tuominen 2011). Ethylene signalling has also been found to be connected to jasmonate signalling in *Arabidopsis*, with elevated jasmonate signalling, causing an increase in secondary growth and stem diameter (Zhu et al. 2011).

Apart from hormones, ions might also play an important role in cambial reactivation. Immediately before the resumption of cell division, a strong temporary increase in calcium concentrations has been observed in the cambium of both beech (Follet-Gueye et al. 1998) and poplar (Arend and Fromm 2000). This increase may be involved in the regulation of cambial reactivation because calcium is known to activate several enzymes such as lipases and amylases which play a role in the hydrolysis of lipids and starch respectively.

## 2.4 Transition to Dormancy

In temperate latitudes, growth cessation occurs in response to shortening day length and coincides with autumnal senescence, leaf shedding, completion of bud set and cambial cessation. Dormancy consists of many interrelated subprocesses that are active during the different periods. For example, prior to the transition to dormancy, E2F phosphorylation is elevated in the cambium of hybrid aspen (Espinosa-Ruiz et al. 2004). After entering dormancy, increase in drought resistance and acquisition of frost resistance are characteristic subprocesses in buds (Rohde and Boerjan 2001; Ruttink et al. 2007). Also, in the stem, genes related to cold hardiness and defence are overrepresented in winter during dormancy, as shown by whole transcriptome analysis in poplar (Ko et al. 2011).

Since cessation of cambial cell division in aspen in mid-August occurs before the temperature becomes suboptimal (Druart et al. 2007), temperature may not play a critical role in growth cessation. In correlation to the cessation of cell division, transcript levels of cell cycle genes decline in autumn, while those for enzymes involved in phospholipid biosynthesis (necessary for the synthesis of new vacuolar membranes), lipid desaturation, dehydrins and cold-regulated proteins are induced (Druart et al. 2007). The development of cold hardiness coincides with the breakdown of starch in autumn, and the generated carbohydrates serve for metabolites such as sucrose, raffinose and cryoprotectants. Interestingly, poplar FT (a RAF-kinase-inhibitor-like protein) and CONSTANS(CO) (a nuclear zinc-finger protein) have been identified as mediators of short-day signals for growth cessation, since growth does not stop upon exposure to short days when FT1 and CO homologues of poplar (*P. trichocarpa*) are overexpressed in transgenic aspen (*P. tremula* × *P. tremuloides*) (Böhlenius et al. 2006).

At the end of the growing season, when trees enter dormancy, hormones also play a role. The sensitivity of the cambium to auxin, which plays a key role in regulating wood formation, is lost (Little and Bonga 1974). The molecular basis of short-day-induced growth cessation and dormancy in the cambial meristem involves a decrease of auxin responsiveness, although basipetal auxin transport remains active (Baba et al. 2011) and cambial auxin levels remain stable (Uggla et al. 1998; Schrader et al. 2003, 2004b). In addition, the timing of the cessation of cambial cell division caused by short days differs between northern and southern genotypes of hybrid poplar and is coincident with the changes in the pattern of expression of the auxin-regulated genes (Resman et al. 2010). In the cambial region of *Eucommia ulmoides* Oliv., the expression of ABP1, one of the putative receptors of auxin, was found to be high, low and remarkably scarce in the active, quiescent and resting stages, respectively (Hou et al. 2006). This would suggest a role for ABP1 in mediating auxin-dependent regulation of cambial activity. Results also show that ABP1 expression is improved by IAA but inhibited by ABA (Hou et al. 2006), indicating a possible role for ABA in the cambial activity-dormancy cycle. In addition, expression of the *Arabidopsis* mutant *abi1* gene alters ABA sensitivity, stomatal development, and growth morphology in poplar, indicating that ABA acts as a negative regulator of shoot growth and, furthermore, has a role in shoot branching by inhibiting lateral bud outgrowth (Arend et al. 2009).

## 2.5 Within-Tree Variations

Cambial age has an important effect on the structure of cambial cells and their derivatives. Wood produced during the early years of cambial growth is called juvenile wood. Juvenile wood is more elastic, thus allowing flexibility, while mature wood is stiffer because it has to carry a greater mechanical load with increasing age. In comparison to mature wood, juvenile wood in poplar is characterised by shorter fusiform cambial cells as well as shorter xylem cells that are derived from these, a lower crystallinity of the fibres, a larger microfibril angle, thinner secondary walls, a higher density of vessels and a lesser amount of latewood (Hejnowicz and Hejnowicz 1958; Kroll et al. 1992). The large S2 microfibril angle in juvenile wood causes increased longitudinal shrinkage as well as decreased transverse shrinkage in sawn lumber during drying. Chemically, juvenile wood shows a lower cellulose content and a higher content of lignins and hemicelluloses. Since earlywood cells predominate, overall wood density decreases in line with lower strength properties (modulus of elasticity and modulus of rupture). Thus, in comparison to mature wood, juvenile wood exhibits marked differences in strength, stability and stiffness and is generally considered to be of inferior quality (Clark et al. 2006; Jordan et al. 2006; Mansfield et al. 2007). In addition, in gymnosperms it tends to contain more compression wood and a higher incidence of spiral grain. Juvenile wood is suitable for the characterisation of the molecular mechanisms controlling MFA and mechanical strength. Thus, transcriptome profiling of juvenile wood with contrasting levels of stiffness from *Pinus radiata* has identified putative candidate genes involved in microfibril orientation and cell wall mechanics (Li et al. 2011). Since the global proportions of construction timber originating from plantations is growing rapidly, the amount of juvenile wood provided for further processing is also increasing.

Depending on the species, the structure of root and branch wood may also differ from stem wood. The pattern of cell divisions in root cambia differs little from those in stem cambia. In root wood cellular changes are most prominent with increasing distance to the stem. In general, root wood cells have wider lùmina and reduced cell wall thickness. Especially in angiosperm trees, the density and size of vessels increase in roots, showing a homogenous distribution throughout the whole growth ring. For instance, in species with ring-porous stem wood, the vessels in the root wood exhibit a diffuse-porous pattern (Fig. 5a, b). In conifers, root tracheids have wider lumina compared to stem tracheids. Also, root tracheids often have bordered pits on radial walls which occur in pairs, whereas in stem wood tracheids, these bordered pits occur only singly in most species. With regard to cell length, tracheids increase in length as their distance from the stem increases and differences between early- and latewood are diminished (Fig. 5c, d). Furthermore, the amount of parenchyma increases in the root wood of both angiosperms and gymnosperms.

Turning now to branches, these have often been a source of cambial studies because sampling is relatively easy. When rates of radial and circumferential branch expansion, age and eccentricity are considered, the patterns and consequences of anticlinal divisions in branch material differ little from those in stem material (Larson 1994). Often, branch wood has an increased density in comparison to stem wood, and generally, xylem elements in branches are smaller than those of stems. Vessels from angiosperm branch wood often have reduced lumina while in gymnosperms the number of resin ducts is greater. Also, changes in chemical composition occur in branch wood, e.g. in spruce branches there are higher concentrations of polyoses, lignin and resin than in the stem.

## 3 Cell Expansion

Following cell division in the cambial zone, xylem mother cells leave the meristem and wood development progresses through the expansion of differentiating cells. By combining symplastic growth—when neighbouring cells differentiate together—and intrusive growth—when they move past each other—wood cells

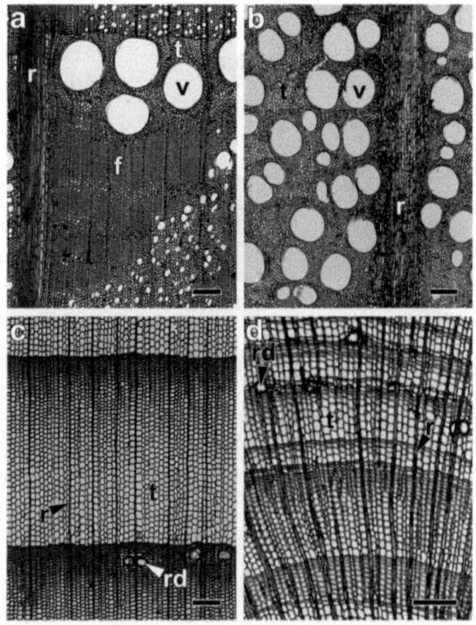


**Fig. 5** (**a**) Transverse view of stem wood from *Quercus robur* showing a ring-porous pattern of vessels. (**b**) Transverse view of root wood from the same species shows a diffuse-porous vessel distribution. (**c**) Transverse view of stem wood from *Picea abies* showing a characteristic tree ring with early- and latewood. (**d**) Transverse view of root wood with diminishing differences between early- and latewood. *f* fibres, *r* ray, *rd* resin duct, *t* tracheids, *v* vessel. Bar 200 μm (Courtesy of Dr. A. Olbrich)

develop to their final shape. Within the xylem elongation zone, all cells undergo expansion but with individual cell types differing in the extent, type and direction of their enlargement.

In regard to the direction of enlargement, for example, while axial parenchyma cells enlarge primarily radially, vessel elements expand tangentially as well as radially, leading to the lateral displacement of neighbouring cells. In regard to the type of growth, meanwhile, both symplastic growth and intrusive growth are involved in the enlargement of a vessel element. Intrusive growth can relocate neighbouring cells out of their position. Ray cells expand mainly in the radial direction whereas fibres grow radially as well as longitudinally through intrusive tip growth. Tip growth can be very extensive, with the length of the fibres sometimes exceeding by severalfold that of fusiform cambial cells. The dimension of expansion, and eventually cell morphology, is controlled by the anisotropic extension of the primary cell wall (Cosgrove 2005), which depends on both the orientation of cellulose microfibrils and the extent of turgor pressure (Tyerman et al. 2002). The orientation of cellulose microfibrils, which determine the direction of cell expansion, is controlled by cortical microtubules. In differentiating conifer tracheids, cortical microtubules have been shown to be randomly orientated when radial primary walls are formed, but then to change their orientation progressively from longitudinal to transverse as the cells expand (Funada et al. 1997). Since the extent of expansion differs between cell types, controlling mechanisms such as turgor pressure and/or cell wall plasticity must be regulated differentially. Also, the incorporation of hemicelluloses, modification of cellulose and subsequent remodelling of the primary wall play a key role in defining cell morphology.

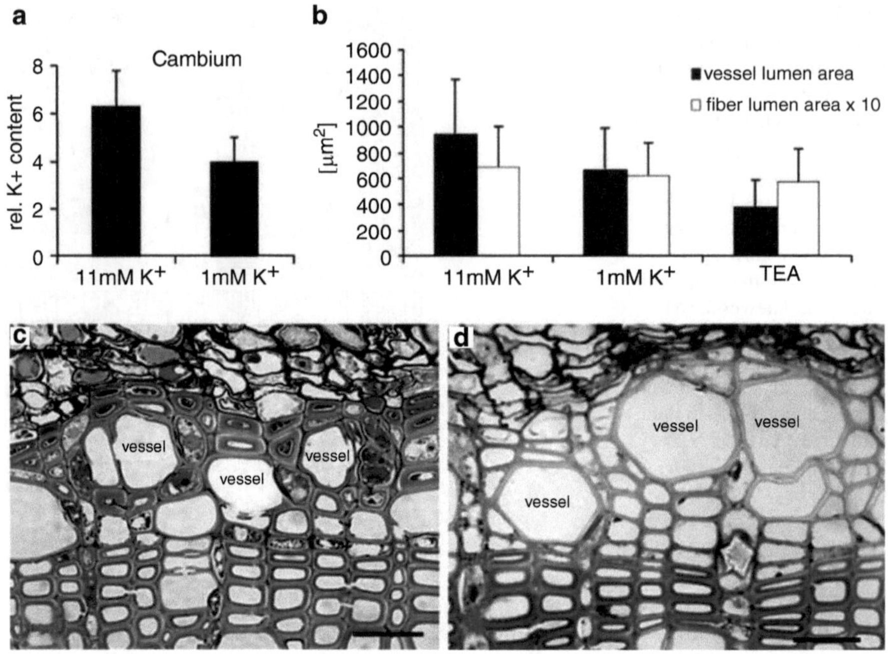


**Fig. 6** Relative cambium $K^+$ content and potassium-dependent vessel lumen. (**a**) The rel. $K^+$ content of the cambium increases with root potassium supply from 1 to 11 mM. (**b**) Plants supplied for 2 weeks with $K^+$ and treated with TEA (5 mM) show reduced vessel size under TEA treatment as well as 1 mM $K^+$. In contrast, fibre lumen does not respond significantly to TEA treatment or different $K^+$ supply. (**c**) TEA-treated twigs have reduced vessel size compared to untreated twigs (**d**), Bar 20 µm (According to Langer et al. 2002)

With regard to turgor regulation, it is well known that potassium is essential for cell expansion during wood formation in trees. In poplar, the $K^+$ concentration in the cambial zone and developing xylem is much higher than in the mature xylem and phloem (Fromm 2010). The $K^+$ concentration is also significantly higher in differentiating vessels in comparison to young fibres (Langer et al. 2002). The pattern of $K^+$ in developing xylem cells also correlates well with the size of the newly formed vessels and fibres. When poplars were grown under non-limiting $K^+$ regimes, $K^+$ content increased within the cambium and also the vessel size clearly increased in contrast to fibre size (Fig. 6, Langer et al. 2002). Furthermore, the developing xylem zone was threefold larger in comparison to plants grown under $K^+$ depletion. When poplars were treated with tetraethylammonium (TEA), a $K^+$ channel blocker, the size of the differentiating vessels was significantly reduced (Fig. 6). In contrast, the size of newly developed fibres was neither influenced by $K^+$ supply nor by TEA treatment. The osmotic function of $K^+$ seems, therefore, to be restricted to vessel and cambial cell enlargement. Moreover, in the poplar cambium, a seasonal variation has been observed, with a high potassium content in spring and summer and a significant reduction in autumn and winter, correlating with the radial width and the osmotic potential of the cambial zone (Wind et al. 2004).

With regard to the molecular analysis of $K^+$-dependent wood formation, so far, ten $K^+$ channels have been identified from the poplar genome (Ache et al. 2010). In close correlation with the increasing cambial $K^+$ concentration in spring is the expression of two ion channels named PTORK (*P. tremula outward rectifying $K^+$ channel*) and PTK2 (*P. tremula $K^+$ channel 2*), which are induced at temperatures >10–15 °C during cambial reactivation. The biophysics of these two poplar $K^+$ channels have been measured by the double-electrode voltage-clamp technique after injection of their gene products' cRNAs into *Xenopus* oocytes. The results indicate that depolarization of the membrane elicits an outward rectifying current. PTORK, therefore, is under the control of the membrane potential, as well as external $K^+$ concentration, and releases $K^+$ in a voltage and K-dependent manner. In contrast to PTORK, PTK2 mediates both the uptake and release of $K^+$ in response to changes in membrane potential, calcium and pH (Langer et al. 2002; Fromm and Hedrich 2007). By using immunofluorescence microscopy PTORK labelling can be localised in the plasmalemma of differentiating fibres and vessel-associated cells (VACs) of mature wood rays (Fig. 7, Arend et al. 2005). Since PTORK was absent in vessels, however, it is assumed that this channel might limit the radial expansion of differentiating fibres by mediating $K^+$ efflux. Moreover, the function of PTORK in VACs of mature wood might be to enable the release of $K^+$ into vessels from where it can be remobilised within the shoot.

In addition to ion channels, the plasma membrane $H^+$-ATPase can also be detected in developing xylem and cambial cells, as well as in VACs during active growth (Arend et al. 2002, 2004). It is assumed that the PM $H^+$-ATPase generates the proton-motive force necessary for the uptake of $K^+$ and other nutrients into developing wood cells, as well as VACs from the xylem stream.

With regard to the control of cell wall extension, there is convincing evidence, in line with the acid growth hypothesis, that auxin plays a significant role by activating a plasma membrane $H^+$-ATPase (Brett and Waldron 1990). By pumping protons from the cytoplasm into the cell wall, the produced acidification causes a loosening of the cell wall structure, and therefore, a turgor-driven extension becomes possible. Cell wall-loosening enzymes also play an important role in cell wall extension, however. The main load-bearing molecules in the primary wall are cellulose microfibrils coated with xyloglucan and linked by xyloglucan bridges (Mellerowicz et al. 2001). Therefore, wall expansion depends mainly on the ability to break these xyloglucan bridges and cut the hydrogen bondings between xyloglucans and cellulose. Of fundamental importance in these processes are cell wall-loosening enzymes such as xylanase, endoglucanases and xyloglucan endotransglucosylase/hydrolase (XET; Mellerowicz et al. 2001; Cosgrove 2005), the last of which is involved in the incorporation of newly synthesised xyloglucan into the existing macromolecular network. Other enzymes that contribute to cell expansion are xyloglucan-specific glucanases, which hydrolyse xyloglucan (Matsumoto et al. 1997) and endoglucanases, that are involved either in the degradation of

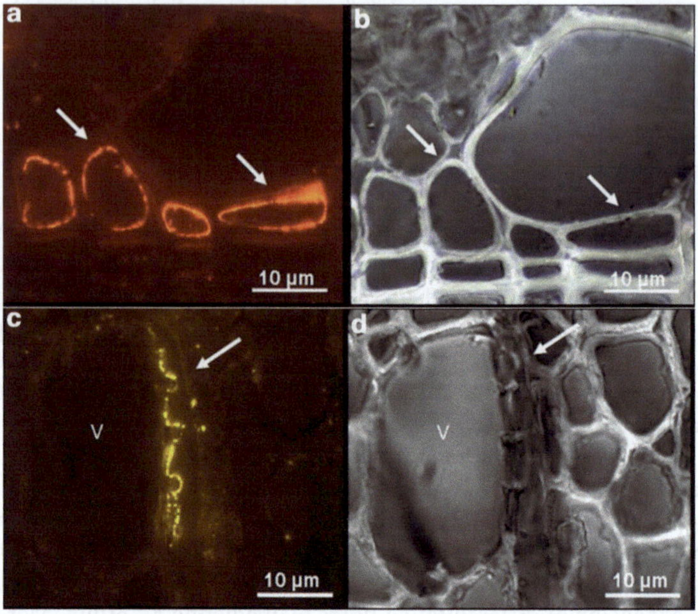


**Fig. 7** Immunolocalisation of PTORK in poplar stems. (**a**) In young fibres the PTORK-specific fluorescence labelling is concentrated along the plasma membrane (arrows). (**b**) Light micrograph of the same section as shown in (**a**). (**c**) A vessel-associated ray cell shows PTORK labelling at the membrane site facing the vessel element. (**d**) Light micrograph of the same section as shown in (**c**) Bar 10 μm (According to Arend et al. 2005)

noncrystalline cellulose (Ohmiya et al. 2000) or in the biosynthesis of cellulose during growth (del Campillo 1999).

Of special interest in mediating cell wall extension are expansins. These are proteins bound to the cell wall that are able to regulate the rheological properties of the cell wall (McQueen-Mason 1997). They bind at the interface between cellulose microfibrils and matrix polysaccharides, showing no hydrolysing activity but reversibly disrupting non-covalent bonds in a pH-dependent manner (McQueen-Mason and Cosgrove 1995). Expansins are supposed to be involved in the radial expansion and tip growth of fusiform cambial cells and developing xylem cells. Tip growth requires the softening of the pectinous middle lamella of neighbouring cells which would otherwise provide resistance to the intruding growing tip. Also, the formation of $Ca^{2+}$-bound pectins in the middle lamella of neighbouring cells can limit the growing fibre tip (Catesson et al. 1994). Thus, the amount and composition of pectins might be assigned an important role in expansion. For example, in poplar cell elongation, tip growth is initiated concomitantly with cell division, and the final morphology of fibres is affected by the function of pectin methyl esterase (PME), which acts on the pattern of methyl-esterification of pectic homogalacturonan in the compound middle lamella, and therefore serves to constrain both intrusive and symplastic cell growth (Siedlecka et al. 2008; Pelloux et al. 2007). Wall plasticity, then, could be controlled by PME through its regulation of the status of pectin methylation. Upregulation of *Populus PME1* induces de-esterification of homogalacturonan and inhibits fibre elongation. Downregulation of this gene, however, stimulates both fibre elongation and the radial expansion of fibres and vessels (Siedlecka et al. 2008). Since spectra of PME isoforms vary across the developmental gradients of wood-forming tissues (Mellerowicz and Sundberg 2008), the sequential expression of different PMEs could play a significant role in the dynamics of cell expansion.

# 4 Cell Wall Thickening

## *4.1 Structure and Composition of the Cell Wall*

Both before and during enlargement, young cells have primary cell walls which are composed of 30–50 % pectins, 20–30 % cellulose, 20–25 % hemicelluloses, up to 10 % proteins and a considerable amount of water. Cellulose microfibrils are embedded within a matrix of hemicelluloses, pectins and glycoproteins. Within the hemicelluloses the main component is xyloglucans, which are closely connected with cellulose microfibrils by hydrogen bonds (Hoson 1991) in order to control cell wall expansion. Pectins are mainly composed of polygalacturonic acid and rhamnogalacturonan, which generate a strongly hydrophilic gel surrounding the cellulose-hemicellulose network. After completion of cell elongation, pectins are combined through $Ca^{2+}$ in order to prevent further cell expansion. In addition, the primary cell wall consists of structural proteins (glycoproteins) which are mainly composed of hydroxyproline-rich proteins (HRGPs), proline-rich proteins (PRPs) and glycine-rich proteins (GRPs). Furthermore, numerous enzymes have been found in the primary wall, such as peroxidases and laccases, which can catalyse lignifications, as well as cellulase and pectinase, which are important during cell wall degradation, e.g. during the formation of perforation plates in differentiating vessels. Phosphatases, invertases, pectinases and pectin methylesterases have also been found in the primary wall.

In terms of chemical composition, distinct differences occur between primary and secondary cell walls. While the main components of primary walls in the developing xylem are pectin, cellulose, hemicelluloses and protein, secondary walls are composed mainly of cellulose, hemicelluloses and lignin. The transition from the primary to the secondary wall is mainly characterised by a decrease in cell wall water content, in association with a decrease in wall porosity caused by a tight accumulation of cellulose microfibrils, as well as by the cross-linking of lignin that displaces water and fills the available interspaces (Fujino and Itoh 1998). The biosynthesis of the secondary cell wall can already start before cell expansion ceases, and this process of formation can be easily identified with polarised light microscopy, since cellulose consists of crystalline zones which cause double

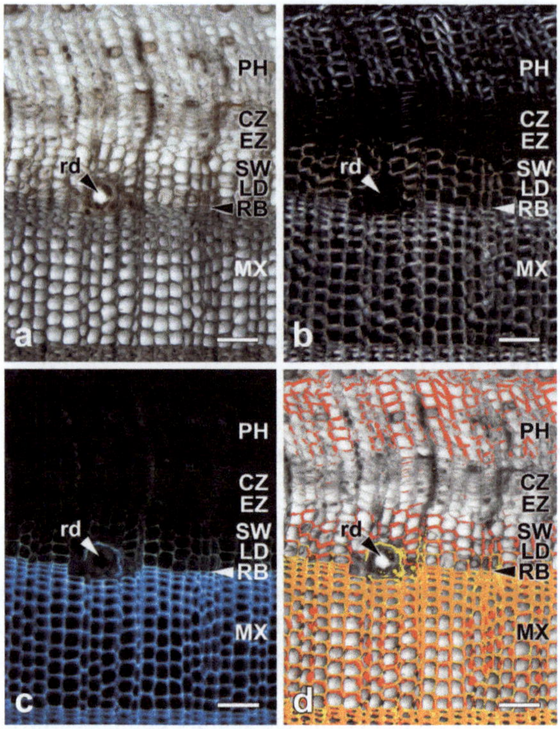


**Fig. 8** Secondary wall formation and lignification in wood from *Pinus sylvestris* grown at a cold location in Kevo (North Finland) where a clear gap occurs between S2 formation and lignification. (**a**) Light micrograph of phloem and wood formation zone. (**b**) The same section viewed under polarised light indicating S2 formation. (**c**) Lignified cell walls shown by lignin autofluorescence. (**d**) Combined photograph showing (in false colours) S2 formation in *red*, lignifications in *yellow* and lignified S2 walls in *orange*. Note that the sieve cells of the phloem also show secondary wall thickening (**b**, **d**), which is characteristic of the secondary phloem of the Pinaceae (Abbe and Crafts 1939). *CZ* cambial zone, *EZ* elongation zone, *LD* lignin deposition, *MX* mature xylem, *PH* phloem, *RB* ring border, *rd* resin duct, *SW* secondary wall formation. Bar 100 μm (Courtesy of Dr. A. Olbrich)

refraction of light and numerous cellulose microfibrils are parallel oriented in the secondary wall (Fig. 8).

During secondary cell wall formation, a reprogramming of wall biosynthesis occurs. Zhong and Ye (2007) identified fibre- and vessel-element-specific master switches that activate transcription factors which induce secondary wall programmes (see also chapter "Transcriptional Regulation of Wood Formation in Tree Species") that at least partially overlap between fibres and vessel elements and coordinate expression of cell wall-related genes. For instance, in poplar, wood-associated NAC domain transcription factors (Ptr WNDs) are master switches activating a series of downstream transcription factors involved in the regulation of secondary wall biosynthesis during wood formation (Zhong et al. 2011). With regard to genes encoding cell wall-related enzymes, various gene families have been characterised in tree species. These include, for example, genes encoding carbohydrate-active enzymes (CAZymes; Aspeborg et al. 2005), cellulose synthases (CesAs; Djerbi et al. 2005), cellulose synthase-like synthases (CSLs; Suzuki et al. 2006), pectin methyl esterases (PMEs; Pelloux et al. 2007), XTH genes encoding xyloglucan (XG) endotransglucosylases and hydrolases (XETs and XEHs; Baumann et al. 2007; Nishikubo et al. 2007) and expansins (Sampedro et al. 2006).

Mature wood cell walls are composed of multiple layers which differ in microfibril angle and ratios of cellulose to lignin, hemicellulose and pectin. The outermost layer is the compound middle lamella (CML) which consists of the middle lamella and the primary walls, followed by three layers of secondary cell walls (S1, S2, S3). Before the secondary wall is formed, the CML is rich in pectin and xyloglucan. In contrast, after completion of secondary wall formation, the CML has a high lignin concentration, e.g. 68 % in poplar. After definition of cell size and shape, cell wall thickening generally consists of the lamination of S1, S2 and S3 layers with distinct microfibril angles, in a spiral orientation either to the right or to the left (Fig. 9).

Each of these wall layers is composed of parallel-arranged cellulose microfibrils, lignin and hemicelluloses. The amount of these macromolecules within the different cell layers can be influenced by abiotic factors such as mechanical stress. The S1 layer is formed first and has a relatively flat microfibril angle (MFA) with regard to the cell axis (60°–80°). Here the microfibrils are arranged in a spiral orientation. Since it is an intermediate layer between the primary wall and the S2, the S1 is only 0.1–0.4 μm thick (Fig. 9). The S2, meanwhile, is the thickest of the secondary wall layers (1–10 μm) and has the lowest microfibril angle (4°–30°). For instance, in the S2 of normal wood fibres from poplar, the MFA is 4° while it is 13° in wood fibres under tension (Lautner et al. 2012). The MFA of the S2 layer varies both longitudinally and radially within the tree stem. The microfibril angle strongly affects the mechanical properties of the wood cell; with an increasing angle, wood becomes less rigid and its swelling and shrinkage properties change. The S2 layer is most important for mechanical stability and the properties of the wood are affected mainly by the S2. Most wood biomass is also in the S2 layers of fibres. Eventually, the microfibrils reorientate themselves to a transverse helix, and the innermost layer of the secondary wall, the S3, is formed. This is only 0.5–1.0 μm thick and the MFA is flat (60°–90°), similar to the S1.

Wood cells of different types show variations in their secondary wall formation. For example, xylem fibres need more time to differentiate than vessels. In poplar, vessel elements and their contact cells are the first to form a secondary wall (Murakami et al. 1999). The S2 layer is proportionally thinner in vessels when compared to fibres (Harada and Coté 1985); however, both fibres and vessels show a three-layered secondary cell wall. Next to vessels, fibres can also show pectin-like fibrillar cell wall deposits lining the lumen-facing side of the cell wall (Arend et al. 2008). To ensure water transport is the most important function of newly formed wood, and the differentiation of the various cell types to achieve this has to occur in

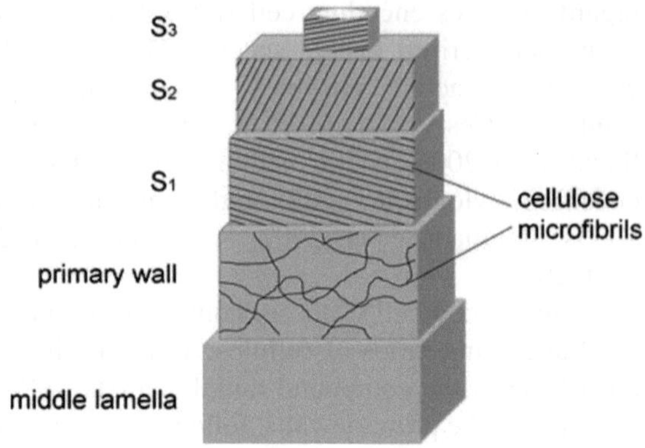


**Fig. 9** Three-dimensional structure of a wood cell wall of a tracheary element showing the different wall layers with their particular cellulose MFA

a coordinated manner. Each cell, therefore, has to receive spatial information in order to develop in a concerted action with neighbouring cells. Numerous vessel elements must be joined together end-to-end to form a functional, often several metre long, vessel. Additionally, pit connections between neighbouring cells have to be developed without any skips. In expanding vessels pits and perforations are free of microtubules (Chaffey et al. 1999), and secondary wall deposition does not occur in the area of pits.

## *4.2 Lignification*

The secondary wall is assumed to be composed of intimate lignin cross-linking between cellulose layers (Kerr and Goring 1975). Lignification is an irreversible process, and lignified cell walls can be identified easily by fluorescence microscopy (Fig. 8) and UV microspectrophotometry (see chapter "Topochemical and Electron Microscopic Analyses on the Lignification of Individual Cell Wall Layers During Wood Formation and Secondary Changes"). Lignin gives compression and bending strength to the wood and imparts hydrophobic qualities on the cell wall, required for the transport of water. Water transportation itself occurs mainly in earlywood cells of the sapwood, as shown by computer tomography (Fromm et al. 2001). Thus, lignin strengthens xylem cells and inhibits collaboration through transpiration-induced negative pressures. It also prevents lateral water diffusion through the cell wall and hence facilitates longitudinal water transport. In addition, lignified cell walls are resistant against microbiotic attack (Nicholson and Hammerschmidt 1992). Lignin occurs at a concentration of 16–31 % in the wood and is derived from the three monolignols *p*-coumaryl alcohol, coniferyl alcohol and sinapyl alcohol, which give rise to H, G and S units, respectively. These units differ from each other in their degree of methoxylation. Several enzymes required for the biosynthesis of monolignols have been localised in membranes of the ER and Golgi-derived vesicles, indicating that monolignols are secreted in order to be incorporated in the cell wall (Mellerowicz et al. 2001; Kenada et al. 2008). During polymerization of lignin, monolignol radicals caused by dehydrogenation reactions undergo radical coupling *in muro* (Boerjan et al. 2003). Vessel elements are lignified first in angiosperms, while fibres lignify later and finally even pit membranes are lignified within the heartwood (Fromm et al. 2003). When the S1 layer is completed during wood formation, lignification starts, particularly at cell corners and then in the middle lamella, and progresses towards the cell lumen after the formation of the S2 and S3 layers (Terashima et al. 1993). Across the gradient of the developing xylem, lignin composition shifts from having more *p*-coumaryl and coniferyl alcohol incorporated within it to later having more sinapyl alcohol (Mellerowicz et al. 2001). The composition and content of lignin vary substantially among different taxa, cell types and wall layers. For example, lignin of angiosperms (hardwood) is composed mainly of coniferyl and sinapyl monolignols, with traces of *p*-coumaryl alcohol, while in gymnosperms (softwood) it is mainly coniferyl alcohol that occurs. Insights into gene function and biosynthesis of lignin have been given by transgenic work focusing on a reduction of lignin content as well as manipulation of the S-to-G ratio and monolignol incorporation (Chiang 2006; Voelker et al. 2010).

### *4.3 Cellulose Formation*

In addition to lignin, wood consists of 40–50 % cellulose which provides the basis for the tensile strength of the cell wall. The cellulose content, as well as the microfibril angle (MFA), has a significant effect on the properties of wood, particularly the modulus of elasticity (MOE) or stiffness. Cellulose appears in the form of microfibrils (MF) which are ~3–10 nm thick (Somerville 2006). Recently, the structure of the microfibrils of spruce wood cellulose has been studied using a range of spectroscopic methods coupled to small-angle neutron and wide-angle X-ray scattering (Fernandes et al. 2011). Results suggest that microfibrils consist of about 24 chains and that these are possibly twisted, with the level of disorder increasing towards the surfaces (Fernandes et al. 2011). The microfibrillar cellulose is largely crystalline; microfibrils can either be accumulated laterally in two crystalline forms, cellulose $I_{alpha}$ or cellulose $I_{beta}$, or they can form an amorphous, paracrystalline structure (Jarvis 2003). Within the cell wall, MFs sum up into macrofibrils which are organised in tangential or random sectors (Donaldson 2007). The size of the macrofibrils depends on the lignin and hemicellulose content, but they vary between 15 and 500 nm in thickness and 4 and 7 μm in length within wood cell walls.

By using freeze-fracture preparations, it can be shown that cellulose is synthesised by plasma membrane-bound rosettes: the cellulose synthase complexes (Brett 2000). Each rosette synthesises cellulose from UDP-glucose and consists of six globules, each of which is assumed to contain six CesA proteins (Mutwil et al. 2008). Cellulose synthase genes carry the type A catalytic domain and are therefore called CesA genes (Holland et al. 2000), involved in both primary and secondary wall biosynthesis. CesA genes have cell-specific expression patterns and have been cloned from the wood-producing tissues of trees, including poplar (Sterky et al. 1998; Wu et al. 2000) and pine (Allona et al. 1998). In the *Populus* genome, 18 CesA genes have been identified (Djerbi et al. 2005; Suzuki et al. 2006). In *Arabidopsis* mutant, expression analyses have demonstrated that three CesA proteins are different from those involved in primary wall biosynthesis and play key roles in secondary wall biosynthesis (Mutwil et al. 2008). Expression analysis indicates that homologues of these three proteins are also the most abundant CesAs during wood formation in *Populus* and *Eucalyptus* (Aspeborg et al. 2005; Djerbi et al. 2004; Prassinos et al. 2005; Bhaudari et al. 2006; Ranik and Myburg 2006). During cellulose synthesis, the moving rosettes release cellulose fibrils into the cell wall. The rosettes are synthesised within the endoplasmic reticulum and transported via Golgi vesicles into the plasma membrane (Haigler and Brown 1986).

Since a parallel correlation exists between cellulose microfibrils and cortical microtubules, a constraint model has been established which describes a rail-like system consisting of microtubules that guide the cellulose synthase complexes in a defined direction (Heath 1974; Baskin 2001). Genetic and pharmacological studies have provided experimental evidence for this model. To visualise CesA, it was fused with a yellow fluorescent protein (YFP) in order to show that the microtubules and microfibrils are pharmacologically disrupted in a way that supports a direct mechanism for the guidance of cellulose deposition by microtubules (Paredez et al. 2006a, b). Furthermore, by using a green fluorescent fusion protein (GFP:CesA), it can be shown that the cortical microtubules also position the delivery of cellulose synthase complexes to the plasma membrane (Paredez et al. 2006a, b). Since microtubules consist of tubulin, some isoforms of the family of α-tubulin and β-tubulin are highly expressed during xylem secondary wall formation in *Populus* (Oakley et al. 2007). Furthermore, a strong association exists between MFA and α-tubulin in *Pinus taeda* (Gonzalez-Martinez et al. 2006).

Two models have been proposed regarding the rail system guiding cellulose synthesis through cortical microtubules (Nick 2008). The monorail model envisages that the cellulose-synthesising complexes are moved along microtubules driven by a microtubule-dependent motor. In contrast, the guardrail model envisages that the cellulose-synthesising complexes are moved by the force from the crystallising cellulose (Nick 2008). Since the practical discrimination between these two models is not easy, further investigations are required in order to get more insights into this process.

## 4.4 Hemicellulose Deposition

Apart from lignin and cellulose, wood is made up of hemicelluloses which account for about 25–35 % of the dry weight of wood and therefore are one of the main wood components. Generally they either occur as heteropolymer-like glucomannan, galactoglucomannan, arabinogalactan and glucuronoxylan or as homopolymer such as galactan, arabinan and ß-1,3-glucan. The biosynthesis of hemicelluloses occurs in the Golgi apparatus in two major steps. Firstly, the formation of the backbone occurs through polysaccharide synthases, and then, secondly, side chain residues catalysed by glycosyltransferases are added (Keegstra and Raikhel 2001). In the secondary walls of dicotyledons 4-*O*-methylglucuronoxylan is the most important hemicellulose (York and O'Neill 2008). Xylans are composed of 1,4-linked ß-D-xylopyranosyl backbone residues, which are nearly always substituted with mono- or disaccharide side chains. Xylan coats microfibrils in secondary walls and seems not to be distributed uniformly within the wood cell wall. For instance, it is associated with the thick cellulose microfibrils in the S1 and S2 layers but not with the thin microfibrils in the S1 layer of beech fibres, as shown by Awano et al. (2000) using immuno-field emission SEM. During secondary wall formation, the biosynthesis of xylan is specifically upregulated (Gregory et al. 1998), and xylan deposition increases in the secondary walls of tracheary elements in comparison to the primary walls (Ingold et al. 1988). It has been demonstrated that in poplar family members, GT43 is involved in the biosynthesis of xylan backbones and that the poplar GT8D is essential for the biosynthesis of the xylan reducing end sequence (Lee et al. 2011). In general, genes for hemicellulose biosynthesis are highly expressed in developing wood of *Populus* (Geisler-Lee et al. 2006) and are especially upregulated in the formation of secondary walls (Aspeborg et al. 2005).

## 5 Programmed Cell Death

Xylem cell death is an important part of the wood formation programme and a prerequisite for the transport of water in vessels and tracheids, commonly known as tracheary elements (TEs). The necessary components for cell death seem to be produced early in xylem differentiation. Inhibitors, as well as storage of hydrolytic enzymes in inactive forms in the vacuole, prevent cell death until vacuolar rupture, which triggers autolytic hydrolysis of the cell contents. There are significant differences between different xylem cell types, however. While vessel elements need only a couple of days to differentiate and die, the lifetime of libriform fibres is estimated to be approximately one month in *Populus tremula* and *Picea abies* (Bollhöner et al. 2012). Dead vessels, therefore, are often bordered by still living fibres, as shown in poplar (Fig. 10). Finally, ray parenchyma cells can live several decades before they die (Nakaba et al. 2006). Another main difference occurs between water-transporting TEs and fibres. While TEs die rapidly after vacuolar bursting and hydrolysis of cell contents, fibres die slower and disintegrate cellular contents well before cell death (Bollhöner et al. 2012). In poplar vessels, membrane characteristics change and the cytoplasm becomes very sparse in appearance while the organelles show increasing degradation. Subsequently, the collapse of the vacuole causes an abrupt loss of the protoplast in order to develop a functioning vessel (Arend and Fromm 2003). In poplar fibres, however, death is accompanied

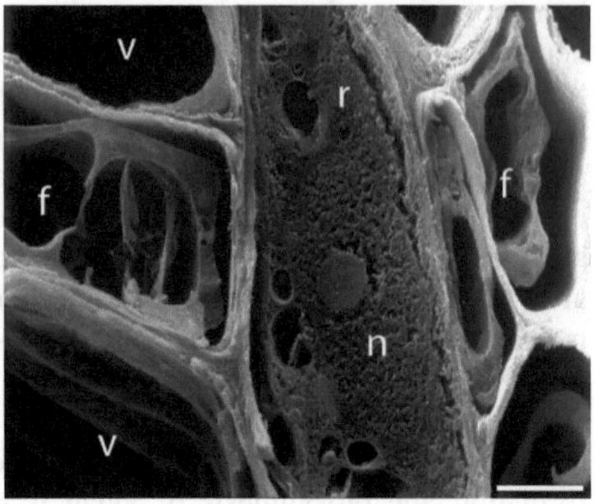


**Fig. 10** Transverse view of xylem cell death in poplar wood. Dead vessels are bordered by still living fibres and ray parenchyma, indicating rapid vessel death. *f* fibre, *n* nucleus, *r* ray parenchyma, *v* vessel (Courtesy of Dr. G. Wanner)

by a diminishing tonoplast and the cytoplasm mixes with the vacuolar content. Long before cell death, fibres show DNA breaks, and the cytoplasmic contents start to be hydrolysed gradually before vacuolar rupture (Courtois-Moreau et al. 2009). The slow death programme of fibres occurs in a more controlled fashion and also indicates autophagy as a degradation process where autophagic bodies enclose cytoplasmic contents and move them to the vacuole for degradation. Lastly, fibres show a turgor reduction and the remaining organelles swell, the vacuole bursts and organelles become autolysed.

With regard to morphological changes, studies of *Zinnia* mesophyll cells, which can be induced to transdifferentiate into TEs in vitro (Fukuda 1996), have given a comprehensive understanding of the cell death of TEs. When *Zinnia* TEs die, the first indication is the swelling of the vacuole, and a subsequent change in tonoplast permeability (Kuriyama 1999) before a rapid vacuolar collapse appears (Groover et al. 1997). This is considered to be the moment of death. After vacuolar collapse cytoplasmic streaming ceases, hydrolytic enzymes are released from the vacuole, cytoplasmic enzymes are activated by acidification and nuclear DNA is degraded rapidly within 10–20 min (Obara et al. 2001). Organelles such as the endoplasmic reticulum (ER) and Golgi vesicles start swelling, and the cellular contents degrade (Fukuda 1997). Following cellular hydrolysis, enzymes that are resistant to the lytic environment modify the cell walls of dying TEs. Moreover, lignin deposition, which has started prior to cell death, continues after dying (Steward 1966; Pesquet et al. 2010). Similar to TEs also in fibres, bulk lignifications occur after cell death in poplar (Courtois-Moreau et al. 2009).

Various signals are related to xylem cell death and play a key role in the correct timing of cellular autolysis. Thermospermine has a protective role against premature xylem maturation and death. In the *Zinnia* cell culture, spermidine treatment prolongs tracheary element differentiation, and the size of the differentiating cells increases (Muniz et al. 2008). Ethylene signalling also plays a role in TE maturation. In addition, cell death is transcriptionally regulated as one part of the wood formation process. Recently identified NAC domain transcription factors induce expression of both cell death and secondary wall-related genes (Ohashi-Ito et al.

2010; Zhong et al. 2007, 2010). Another NAC transcription factor seems to inhibit TE cell death (Yamaguchi et al. 2010). Also calcium seems to play an important role in vacuolar rupture. It increases during cell death, and the blocking of $Ca^{2+}$ inward channels has been shown to suppress vacuolar rupture in *Zinnia* TEs (Groover and Jones 1999). Furthermore, numerous proteases, lipases and nucleases are stored in the vacuole until release and cause hydrolysis of cellular contents. For example, cysteine proteases XCP1 and XCP2 aid micro-autolysis within the intact central vacuole and mega-autolysis of cellular contents after vacuolar rupture during xylogenesis in *Arabidopsis* roots (Avci et al. 2008). Also several different genes of vacuolar-processing enzymes (VPEs) are expressed during maturation of fibres in the *Populus* stem (Moreau et al. 2005; Courtois-Moreau et al. 2009), indicating a possible role in cell death. Furthermore, structurally related proteins called metacaspases seem to play a regulative role in xylem cell death because their genes are expressed at the very last stages of wood formation (Courtois-Moreau et al. 2009). Other important enzymes are nucleases that are responsible for degradation of genomic DNA. In *Zinnia* cell cultures, three main nucleases have been shown that are involved in differentiating TEs (Ito and Fukuda 2002). The control of the timing of vacuolar rupture is very important but little is known about this process.

# 6 Conclusions

In the last decades an extensive amount of literature addressing wood formation has been published. Most of the data were obtained on woody models such as *Populus, Eucalyptus* and *Pinus*. For a deeper understanding of the molecular mechanisms of cell wall biosynthesis, however, herbaceous plant species such as *Arabidopsis thaliana* and *Zinnia elegans* are also valuable systems. Nevertheless, trees are of course the main model system for wood formation partly because of the complicated functioning of the cambium through their seasonal cycle and partly because the different maturation of various wood cell types, as well as heartwood formation, can only be studied in trees. Recent advances in cell and molecular biological techniques have broadened our knowledge of the fascinating process of wood production. An increasing amount of genomic data has become available for trees, and transgenic plants have been designed to either overexpress or silence a gene of interest. By combining various disciplines such as genomics, cell biology, biochemistry, anatomy and electrophysiology, it has become possible to obtain information on gene action in the context of structural cellular components.

## References

Abbe LB, Crafts AS (1939) Phloem of white pine and other coniferous species. Bot Gaz 100:695–722

Ache P, Fromm J, Hedrich R (2010) Potassium-dependent wood formation in poplar: seasonal aspects and environmental limitations. Plant Biol 12:259–267

Allona I, Quinn M, Shoop E, Swope K, St. Cyr S, Carlis J, Riedl J, Retzel E, Campbell MM, Sederoff R, Whetten RW (1998) Analysis of xylem formation in pine by cDNA sequencing. Proc Natl Acad Sci USA 95:9693–9698

Alves ES, Angyalossy-Alfonso V (2000) Ecological trends in the wood anatomy of some Brazilian species. 1. Growth rings and vessels. IAWA J 21:3–30

Arend M, Fromm J (2000) Seasonal variation in the K, Ca and P content and distribution of plasma membrane $H^+$-ATPase in the cambium of *Populus trichocarpa*. In: Savidge RA, Barnett JR, Napier R (eds) Cell and molecular biology of wood formation. BIOS Scientific Publishers, Oxford, pp 67–70

Arend M, Fromm J (2003) Ultrastructural changes in cambial cell derivatives during xylem differentiation in poplar. Plant Biol 5:255–264

Arend M, Fromm J (2004) Die Rolle von Kalium und $H^+$-ATPasen bei der Holzbildung. AFZ-DerWald 19:1032–1033

Arend M, Weisenseel MH, Brummer M, Osswald W, Fromm J (2002) Seasonal changes of plasma membrane $H^+$-ATPase and endogenous ion current during cambial growth in poplar plants. Plant Phys 129:1651–1663

Arend M, Monshausen G, Wind C, Weisenseel MH, Fromm J (2004) Effect of potassium deficiency on the plasma membrane $H^+$-ATPase of the wood ray parenchyma in poplar. Plant Cell Environ 27:1288–1296

Arend M, Stinzing A, Wind C, Langer K, Latz A, Ache P, Fromm J, Hedrich R (2005) Polar-localised poplar $K^+$ channel capable of controlling electrical properties of wood-forming cells. Planta 223:140–148

Arend M, Muninger M, Fromm J (2008) Unique occurrence of pectin-like fibrillar cell wall deposits in xylem fibres of poplar. Plant Biol 10:763–770

Arend M, Schnitzler J-P, Ehlting B, Hänsch R, Lange T, Rennenberg H, Himmelbach A, Grill E, Fromm J (2009) Expression of the Arabidopsis mutant *abi1* gene alters abscisic acid sensitivity, stomatal development, and growth morphology in gray poplars. Plant Phys 151:2110–2119

Aspeborg H, Schrader J, Coutinho PM, Stam M, Kallas Å, Nilsson P, Denman S, Amini B, Sterky F, Master E et al (2005) Carbohydrate-active enzymes involved in the secondary cell wall biogenesis in hybrid aspen. Plant Physiol 137:983–997

Avci U, Earl Petzold H, Ismail IO, Beers EP, Haigler CH (2008) Cysteine proteases XCP1 and XCP2 aid micro-autolysis within the intact central vacuole during xylogenesis in *Arabidopsis* roots. Plant J 56:303–315

Avery GS, Burkholder PR, Creighton HB (1937) Production and distribution of growth hormone in shoots of *Aesculus* and *Malus*, and its probable role in stimulating cambial activity. Am J Bot 24:51–58

Awano T, Takabe K, Fujita M, Daniel G (2000) Deposition of glucuronoxylans on the secondary cell wall of Japanese beech as observed by immune-scanning electron microscopy. Protoplasma 212:72–79

Baba K, Karlberg A, Schmidt J, Schrader J, Hvidsten TR, Bako L, Bhalerao RP (2011) Activity-dormancy transition in the cambial meristem involves stage-specific modulation of auxin response in hybrid aspen. Proc Nat Acad Sci USA 108:3418–3423

Baier M, Goldberg R, Catesson A-M, Liberman M, Bouchemal N, Michon V, Herve du Penhoat C (1994) Pectin changes in samples containing poplar cambium and inner bark in relation to the seasonal cycle. Planta 193:446–454

Bannan MW (1955) The vascular cambium and radial growth in *Thuja occidentalis* L. Can J Bot 33:113–138

Baskin TI (2001) On the alignment of cellulose microfibrils by cortical microtubules: a review and a model. Protoplasma 215:150–171

Baumann MJ, Eklof JM, Michel G, Kallas AM, Teeri TT, Czjzek M, Brumer H (2007) Structural evidence for the evolution of xyloglucanase activity from xyloglucan endotransglycosylases: biological implications for cell wall metabolism. Plant Cell 19:1947–1963

Begum S, Nakaba S, Oribe Y, Kubo T, Funada R (2007) Induction of cambial reactivation by localized heating in a deciduous hardwood hybrid poplar (*Populus sieboldii* × *P. grandidentata*). Ann Bot 100:439–447

Begum S, Nakaba S, Oribe Y, Kubo T, Funada R (2010) Changes in the localization and levels of starch and lipids in cambium and phloem during cambial reactivation by artificial heating of main stems of *Cryptomeria japonica* trees. Ann Bot 106:885–895

Bertaud F, Holmbom B (2004) Chemical composition of earlywood and latewood in Norway spruce heartwood, sapwood and transition zone wood. Wood Sci Tech 38:245–256

Bhaudari S, Fujino T, Thammanagowda S, Zhang DY, Xu FY, Joshi CP (2006) Xylem-specific and tension stress-responsive coexpression of KORRIGAN endoglucanase and three secondary wall-associated cellulose synthase genes in aspen trees. Planta 224:828–837

Bishopp A, Lehesranta S, Vatén A, Help H, El-Showk S, Scheres B, Helariutta K, Mähönen AP, Sakakibara H, Helariutta Y (2011) Phloem-transported cytokinin regulates polar auxin transport and maintains vascular pattern in the root meristem. Curr Biol 21:927–932

Bissett IJW, Dadswell HE (1950) The variation in cell length within one growth ring of certain angiosperms and gymnosperms. Aust For 14:17–29

Björklund S, Antti H, Uddestrand I, Moritz T, Sundberg B (2007) Cross-talk between gibberellin and auxin in development of Populus wood: gibberellin stimulate polar auxin transport and has a common transcriptome with auxin. Plant J 52:499–511

Boerjan W, Ralph J, Baucher M (2003) Lignin biosynthesis. Annu Rev Plant Biol 54(1):519–546

Böhlenius H, Huang T, Charbonnel-Campaa L, Brunner AM, Jansson S, Strauss SH, Nilsson O (2006) CO/FT regulatory module controls timing of flowering and seasonal growth cessation in trees. Science 312:1040–1043

Bollhöner B, Prestele J, Tuominen H (2012) Xylem cell death: emerging understanding of regulation and function. J Exp Bot 63(3):1081–1094

Brett CT (2000) Cellulose microfibrils in plants: biosynthesis, deposition, and integration into the cell wall. Int Rev Cytol 199:161–199

Brett C, Waldron K (1990) Physiology and biochemistry of plant cell walls. Unwin Hyman, London

Brienen RJW, Zuidema PA (2005) Relating tree growth to rainfall in Bolivian rain forests: a test for six species using tree ring analysis. Oecologia 146:1–12

Catesson A-M (1989) Specific characters of vessel primary walls during the early stages of wood differentiation. Biol Cell 67:221–226

Catesson A-M (1990) Cambial cytology and biochemistry. In: Iqbal M (ed) The vascular cambium, vol 7, Research studies in botany and related applied fields. Research Studies Press, Taunton, MA, pp 63–112

Catesson A-M (1994) Cambial ultrastructure and biochemistry: changes in relation to vascular tissue differentiation and the seasonal cycle. Int J Plant Sci 155:251–261

Catesson A-M, Roland JC (1981) Sequential changes associated with cell wall formation and fusion in the vascular cambium. IAWA Bull 2:151–162

Catesson A-M, Funada R, Robertbaby D, Quinetszely M, Chuba J, Goldberg R (1994) Biochemical and cytochemical cell-wall changes across the cambial zone. IAWA J 15:91–101

Chaffey N (1999) Cambium: old challenges – new opportunities. Trees 13:138–151

Chaffey NF, Barnett JR, Barlow PW (1998) A cycle of microtubule rearrangement accompanies the seasonal cycle of wall thickening within the vascular cambium of *Aesculus hippocastanum* L. taproots. New Phytol 139:623–635

Chaffey N, Barnett J, Barlow P (1999) A cytoskeletal basis for wood formation in angiosperm trees: the involvement of cortical microtubules. Planta 208:19–30

Chaffey N, Cholewa E, Regan S, Sundberg B (2002) Secondary xylem development in Arabidopsis: a model for wood formation. Physiol Plant 114:594–600

Chiang VL (2006) Monolignol biosynthesis and genetic engineering of lignin in trees, a review. Environ Chem Lett 4(3):143–146

Clark A, Daniels RF, Jordan L (2006) Juvenile/mature wood transition in loblolly pine as defined by annual ring specific gravity, proportion of latewood, and microfibril angle. Wood Fiber Sci 38:292–299

Cosgrove DJ (2005) Growth of the plant cell wall. Nat Rev Mol Cell Biol 6(11):850–861

Courtois-Moreau CL, Pesquet E, Sjodin A, Muniz L, Bollhöner B, Kaneda M, Samuels L, Jansson S, Tuominen H (2009) A unique program for cell death in xylem fibers of *Populus* stem. Plant J 58:260–274

Del Campillo E (1999) Multiple endo-1,4-D-glucanase (cellulase) genes in *Arabidopsis*. Curr Top Dev Biol 46:39–61

Djerbi S, Aspeborg H, Nilsson P, Sundberg B, Mellerowicz E, Blomqvist K, Teeri TT (2004) Identification and expression analysis of genes encoding putative cellulose synthases (CesA) in the hybrid aspen, *Populus tremula* (L.) × *P. tremuloides* (Michx.). Cellulose 11:301–312

Djerbi S, Lindskog M, Arvestad L, Sterky F, Teeri TT (2005) The genome sequence of black cottonwood (*Populus trichocarpa*) reveals 18 conserved cellulose synthase (CesA) genes. Planta 221:739–746

Donaldson L (2007) Cellulose microfibril aggregates and their size variation with cell wall type. Wood Sci Technol 41:443–460

Druart N, Johansson A, Baba K, Schrader J, Sjödin A, Bhalerao RR, Resman L, Trygg J, Moritz T, Bhalerao RP (2007) Environmental and hormonal regulation of the activity-dormancy cycle in the cambial meristem involves stage-specific modulation of transcriptional and metabolic networks. Plant J 50:557–573

Endo S, Pesquet E, Tashiro G, Kuriyama H, Goffner D, Fukuda H, Demura T (2008) Transient transformation and RNA silencing in *Zinnia* tracheary element differentiating cell cultures. Plant J 53:864–875

Eriksson ME, Israelsson M, Olsson O, Moritz T (2000) Increased gibberellin biosynthesis in transgenic trees promotes growth, biomass production and xylem fiber length. Nat Biotechnol 18:784–788

Ermel FF, Follet-Gueye M-L, Cibert C, Vian B, Morvan C, Catesson A-M, Goldberg R (2000) Differential localization of arabinan and galactan side chains of rhamnogalacturonan 1 in cambial derivatives. Planta 210:732–740

Espinosa-Ruiz A, Saxena S, Schmidt J, Mellerowicz E, Miskolczi P, Bako L, Bhalerao R (2004) Differential stage-specific regulation of cyclin-dependent kinases during cambial dormancy in hybrid aspen. Plant J 38:603–615

Farrar JJ, Evert RF (1997) Seasonal changes in the ultrastructure of the vascular cambium of *Robinia pseudoacacia*. Trees 11:191–202

Fernandes AN, Thomas LH, Altaner CM, Callow P, Forsyth VT, Apperley DC, Kennedy CJ, Jarvis MC (2011) Nanostructure of cellulose microfibrils in spruce wood. Proc Natl Acad Sci USA 108:E1195–E1203

Fisher K, Turner S (2007) PXY, a receptor–like kinase essential for maintaining polarity during plant vascular-tissue development. Curr Biol 17:1061–1066

Follet-Gueye ML, Verdus MC, Demarty M, Thellier M, Ripoll C (1998) Cambium pre-activation in beech correlates with a strong temporary increase of calcium in cambium and phloem but not in xylem cells. Cell Calcium 24(3):205–211

Follet-Gueye ML, Ermel FF, Vian B, Catesson A-M, Goldberg R (2000) Pectin remodeling during cambial derivative differentiation. In: Savidge RA, Barnett JR, Napier R (eds) Cell and molecular biology of wood formation. BIOS Scientific Publishers, Oxford, pp 289–294

Foucart C, Paux E, Ladouce N, San-Clemente H, Grima-Pettenati J, Sivadon P (2006) Transcript profiling of a xylem vs phloem cDNA subtractive library identifies new genes expressed during xylogenesis in *Eucalyptus*. New Phytol 170:739–752

Fromm J (1997) Hormonal physiology of wood growth in willow (*Salix viminalis* L.): effects of spermine and abscisic acid. Wood Sci Technol 31:119–130

Fromm J (2010) Wood formation of trees in relation to potassium and calcium nutrition. Tree Phys 30:1140–1147

Fromm J, Hedrich R (2007) The role of potassium in wood formation of poplar. In: Sattelmacher B, Horst WJ (eds) The Apoplast of higher plants: compartment of storage, transport and reactions. Springer, New York, pp 137–149

Fromm J, Sautter I, Matthies D, Kremer J, Schumacher P, Ganter C (2001) Xylem water content and wood density in spruce and oak trees detected by high-resolution computed tomography. Plant Phys 127:416–425

Fromm J, Rockel B, Lautner S, Windeisen E, Wanner G (2003) Lignin distribution in wood cell walls determined by TEM and backscattered SEM techniques. J Struct Biol 143:77–84

Fuchs M, Ehlers K, Will T, van Bel AJE (2010a) Immunolocalization indicates plasmodesmal trafficking of storage proteins during cambial reactivation in *Populus nigra*. Ann Bot 106:385–394

Fuchs M, van Bel AJE, Ehlers K (2010b) Season-associated modifications in symplasmic organization of the cambium in *Populus nigra*. Ann Bot 105:375–387

Fujino T, Itoh T (1998) Changes in the three dimensional architecture of the cell wall during lignifications of xylem cells in *Eucalyptus tereticornis*. Holzforschung 52:111–116

Fukuda H (1996) Xylogenesis: initiation, progression and cell death. Annu Rev Plant Physiol Plant Mol Biol 47:299–325

Fukuda H (1997) Tracheary element differentiation. Plant Cell 9:1147–1156

Fukuda H (2004) Signals that control plant vascular cell differentiation. Nat Rev 5:379–391

Funada R, Catesson A-M (1991) Partial cell wall lysis and the resumption of meristematic activity in *Fraxinus excelsior* cambium. IAWA Bull ns 12:439–444

Funada R, Abe H, Furusawa O, Imaizumi H, Fukuzawa K, Ohtani J (1997) The orientation and localization of cortical microtubules in differentiating conifer tracheids during cell expansion. Plant Cell Physiol 38(2):210–212

Funada R, Kubo T, Tabuchi M, Sugiyama T, Fushitani M (2001) Seasonal variations in endogenous indole-3-acetic acid and abscisic acid in the cambial region of *Pinus densiflora* Sieb. Et Zucc. stems in relation to earlywood-latewood transition and cessation of tracheid production. Holzforschung 55:128–134

Geisler-Lee J, Geisler M, Coutinho PM, Segerman B, Nishikubo N, Takahashi J, Aspeborg H, Djerbi S, Master E, Andersson-Gunnerås S et al (2006) Poplar carbohydrate-active enzymes (CAZymes). Gene identification and expression analyses. Plant Physiol 140:946–962

Ghouse AKM, Hashmi S (1983) Periodicity of cambium and of the formation of xylem and phloem in *Mimusops elengi* L., an evergreen member of tropical India. Flora 173:479–487

Gion JM, Carouché A, Deweer S, Bedon F, Pichavant F, Charpentier JP, Baillères H, Rozenberg P, Carocha V, Ognouabi N et al (2011) Comprehensive genetic dissection of wood properties in a widely-grown tropical tree: Eucalyptus. BMC Genomics 12:301

González-Martínez SC, Wheeler NC, Ersoz E, Nelson CD, Neale DB (2006) Association genetics in *Pinus taeda* L. I. Wood property traits. Genetics 175:399–409

Goué N, Lesage-Descauses M C, Mellerowicz EJ, Magel E, Label P, Sundberg B (2008) Microgenomic analysis reveals cell type-specific gene expression patterns between ray and fusiform initials within the cambial meristem of *Populus*. New Phytol 180:45–56

Gregory RA (1971) Cambial activity in Alaskan white spruce. Am J Bot 58:160–171

Gregory ACE, O'Connell AP, Bolwell GP (1998) Xylans. Biotechnol Genet Eng Rev 15:439–455

Groover A, Jones AM (1999) Tracheary element differentiation uses a novel mechanism coordinating programmed cell death and secondary cell wall synthesis. Plant Physiol 119:375–384

Groover A, DeWitt N, Heidel A, Jones A (1997) Programmed cell death of plant tracheary elements differentiating *in vitro*. Protoplasma 196:197–211

Haigler CH, Brown RM Jr (1986) Transport of rosettes from the Golgi apparatus to the plasma membrane in isolated mesophyll cells of *Zinnia elegans* during differentiation to tracheary elements in suspension culture. Protoplasma 134:111–120

Harada H, Coté WA Jr (1985) Structure of wood. In: Biosynthesis and biodegradation of wood components. Academic, Orlando, FL, pp 1–44

Heath IB (1974) A unified hypothesis for the role of membrane bound enzyme complexes and microtubules in plant cell wall synthesis. J Theor Biol 48:445–449

Heide O (1993) Daylength and thermal time responses of bud burst during dormancy release in some northern deciduous trees. Physiol Plant 88:531–540

Hejnowicz A, Hejnowicz Z (1958) Variation of length of vessel members and fibres in the trunk of *Populus tremula* L. Acta Soc Bot Pol 27:131–159

Hirakawa Y, Shinohara H, Kondo Y, Inoue A, Nakanomyo I, Ogawa M, Sawa S, Ohashi-Ito K, Matsubayashi Y, Fukuda H (2008) Non-cell autonomous control of vascular stem cell fate by a CLE peptide/receptor system. Proc Natl Acad Sci USA 105:15208–15213

Hoenicka H, Lautner S, Klingberg A, Koch G, El-Sherif F, Lehnhardt D, Zhang B, Burgert I, Odermatt J, Melzer S, Fromm J, Fladung M (2012) Influence of over-expression of the *FLOWERING PROMOTING FACTOR 1* gene (*FPF1*) from Arabidopsis on wood formation in hybrid poplar (*Populus tremula L.* × *P. tremuloides Michx.*). Planta 235:359–373

Holland N, Holland D, Helentjaris T, Dhugga KS, Xoconostle-Cazares B, Delmer DP (2000) A comparative analysis of the plant cellulose synthase (CesA) gene family. Plant Phys 123:1313–1323

Hoson T (1991) Structure and function of plant cell walls: immunological approaches. Int Rev Cytol 130:233–268

Hou H-W, Zhou Y-T, Mwange K-N, Li W-F, He X-Q, Cui K-M (2006) ABP1 expression regulated by IAA and ABA is associated with the cambium periodicity in *Eucommia ulmoides* Oliv. J Exp Bot 57(14):3857–3867

Ingold E, Sugiyama M, Komamine A (1988) Secondary cell wall formation: changes in cell wall constituents during the differentiation of isolated mesophyll cells of *Zinnia elegans* to tracheary elements. Plant Cell Physiol 29:295–303

Ito J, Fukuda H (2002) ZEN1 is a key enzyme in the degradation of nuclear DNA during programmed cell death of tracheary elements. Plant Cell 14:3201–3211

Jarvis M (2003) Chemistry: cellulose stacks up. Nature 426:611–612

Jenkins PA, Shepherd KR (1974) Seasonal changes in levels of indoleacetic acid and abscisic acid in stem tissues of *Pinus radiata*. N Z J For Sci 4:511–519

Jordan L, He R, Hall DB, Clark A, Daniels RF (2006) Variation in loblolly pine cross-sectional microfibril angle with tree height and physiographic region. Wood Fiber Sci 38:390–398

Keegstra K, Raikhel N (2001) Plant glycosyltransferases. Curr Opin Plant Biol 4:219–224

Kenada M, Rensing KH, Wong JCT, Banno B, Mansfield SD, Samuels AL (2008) Tracking monolignols during wood development in lodgepole pine. Plant Phys 147:1750–1760

Kerr AJ, Goring DAI (1975) The ultrastructural arrangement of the wood cell wall. Cellulose Chem Technol 9:563–573

Ko JH, Prassinos C, Keathley D, Han KH (2011) Novel aspects of transcriptional regulation in the winter survival and maintenance mechanism of poplar. Tree Phys 31:208–225

Kroll RE, Ritter DC, Gertjejansen RO, Au KC (1992) Anatomical and physical properties of balsam poplar (*Populus balsamifera* L.) in Minnesota. Wood Fiber Sci 24:13–24

Kuriyama H (1999) Loss of tonoplast integrity programmed in tracheary element differentiation. Plant Physiol 121:763–774

Lachaud S (1989) Participation of auxin and abscisic acid in the regulation of seasonal variations in cambial activity and xylogenesis. Trees 3:125–137

Lang GA (1987) Dormancy: a new universal terminology. Hort Sci 22:817–820

Langer K, Ache P, Geiger D, Stinzing A, Arend M, Wind C, Regan S, Fromm J, Hedrich R (2002) Poplar potassium transporters capable of controlling $K^+$ homeostasis and $K^+$-dependent xylogenesis. Plant J 32:997–1009

Larisch C, Dittrich M, Wildhagen H, Lautner S, Fromm J, Polle A, Hedrich R, Rennenberg H, Müller T, Ache P (2012) Poplar wood rays are involved in seasonal remodeling of tree physiology. Plant Physiol 160(3):1515–1529

Larson PR (1994) The cambium: development and structure. Springer, Berlin

Lautner S, Zollfrank C, Fromm J (2012) Microfibril angle distribution of poplar tension wood. IAWA J 33(4):431–439

Lavigne MB, Little CHA, Riding RT (2004) Changes in stem respiration rate during cambial reactivation can be used to refine estimates of growth and maintenance respiration. New Phytol 162:81–93

Lee C, Teng Q, Zhong R, Ye ZH (2011) Molecular dissection of xylan biosynthesis during wood formation in poplar. Mol Plant 4:730–747

Li X, Wu HX, Southerton SG (2011) Transcriptome profiling of *Pinus radiata* juvenile wood with contrasting stiffness identifies putative candidate genes involved in microfibril orientation and cell wall mechanics. BMC Genomics 12:480–496

Little CHA, Bonga JM (1974) Rest in the cambium of *Abies balsamea*. Can J Bot 52:1723–1730

Little CHA, Savidge RA (1987) The role of plant growth regulators in forest tree cambial growth. Plant Growth Regul 6:137–169

Love J, Björklund S, Vahala J, Hertzberg M, Kangasjärvi J, Sundberg B (2009) Ethylene is an endogenous stimulator of cell division in the cambial meristem of Populus. Proc Natl Acad Sci USA 106:5984–5989

Magel EA, Monties B, Drouet A, Jay-Allemand C, Ziegler H (1995) Heartwood formation: biosynthesis of heartwood extractives and "secondary" lignifications. In: Sandermann H Jr, Bonnet-Masimbert M (eds) Eurosilva – contribution to forest tree physiology. INRA, Versailles, pp 35–56

Mansfield SD, Iliadis L, Avramidis S (2007) Neural network prediction of bending strength and stiffness in western hemlock (*Tsuga heterophylla* Raf.). Holzforschung 61:707–716

Matsumoto T, Sakai F, Hayashi T (1997) A xyloglucan-specific endo-1,4-ß-glucanase isolated from auxin-treated pea stems. Plant Phys 114:661–667

Matsumoto-Kitano M, Kusumoto T, Tarkowski P, Kinoshita-Tsujimura K, Václavíková K, Miyawaki K, Kakimoto T (2008) Cytokinins are central regulators of cambial activity. Proc Natl Acad Sci USA 105:20027–20031

Mauriat M, Moritz T (2009) Analyses of GA20ox-and GID1-over-expressing aspen suggest that gibberellins play two distinct roles in wood formation. Plant J 58:989–1003

McQueen-Mason S (1997) Plant cell walls and the control of growth. Biochem Soc Trans 25:204–214

McQueen-Mason SJ, Cosgrove DJ (1995) Expansin mode of action on cell walls (analysis of wall hydrolysis, stress relaxation, and binding). Plant Phys 107(1):87–100

Mellerowicz EJ, Sundberg B (2008) Wood cell walls: biosynthesis, developmental dynamics and their implications for wood properties. Curr Opin Plant Biol 11:293–300

Mellerowicz EJ, Baucher M, Sundberg B, Boerjan W (2001) Unravelling cell wall formation in the woody dicot stem. Plant Mol Biol 47:239–274

Moreau C, Aksenov N, Lorenzo M, Segerman B, Funk C, Nilsson P, Jansson S, Tuominen H (2005) A genomic approach to investigate developmental cell death in woody tissues of *Populus* trees. Genome Biol 6:R34

Moyle R, Schrader J, Stenberg A, Olsson O, Saxena S, Sandberg G, Bhalerao RP (2002) Environmental and auxin regulation of wood formation involves members of the *Aux/IAA* gene family in hybrid aspen. Plant J 31(6):675–685

Muñiz L, Minguet EG, Singh SK, Pesquet E, Vera-Sirera F, Moreau-Courtois CL, Carbonell J, Blázquez MA, Tuominen H (2008) ACAULIS5 controls *Arabidopsis* xylem specification through the prevention of premature cell death. Development 135:2573–2582

Murakami Y, Funada R, Sano Y, Ohtani J (1999) The differentiation of contact cells and isolation cells in the xylem ray parenchyma of *Populus maximowiczii*. Ann Bot 84:429–435

Mutwil M, Debold S, Persson S (2008) Cellulose synthesis: a complex. Curr Opin Plant Biol 11:252–257

Nakaba S, Sano Y, Kubo T, Funada R (2006) The positional distribution of cell death of ray parenchyma in a conifer, *Abies sachalinensis*. Plant Cell Rep 25:1143–1148

Nicholson RL, Hammerschmidt R (1992) Phenolic compounds and their role in disease resistance. Annu Rev Phytopathol 30:369–389

Nick P (2008) Control of cell axis. In: Nick P (ed) Plant microtubules – development and flexibility. Springer, Berlin, pp 3–46

Nieminen K, Immanen J, Laxell M, Kauppinen L, Tarkowski P, Dolezal K, Tähtiharju S, Elo A, Decourteix M, Ljung K et al (2008) Cytokinin signaling regulates cambial development in poplar. Proc Natl Acad Sci USA 105:20032–20037

Nieminen K, Robischon M, Immanen J, Helariutta Y (2012) Towards optimizing wood development in bioenergy trees. New Phytol 194:46–53

Nilsson J, Karlberg A, Antti H, Lopes-Vernaza M, Mellerowicz E, Perrot-Rechenmann C, Sandberg G, Bhalerao RP (2008) Dissecting the molecular basis of the regulation of wood formation by auxin in hybrid aspen. Plant Cell 20:843–855

Nishikubo N, Awano T, Banasiak A, Bourquin V, Ibatullin F, Funada R, Brumer H, Teeri TT, Hayashi T, Sundberg B, Mellerowicz EJ (2007) Xyloglucan endotransglycosylase (XET) functions in gelatinous layers of tension wood fibers in poplar – a glimpse into the mechanism of the balancing act of trees. Plant Cell Physiol 48:843–855

Oakley RV, Wang YS, Ramakrishna W, Harding SA, Tsai CJ (2007) Differential expansion and expression of α- and β-tubulin gene families in *Populus*. Plant Physiol 145:961–973

Obara K, Kuriyama H, Fukuda H (2001) Direct evidence of active and rapid nuclear degradation triggered by vacuole rupture during programmed cell death in *Zinnia*. Plant Physiol 125:615–626

Oda H, Fukuda H (2012) Secondary cell wall patterning during xylem differentiation. Curr Opin Plant Biol 15:38–44

Ohashi-Ito K, Oda Y, Fukuda H (2010) *Arabidopsis* VASCULAR-RELATED NAC-DOMAIN6 directly regulates the genes that govern programmed cell death and secondary wall formation during xylem differentiation. Plant Cell 22:3461–3473

Ohmiya Y, Samejima M, Shiroishi M, Amano Y, Kanda T, Sakai F, Hayashi T (2000) Evidence that endo-1,4-ß-glucanases act on cellulose in suspension-cultured poplar cells. Plant J 24:147–158

Oribe Y, Funada R, Shibagaki M, Kubo T (2001) Cambial reactivation in the partially heated stem in an evergreen conifer *Abies sachalinensis*. Planta 212:684–691

Oribe Y, Funada R, Kubo T (2003) Relationships between cambial activity, cell differentiation and the localization of starch in storage tissues around the cambium in locally heated stems of *Abies sachalinensis* (Schmidt) Masters. Trees 17:185–192

Paredez A, Wright A, Ehrhardt DW (2006a) Microtubule cortical array organization and plant cell morphogenesis. Curr Opin Plant Biol 9:571–578

Paredez AR, Somerville CR, Ehrhardt DW (2006b) Visualization of cellulose synthase demonstrates functional association with microtubules. Science 312:1491–1495

Pelloux J, Rustérucci C, Mellerowicz EJ (2007) New insights into pectin methylesterase (PME) structure and function. Trends Plant Sci 12:267–277

Pesquet E, Tuominen H (2011) Ethylene stimulates tracheary element differentiation in *Zinnia elegans* cell cultures. New Phytol 190:138–149

Pesquet E, Korolev AV, Calder G, Lloyd CW (2010) The microtubule-associated protein AtMAP70-5 regulates secondary wall patterning in *Arabidopsis* wood cells. Curr Biol 20:744–749

Prassinos C, Ko JH, Han KH (2005) Transcriptome profiling of vertical stem segments provides insights into the genetic regulation of secondary growth in hybrid aspen trees. Plant Cell Physiol 46:1213–1225

Ragni L, Nieminen K, Pacheco-Villalobos D, Sibout R, Schwechheimer C, Hardtke CS (2011) Mobile gibberellin directly stimulates Arabidopsis hypocotyl xylem expansion. Plant Cell 23:1322–1336

Ralph J, MacKay JJ, Hatfield RD, O'Malley DM, Whetten RW, Sederoff RR (1997) Abnormal lignin in a loblolly pine. Science 277:235–239

Ranik M, Myburg AA (2006) Six new cellulose synthase genes from *Eucalyptus* are associated with primary and secondary cell wall biosynthesis. Tree Physiol 26:545–556

Rathgeber CBK, Rossi S, Bontemps JD (2011) Cambial activity related to tree size in a mature silver-fir plantation. Ann Bot 108:429–438

Resman L, Howe G, Jonsen D, Englund M, Druart N, Schrader J, Antti H, Skinner J, Sjodin A, Chen T, Bhalerao RP (2010) Components acting downstream of short day perception regulate differential cessation of cambial activity and associated responses in early and late clones of hybrid poplar. Plant Phys 154:1294–1303

Rinne PLH, Welling A, Vahala J, Ripel L, Ruonala R, Kangasjarvi J, van der Schoot C (2011) Chilling of dormant buds hyperinduces FLOWERING LOCUS T and recruits GA-inducible 1,3-beta-glucanases to reopen signal conduits and release dormancy in *Populus*. Plant Cell 23:130–146

Robards AW, Kidway P (1969) A comparative study of the ultrastructure of resting and active cambium of *Salix fragilis* L. Planta 84:239–249

Robischon M, Du J, Miura E, Groover A (2011) The Populus class III HD ZIP, popREVOLUTA, influences cambium initiation and patterning of woody stems. Plant Physiol 155:1214–1225

Rohde A, Boerjan W (2001) Insights into bud development and dormancy in poplar. In: Huttunen S, Heikkilä H, Bucher J, Sundberg B, Jarvis P, Matyssek R (eds) Trends in European tree physiology research. Kluwer, The Netherlands, pp 33–52

Ruttink T, Arend M, Morreel K, Strome V, Rombauts S, Fromm J, Bhalerao R, Boerjan W, Rohde A (2007) A molecular timetable for apical bud formation and dormancy induction in poplar. Plant Cell 19:2370–2390

Sampedro J, Carey RE, Cosgrove DJ (2006) Genome histories clarify evolution of the expansin superfamily: new insights from the poplar genome and pine ESTs. J Plant Res 119:11–21

Schrader J, Baba K, May ST, Palme K, Bennett M, Bhalerao RP (2003) Polar auxin transport in the wood forming tissues of hybrid aspen is under simultaneous control of developmental and environmental signals. Proc Natl Acad Sci USA 100:10096–10101

Schrader J, Moyle R, Bhalerao R, Hertzberg M, Lundeberg J, Nilsson P, Bhalerao RP (2004a) Cambial meristem dormancy in trees involves extensive remodelling of the transcriptome. Plant J 40:173–187

Schrader J, Nilsson J, Mellerowicz E, Berglund A, Nilsson P, Hertzberg M, Sandberg G (2004b) A high-resolution transcript profile across the wood-forming meristem of poplar identifies potential regulators of cambial stem cell identity. Plant Cell 16:2278–2292

Sennerby-Forsse L (1986) Seasonal variation in the ultrastructure of the cambium in young stems of willow (*Salix viminalis*) in relation to phenology. Plant Physiol 67:529–537

Sexton TR, Henry RJ, Harwood CE, Thomas DS, McManus LJ, Raymond C, Henson M, Shepherd M (2012) Pectin methylesterase genes influence solid wood properties of *Eucalyptus pilularis*. Plant Physiol 158:531–541

Siedlecka A, Wiklund S, Péronne MA, Micheli F, Lésniewska J, Sethson I, Edlund U, Richard L, Sundberg B, Mellerowicsz EJ (2008) Pectin methyl esterase inhibits intrusive and symplastic cell growth in developing wood cells of *Populus*. Plant Physiol 146:554–565

Söding H (1937) Wuchsstoff und Kambiumtätigkeit der Bäume. Jahrb Wiss Bot 84:639–670

Somerville C (2006) Cellulose synthesis in higher plants. Annu Rev Cell Dev Biol 22:53–78

Sterky F, Regan S, Karlsson J, Hertzberg M, Rohde A, Holmberg A, Amini B, Bhalerao R, Larsson M, Villarroel R, Van Montagu M, Sandberg G, Olsson O, Teeri TT, Boerjan W, Gustafsson P, Uhlén M, Sundberg B, Lundeberg J (1998) Gene discovery in the wood-forming tissues of poplar: analysis of 5692 expressed sequence tags. Proc Natl Acad Sci USA 95:13330–13335

Sterky F, Bhalerao RR, Unneberg P, Segerman B, Nilsson P, Brunner AM, Campaa LC, Lindvall JJ, Tandre K, Strauss SH, Sundberg B, Gustafsson P, Uhlen M, Bhalerao RP, Nilsson O, Sandberg G, Karlsson J, Lundeberg J, Jansson S (2004) A Populus EST resource for plant functional genomics. Proc Natl Acad Sci USA 101:13951–13956

Steward CM (1966) Excretion and heartwood formation in living trees. Science 153:1068–1074

Sundberg B, Uggla C (1997) Origin and dynamics of indoleacetic acid under polar transport in *Pinus sylvestris*. Physiol Plant 104:22–29

Sundberg B, Tuominen H, Little C (1994) Effects of the indole-3-acetic acid (IAA) transport inhibitors N-1-naphthylphthalamic acid and morphactin on endogenous IAA dynamics in relation to compression wood formation in 1-year-old *Pinus sylvestris* (L) shoots. Plant Physiol 106:469–476

Sundberg B, Uggla C, Tuominen H (2000) Cambial growth and auxin gradients. In: Savidge RA, Barnett JR, Napier R (eds) Cell and molecular biology of wood formation. BIOS Scientific Publishers, Oxford, pp 169–188

Suzuki S, Li LG, Sun Y, Chiang VL (2006) The cellulose synthase gene superfamily and biochemical functions of xylem-specific cellulose synthase-like genes in *Populus trichocarpa*. Plant Physiol 142:1233–1245

Tanino K (2004) Hormones and endodormancy induction in woody plants. J Crop Improv 10:157–199

Terashima N, Fukushima K, He LF, Takabe K (1993) Comprehensive model of the lignified plant cell wall. In: Jung HG (ed) Forage cell wall structure and digestibility. ASA-CSSA-SSSA, Madison, WI, pp 247–270

Tuominen H, Sitbon F, Jacobsson C, Sandberg G, Olsson O, Sundberg B (1995) Altered growth and wood characteristics in transgenic hybrid aspen expressing *Agrobacterium tumefaciens* T-DNA indoleacetic-acid-biosynthetic genes. Plant Physiol 109:1179–1189

Tuskan GA et al (2006) The genome of black cottonwood, *Populus trichocarpa* (Torr. & Gray). Science 313:1596–1604

Tyerman SD, Niemietz CM et al (2002) Plant aquaporins: multifunctional water and solute channels with expanding roles. Plant Cell Environ 25(2):173–194

Uggla C, Mellerowicz EJ, Sundberg B (1998) Indole-3-acetic acid controls cambial growth in Scots pine by positional signaling. Plant Physiol 117:113–121

Uggla C, Magel E, Moritz T, Sundberg B (2001) Function and dynamics of auxin and carbohydrates during earlywood/latewood transition in Scots pine. Plant Phys 125 (4):2029–2039

Vetter RE, Botosso PC (1989) Remarks on age and growth rate determination of Amazonian trees. IAWA Bull ns 10:133–145

Voelker SL, Lachenbruch B, Meinzer FC, Jourdes M, Ki C, Patten AM, Davin LB, Lewis NG, Tuskan GA, Gunter L, Decker SR, Selig MJ, Sykes R, Himmel ME, Kitin P, Shevchenko O, Strauss SH (2010) Antisense down-regulation of 4CL expression alters lignification, tree growth and saccharification potential of field-grown poplar. Plant Physiol 154:874–886

Wind C, Arend M, Fromm J (2004) Potassium-dependent cambial growth in poplar. Plant Biol 6:30–37

Wodzicki TJ, Wodzicki AB (1980) Seasonal abscisic acid accumulation in stem and cambial region of *Pinus sylvestris* and its contribution to the hypothesis of a late-wood control system in conifers. Physiol Plant 48:443–447

Worbes M (1985) Structural and other adaptations to long-term flooding by trees in Central Amazonia. Amazoniana 9:459–484

Worbes M (1995) How to measure growth dynamics in tropical trees – a review. IAWA J 16:337–351

Wu L, Joshi CP, Chiang VL (2000) A xylem-specific cellulose synthase gene from aspen (*Populus tremuloides*) is responsive to mechanical stress. Plant J 22:495–502

Yamaguchi M, Kubo M, Fukuda H, Demura T (2008) Vascular-related NAC-DOMAIN7 is involved in the differentiation of all types of xylem vessels in Arabidopsis roots and shoots. Plant J 55:652–664

Yamaguchi M, Ohtani M, Mitsuda N, Kubo M, Ohme-Takagi M, Fukuda H, Demura T (2010) VND-INTERACTING2, a NAC domain transcription factor, negatively regulates xylem vessel formation in *Arabidopsis*. Plant Cell 22:1249–1263

Yordanov YS, Regan S, Busov V (2010) Members of the LATERAL ORGAN BOUNDARIES DOMAIN transcription factor family are involved in the regulation of secondary growth in *Populus*. Plant Cell 22:3662–3677

York SW, O'Neill MA (2008) Biochemical control of xylan biosynthesis – which end is up? Curr Opin Plant Biol 11:258–265

Zagórska-Marek B (1995) Morphogenetic waves in cambium and figured wood formation. In: Iqbal M (Hrgs) Encyclopedia of plant anatomy, Band 9, Teil 4, the cambial derivatives. Gebrüder Borntraeger, Berlin, pp 69–69

Zhang J, Elo A, Helariutta Y (2011) Arabidopsis as a model for wood formation. Curr Opin Biotechnol 22:293–299

Zhong R, Ye Z-H (2007) Regulation of cell wall biosynthesis. Curr Opin Plant Biol 10:564–572

Zhong R, Richardson EA, Ye ZH (2007) Two NAC domain transcription factors, SND1 and NST1, function redundantly in regulation of secondary wall synthesis in fibers of *Arabidopsis*. Planta 225:1603–1611

Zhong R, Lee C, Ye Z-H (2010) Global analysis of direct targets of secondary wall NAC master switches in *Arabidopsis*. Mol Plant 3:1087–1103

Zhong R, McCarthy RL, Lee C, Ye ZH (2011) Dissection of the transcriptional program regulating secondary walls biosynthesis during wood formation in Poplar. Plant Physiol 157:1452–1468

Zhu Z, An F, Feng Y, Li P, Xue L, Mu A, Jiang Z, Kim JM, To TK, Li W et al (2011) Derepression of ethylene-stabilized transcription factors (EIN3/EIL1) mediates jasmonate and ethylene signaling synergy in Arabidopsis. Proc Natl Acad Sci USA 108:12539–12544

# Topochemical and Electron Microscopic Analyses on the Lignification of Individual Cell Wall Layers During Wood Formation and Secondary Changes

**Gerald Koch and Uwe Schmitt**

**Abstract** The topochemical distribution of lignin and phenolic extractives in wood cell walls is determined on a cellular level by using scanning UV microspectrophotometry (UMSP) and transmission electron microscopy (TEM). These improved cellular analytical techniques enable direct imaging of the lignin distribution within individual cell wall layers during wood formation and secondary changes. The UMSP technique is based on the ultraviolet illumination of semithin transverse sections which can be related semiquantitatively to the concentration of lignin. Electron microscopy is variously used to obtain high-resolution information on the lignin distribution in wood cell walls which can be visualised by staining with potassium permanganate ($KMnO_4$). By applying these improved techniques, (1) the topochemistry of lignification in developing xylem and wood tissue after wounding, (2) the topochemical detection of phenolic extractives, and (3) the lignin distribution in tropical bamboo species are presented and illustrated in detail. The described methods and presented results demonstrate that cellular UV microspectrophotometry and electron microscopy are ideally suited to study the topochemical distribution of lignin and phenolic extractives on a subcellular level. In particular, the application of the UV-scanning technique enables a direct imaging of lignin distribution (geometrical resolution of 0.25 μm × 0.25 μm) and provides fundamental information on the topochemistry of lignification. The techniques can be used for a wide range of applications in wood biology and wood topochemistry.

G. Koch (✉) • U. Schmitt
Thünen Institute of Wood Research, Leuschnerstr. 91, 21031 Hamburg, Germany
e-mail: gerald.koch@ti.bund.de; uwe.schmitt@ti.bund.de

J. Fromm (ed.), *Cellular Aspects of Wood Formation*, Plant Cell Monographs 20,
DOI 10.1007/978-3-642-36491-4_2, 

# 1 Introduction

## *1.1 Lignin: Significance and Occurrence*

Next to cellulose, lignin is the most abundant and important polymeric organic substance in plants. It is a characteristic chemical and morphological component of the tissue of higher plants such as pteridophytes and spermatophytes (gymnosperms and angiosperms), where it typically occurs in the vascular tissue, specialised for liquid transport and mechanical strength (e.g. Fengel and Wegener 1989). The lignin content in different plants is quite variable (Table 1). While in wood species the lignin content ranges from 20 to 40 %, aquatic and herbaceous angiosperms as well as many monocotyledons are less lignified (Sarkanen and Hergert 1971). Additionally, the distribution of lignin within the cell wall and the lignin content of different parts of a tree is not uniform. For example, high lignin amounts are characteristic for softwood branches and compression wood (Timell 1986), whereas the so-called gelatinous layers of tension wood fibres in hardwoods may be almost devoid of lignin (Timell 1969; Novaes et al. 2010).

## *1.2 Chemical Structure of Lignin*

The topochemical and electron microscopic study of cellular lignin distribution generally requires basic information on the chemical structure and composition of lignin.

Lignin is a complex phenolic polymer formed by radical coupling reactions of mainly three hydroxycinnamyl alcohols or monolignols (Fig. 1): *p*-coumaryl (4-hydroxycinnamyl), coniferyl (3-methoxy-4-hydroxycinnamyl) and sinapyl (3,5-dimethoxy-4-hydroxycinnamyl) alcohol, which are synthesised via the phenylpropanoid pathway (e.g. Nimz 1981; Boerjan et al. 2003). The complex structure of lignin arises from its biosynthesis in which the last step is a nonenzymatic, random recombination of phenoxy radicals of coniferyl, sinapyl and *p*-coumaryl alcohols. The synthesis of the monolignols (precursors) and the formation of lignin macromolecules comprise complicated biochemical and chemical reactions which have been extensively studied and repeatedly reviewed (e.g. Sarkanen and Hergert 1971; Glasser 1980; Terashima et al. 1993; Boerjan et al. 2003). According to Terashima (2000), the polymerisation of the monolignols is considered to proceed primarily via the following steps: (1) the formation of monolignol radicals by hydrogen peroxide and peroxidase or laccase and oxygen (Dean and Eriksson 1994), (2) the production of dilignols and dilignol quinone methides by coupling of the radicals, (3) the addition of water, lignol or carbohydrates to the quinone methides (Sarkanen and Hergert 1971; Higuchi 1997), (4) formation of phenyl radicals on oligo- and polylignols, and coupling with monolignol radicals to synthesise the polylignol. The mode of polymerisation

**Table 1** Chemical composition of different plants (adopted from Klemm et al. 2002)

| | Composition (%) | | | |
|---|---|---|---|---|
| Source | Cellulose | Hemicelluloses | Lignin | Extract |
| Hardwood | 43–47 | 25–35 | 16–24 | 2–8 |
| Softwood | 40–44 | 25–29 | 25–31 | 1–5 |
| Bagasse | 40 | 30 | 20 | 10 |
| Coir | 32–43 | 10–20 | 43–49 | 4 |
| Cotton | 95 | 2 | 1 | 0.4 |
| Hemp | 70 | 22 | 6 | 2 |
| Jute | 71 | 14 | 13 | 2 |
| Sisal | 73 | 14 | 11 | 2 |
| Wheat straw | 30 | 50 | 15 | 5 |

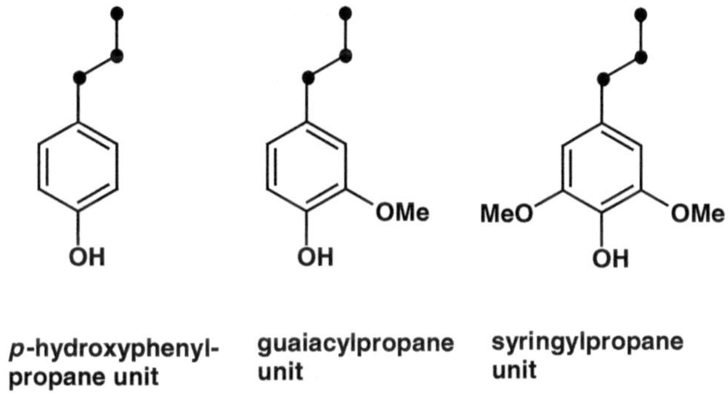


**Fig. 1** Molecular structure of the basic phenylpropane units of lignin

can be affected by relative amounts of radical types participating in the reaction. The first products of the coupling of monolignol radicals are β-O-4, β-5, and β–β dimers (Fig. 2), whereas in the next stages of polymerisation, bulk type oligomers and endwise polymers are formed. As a result, a globular macromolecule is formed which is composed of bulk polymers inside and endwise polymers in the outer part (Terashima et al. 1998).

Variations in the chemical reactivity of lignin are based on the proportions of the three monolignol structural units (guaiacyl (G), syringyl (S), and *p*-hydroxyphenyl (H) units) in different wood species as well as in different tissues and even cell wall layers. Whereas softwood lignin consists mainly of guaiacylpropane (4-hydroxy-3-methoxyphenylpropane) units (G), hardwood lignins also contain up to 50 % syringyl (3,5-dimethoxy-4-hydroxyphenyl) groups (S). Guaiacyl lignins (G lignins) are predominantly polymerisates of coniferyl alcohol, while guaiacyl–syringyl–lignins (GS lignins) are composed of varying parts of the aromatic nuclei guaiacyl and syringyl in addition to small amounts of *p*-hydroxyphenyl units.

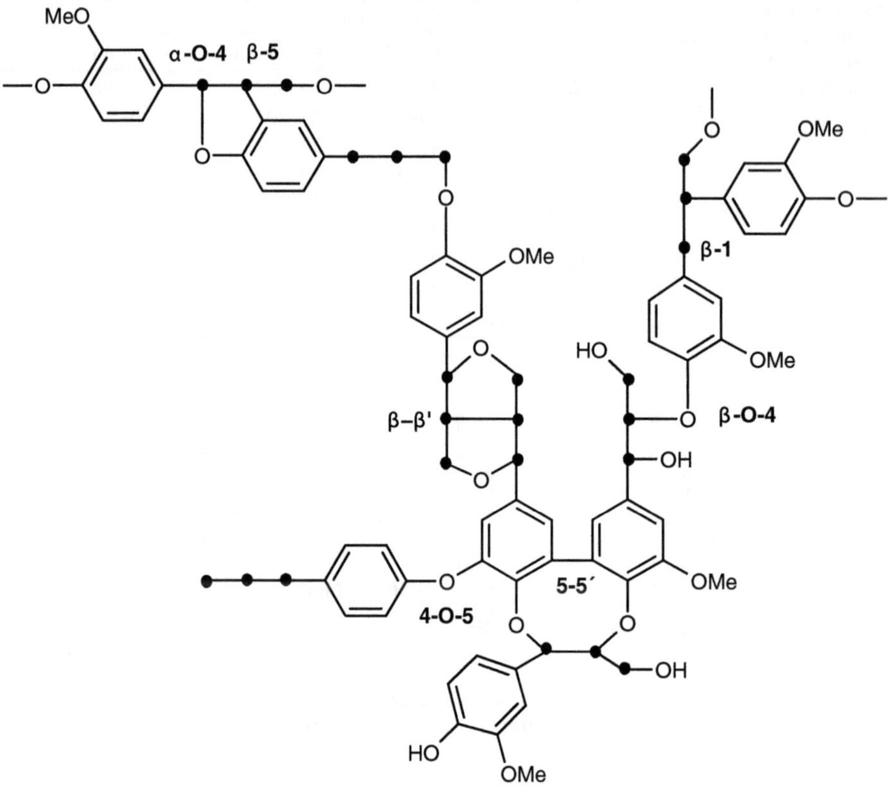


**Fig. 2** Model of important lignin linkages (modified model adopted from Kindl 1991)

# 2 Deposition of Lignin Within the Polysaccharide Cell Wall Framework

## *2.1 Lignification of the Wood Cell Wall*

Wood cell walls are strictly concentrically layered which is very well visible with the electron microscope especially in softwood tracheids. According to numerous wall models (review, see Brändström 2001), the individual wall layers are named in a common way (Fig. 3). The outermost wall portion connecting two neighbouring cells is called middle lamella (ML), followed by the primary wall (P) on both sides. ML and P form the so-called compound middle lamella (CML) because of their uniform appearance in the electron microscope. Well discernable from the CML is the secondary wall which is divided in the narrow and outermost S1 layer, the broad S2 layer, and the tiny, sometimes hardly visible innermost S3 layer. Some species (soft- and hardwoods, such as pine, fir and beech) additionally deposit a warty layer as the innermost wall portion. The middle lamella becomes distinctly broadened

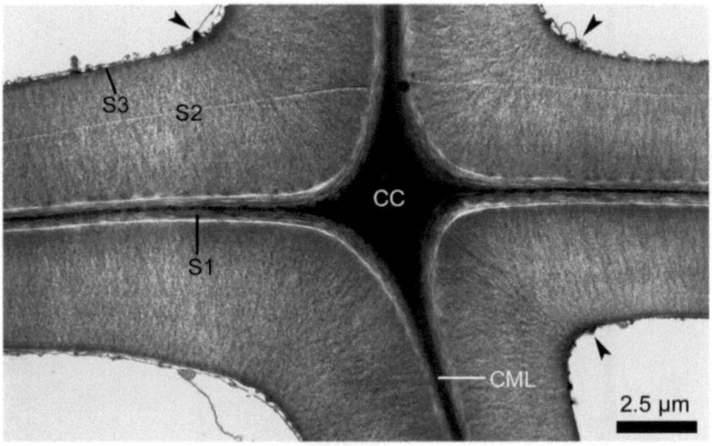


**Fig. 3** TEM micrograph of fir (*Abies alba*) xylem after potassium permanganate staining. Cell corner region (CC) of four neighbouring tracheids. Typical cell wall layering with compound middle lamella (CML), secondary wall layers S1, S2 and S3 as well as a warty layer (*arrowheads*) deposited onto the S3

where more than two cells are in direct contact to each other; the name for this region is cell corner (CC). During wood formation, the incorporation of lignin within the polysaccharide cell wall framework is generally regarded as the final stage of the differentiating process of the typical secondary xylem cell wall. Results from ultraviolet, fluorescence and light autoradiographic microscopic studies as well as from electron microscopic studies have confirmed that lignin is most probably deposited initially in the cell corners when the surface enlargement of the cell is completed, and just before the S1 starts thickening. Lignification proceeds in the ML and the primary wall, starting at the tangential walls and spreading centripetally (Takabe et al. 1981). Lignification of the CML continues during the differentiation of the S1 and S2 layers, and even until the formation of the S3. Lignification of the secondary wall layers proceeds slowly in a first stage but becomes more rapid after formation of the S3 wall has been completed (Takabe et al. 1981). These findings indicate an ongoing lignification process throughout the entire period of cell wall differentiation, with a considerable delay as regards the synthesis of cellulose and hemicelluloses (Fig. 4).

On the ultrastructural level, Ruel et al. (1999) and Ruel (2004) added details using the immunogold-labelling technique to differentiate between condensed and non-condensed lignin subunits during deposition. The authors clearly demonstrated the micro-heterogeneity of lignification within developing wood cell walls. For fibres of hardwoods, they showed a preferable deposition of non-condensed GS lignins in the S1 and outer S2 layer and the incorporation of condensed GS subunits with a rather weak labelling in the S1 and within the entire S2 layer. According to Joseleau et al. (2004) as well as Lehringer et al. (2007, 2009) and in contrast to textbook knowledge, which is describing the G-layer consisting of only

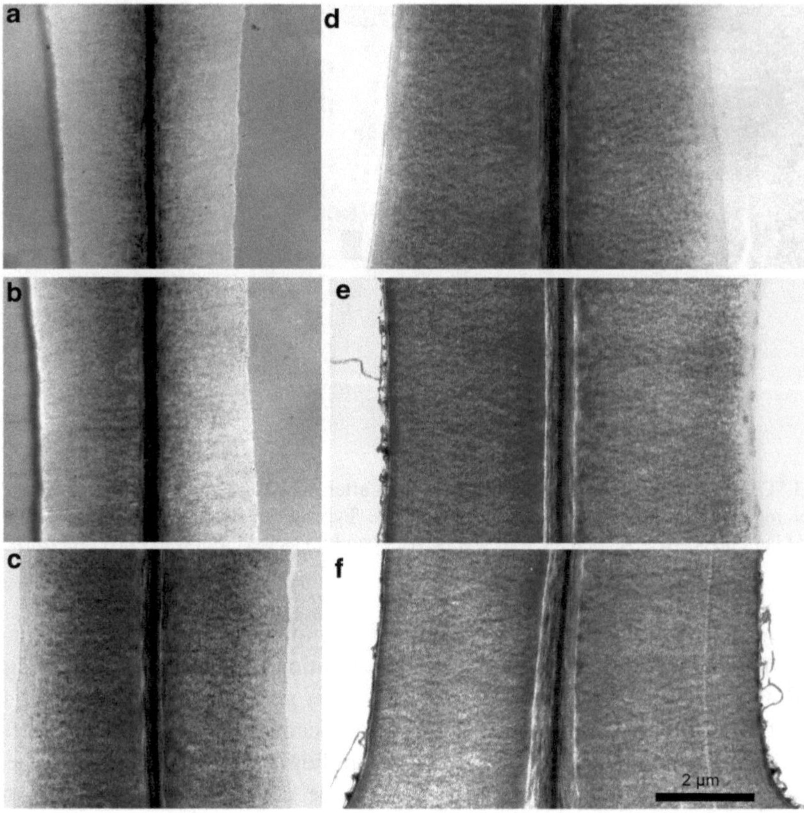


**Fig. 4** TEM micrographs of developing tracheid walls in fir (*Abies alba*), potassium permanganate staining. (**a**) Secondary wall still developing, lignification of middle lamella portions visible, outer secondary wall with early stage of lignification, inner secondary wall unlignified. (**b**) About one-third of the secondary wall lignified. (**c**) Advanced stage of the secondary wall lignification. (**d**) Lignification of the secondary wall nearly completed. (**e**) Lignification of the left wall appears completed, whereas the neighbouring right wall is still lignifying. (**f**) Completed lignification

polysaccharides, a low amount of lignin or lignin-like substances may be occasionally incorporated also in the gelatinous layer of tension wood fibres in hardwoods.

## *2.2 Deposition of Phenolic Extractives*

In addition to its major structural components, i.e. cellulose, hemicelluloses and lignin, wood contains also an exceedingly large number of other low- and high-molecular organic compounds, the so-called accessory compounds or extractives. A major group of extractives (phenolic compounds) in hardwoods are the tannins and flavonoids ranging from simple phenols to condensed flavonoid systems which can

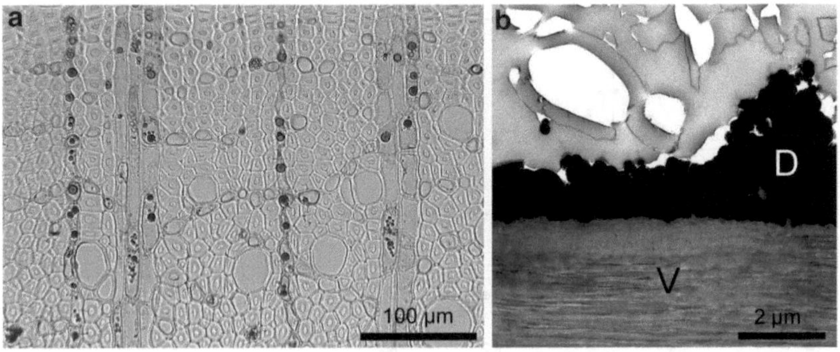


**Fig. 5** (**a**) Light micrograph of beech wood tissue (*Fagus sylvatica*) showing the deposition of high condensed phenolic extractives in the parenchyma cells (axial parenchyma and rays). (**b**) Deposition of crystalline extractives along the vessel wall of *Intsia bijuga* (*V* vessel wall, *D* deposits), TEM micrograph

also be topochemically analysed by UV-spectroscopy. The tannins are subdivided into hydrolysable tannins and non-hydrolysable or condensed tannins (phlobaphenes). The main components of the condensed tannins are the catechins (flavan-3-ols) and the leucoanthocyanidins (flavan-3,4-diols). They are often present in wood as colourless leuco-compounds and the colour will be secondarily developed by biochemical and chemical reactions (Koch 2003b). On the cellular level, the extractives are concentrated in the parenchyma cells (Fig. 5) and the resin canals; lower amounts are also found in the middle lamellae, intercellular spaces and cell walls of tracheids and libriform fibres (e.g. Grosser et al. 1974; Koch et al. 2003b) as well as occasionally also in vessel walls (Kleist and Schmitt 1999).

# 3 Topochemical Detection of Lignin by Using UMSP Technique and Electron Microscopy

## 3.1 Application of UV Microspectrophotometry

Scanning UV microspectrophotometry (UMSP) has been established as a useful technique for the topochemical detection of lignin *in situ* and for its semiquantitative determination in the various layers of wood cell walls (Lange 1954; Fergus et al. 1969; Scott et al. 1969; Saka et al. 1982; Fukazawa 1992; Takabe 2002; Koch and Grünwald 2004). The technique is based on the ultraviolet illumination of semithin transverse sections of woody tissue. Lignin displays a characteristic ultraviolet absorbance spectrum with maxima around 212 nm and 280 nm due to the presence of associated phenylpropane groups with several chromophoric structural elements (Jaffé and Orchin 1962; Sarkanen and Hergert 1971; Hesse et al. 1991). No other component of the mature wood cell wall displays ultraviolet

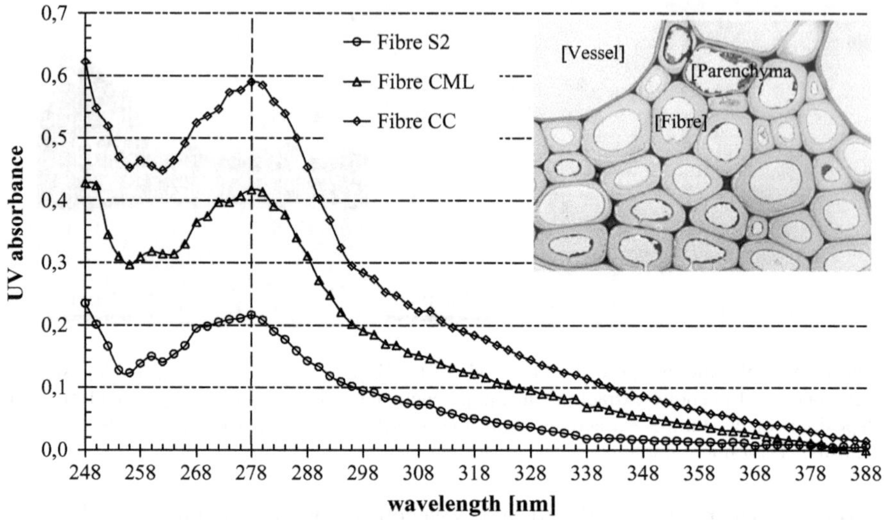


**Fig. 6** Representative UV absorbance spectra of individual cell wall layers in the woody tissue of *Fagus sylvatica* (fibre S2-secondary wall, CML-compound middle lamella, CC-cell corner)

absorbance properties in the same spectral region, and the intensity of absorbance may therefore be related to the concentration of lignin across the cell wall. Furthermore, the UV absorbance maximum is sensitive to structural differences of the lignin allowing discrimination between hard- and softwood lignins due to different ratios of their guaiacyl- and syringylpropane units (Fergus and Goring 1970a, b; Takabe et al. 1992). Softwood lignin is mainly composed of guaiacylpropane units with an absorbance maximum at 280 nm, and the hardwood lignin consists of guaiacyl- and syringylpropane units in varying ratios characterised by a shifting maximum between 270 and 278 nm (Fig. 6) with lower wavelengths representing more syringyl units.

## 3.2 *Application of Electron Microscopy*

Electron microscopy was variously used to obtain high-resolution information on the lignin distribution in wood cell walls. Since the early 1980s, advanced electron microscopic techniques were developed to improve our knowledge on the lignin topochemistry. The bromination technique revealed lignin distribution by combination of electron dispersive X-ray analysis (EDXA) with transmission (TEM) or scanning electron microscopy (SEM) (e.g. Saka and Thomas 1982;Westermark 1985; Donaldson and Ryan 1987). Either sectioned material or isolated wall fractions were analysed. This technique confirmed on a high-resolution level that middle lamella portions contain distinctly more lignin than secondary walls, whereby ratios between middle lamella and secondary wall lignin were determined.

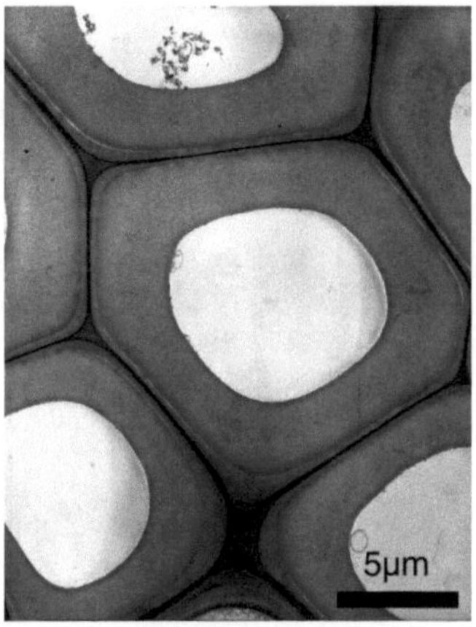


**Fig. 7** TEM micrograph of oak (*Quercus* sp.) xylem showing several fibres with distinct wall layering after potassium permanganate staining (from Čufar et al. 2008)

Similar results were obtained with the mercurisation technique and subsequent SEM/TEM–EDXA analyses which is also based on chemical reactions between lignin and inorganic compounds (Eriksson et al. 1988; Westermark et al. 1988). For example, lignin contents for middle lamella portions of spruce wood were measured obtaining percentage values of 50–60 %.

According to the methods mainly developed by Hepler and Newcomb (1963), Parham (1974), Paszner and Behera (1989), as well as Michalowicz and Robert (1990), potassium permanganate became a very useful general staining agent for cell walls of soft- and hardwoods (Figs. 3, 4 and 7). Later on potassium permanganate was regularly applied especially for lignin localisation in walls of mostly softwood species (e.g. Maurer and Fengel 1991; Donaldson 1992; Singh and Donaldson 1999). However, potassium permanganate does not differentiate between guaiacyl- and syringylpropane units. Since a few years also field-emission scanning electron microscopy (FESEM) was sometimes used to reveal fine structural aspects of the cell wall architecture (Daniel et al. 2004; Lehringer et al. 2009). Fromm et al. (2003) succeeded in obtaining high-resolution micrographs of mercurised softwood tracheids; lignin complexes could be shown within the polysaccharide network using the backscattering mode of a FESEM.

## *3.3 Application of Immunogold Labelling*

Another technique for ultrastructural lignin localisation in wood cell walls is the immunogold labelling with lignin-specific markers. Ruel et al. (1994) raised polyclonal antibodies against synthetic dehydrogenative polymers (DHPs) in rabbits for immunological lignin localisation. Visualisation of antibody binding sites by electron microscopy was achieved with a secondary marker coupled onto small gold particles (usually a protein A-gold complex) binding to the first marker. Those treatments resulted in good labellings of lignin in various cell wall types. Beside a

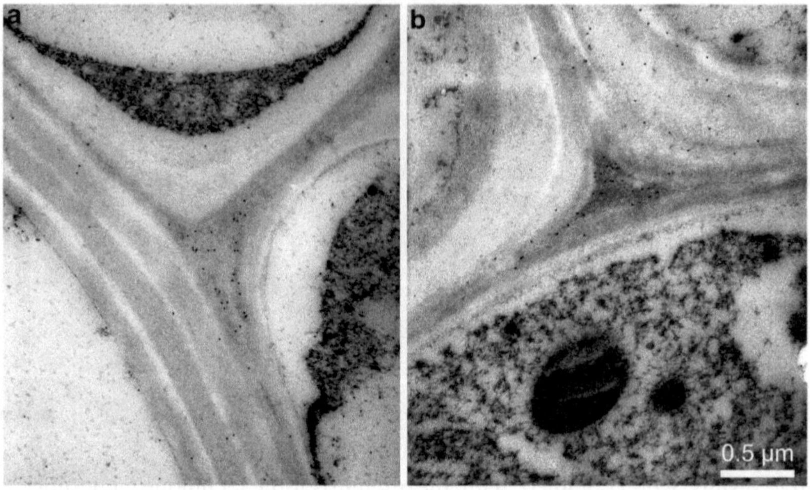


**Fig. 8** TEM micrographs of cell walls in developing xylem of poplar after immunogold labelling with antibodies against peroxidases. Early stages of fibre wall development (**a**) show a strong labelling with gold particles in cell corner regions, whereas later stages are characterised by a weak and more uniform labelling across the entire secondary wall (**b**)

direct lignin labelling, enzymes, such as peroxidases, laccases, or cinnamicacid-4-hydroxylases as key enzymes involved in the lignification of plant cell walls, were localised in developing walls with specific antibodies (Smith et al. 1994; Kim et al. 2002). Kim et al. (2002) localised peroxidases in developing xylem cells of *Populus*. Strong labelling of cell corner regions during early developing stages is probably confirming the onset of lignification in these wall portions (Fig. 8a). On the other hand, advanced stages of cell wall formation are characterised by a weak but more uniform labelling across the entire developing secondary wall (Fig. 8b) which is interpreted as evidence for an incorporation of peroxidases into a young secondary wall simultaneously with the deposition of the polysaccharide matrix.

Recently raised antibodies enabled a discrimination between non-condensed (O-4 aryl–alkyl bonds) and condensed (mainly C–C bonds) lignin interunit linkages (Joseleau and Ruel 1997). Kukkola et al. (2003, 2008) raised a polyclonal antibody against a specific condensed lignin substructure, i.e. the 8-ring dibenzodioxocin, for analysing details on the lignin composition in cell walls of softwood xylem. In addition to electron microscopy, Kukkola et al. (2004) applied confocal laser-scanning fluorescence microscopy and visualised dibenzodioxocin with antibodies coupled to a fluorescent dye. They found out that in mature xylem cells of birch and spruce, dibenzodioxocin was preferably located in the S3 layer of fibres and tracheids as well as in the entire secondary wall of birch vessels. Developing cells did not show any labelling. These results support the assumption that more condensed lignin subunits contribute to a higher stability particularly in vessel walls.

# 4 Experimental Approach of UV Microspectrophotometry and Electron Microscopy

## 4.1 Sampling and Preparation

The preparation and sectioning followed procedures normally employed for electron microscopy. First, small blocks (measuring about $1 \times 1 \times 5\ mm^3$) of selected species are dehydrated in a graded series of acetone, impregnated with Spurr's epoxy resin (Spurr 1969) through a series of acetone/resin mixtures, followed by immersion in pure resin and polymerisation at 70 °C for 24 h. For the investigation of low molecular phenolic extractives, a direct impregnation with pure Spurr's resin under vacuum was additionally developed. For this special impregnation, the specimens are freeze-dried and immediately embedded with Spurr's epoxy resin under mild vacuum with several cycles of evacuation and ventilation as described by Kleist and Schmitt (1999). The embedded blocks are trimmed to provide a face of approximately 0.5 $mm^2$. Semithin sections (1 μm) are cut with an ultramicrotome using a diamond knife. The sections are then transferred to quartz microscope slides, immersed in a drop of non-UV absorbing glycerine and covered with a quartz cover slip. The UV microscopic investigations are carried out using the immersion ultrafluar objectives 32:1 and 100:1. All lenses are completely achromatic in the range of 200...700 nm. The immersion oil used consisted of a glycerine/water mixture $n_D = 1.46$.

## 4.2 Measuring Technique of Scanning UV Microspectrophotometry

The topochemical analyses are carried out using a UV microspectrophotometer (UMSP 80, Zeiss) equipped with a scanning stage enabling the determination of image profiles at defined wavelengths using the scan program APAMOS® (Automatic-Photometric-Analysis of Microscopic Objects by Scanning (Zeiss)). For the detection of the lignin distribution of softwoods and hardwoods, wavelengths of 280 nm and 278 nm, respectively, are selected. The scan programme digitises rectangular fields with a local geometrical resolution of 0.25 μm × 0.25 μm and a photometrical resolution of 4,096 grey scale levels which are converted in 14 basic colours to visualise the absorbance intensities. The scans can be depicted as two- or three-dimensional image profiles including a statistical evaluation (histogram) of the semiquantitative lignin distribution (Fig. 9).

The sections are also conventionally subjected to point measurements with a spot size of 1 $\mu m^2$ with varying wavelengths between 240 and 400 nm (comp. Fig. 6) using the program LAMBDA SCAN® (Zeiss). This program evaluates the UV absorbance spectra of the lignified cell walls and accessory compounds in tissues.

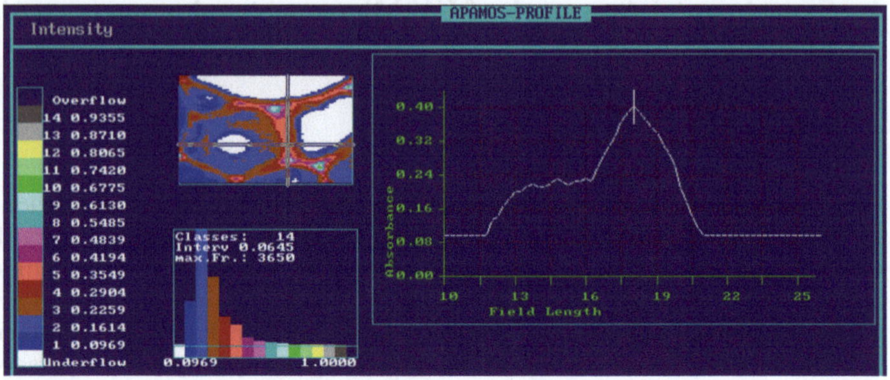


**Fig. 9** UV microscopic scanning profile of lignin distribution in beech wood tissue (cross section of individual cell wall layers). The colour pixels display the UV absorbance at 278 nm wavelength with a resolution of 0.25 μm × 0.25 μm (Röder et al. 2004)

The lignin concentrations are estimated according to the Lambert–Beer's law:

$$\text{UV absorbance} = \varepsilon \cdot \mathrm{C} \cdot \mathrm{d},$$

where $\varepsilon$ is the extinction coefficient, $C$ the volume concentration and d the thickness of the absorbing layer. Considering the cell wall in a cross section of 1 μm thickness, the incident UV light intensity $I_0$ is reduced to the intensity $I_{\text{cell wall}}$ emerging from the cell wall due to the absorbance by the constituent lignin. Measurement of $I_0$ is facilitated by the unchanged passage of the incident radiation through the embedding medium in the cell lumen. $I_0$ may therefore be replaced by $I_{\text{lumen}}$, the intensity of UV light emerging from the lumen:

$$\text{UV absorbance} = \log \mathrm{I}_{\text{lumen}}/\mathrm{I}_{\text{cell wall}}.$$

## 4.3 *Principle of Preparation for Transmission Electron Microscopy*

For transmission electron microscopy (TEM) of woody material, in most cases, a low viscosity embedding resin is used. Spurr (1969) introduced such an epoxy resin with nonenyl succinic anhydride as the main component. Due to its excellent penetration also into woody tissue, it is still in use in many laboratories, although some components may nowadays be replaced by less harmful chemicals. Prior to embedding, the standard preparation of living plant tissue for TEM is an initial fixation with aldehydes followed by a second fixation with osmium tetroxide, which simultaneously acts as a staining agent for mainly membranes and cell walls.

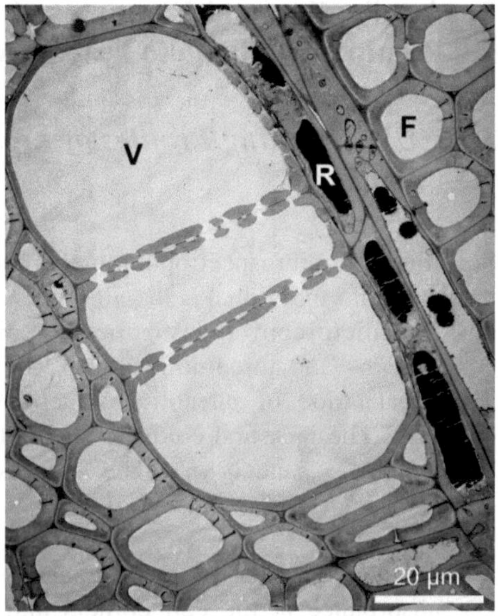


**Fig. 10** TEM micrograph of mature xylem of birch (*Betula pendula* Roth) embedded in Spurr's epoxy resin and en bloc staining with osmium tetroxide. The entire tissue is well prepared and with an excellent contrast after uranyl acetate and lead citrate staining (F fibre, R ray, V vessel)

$$\text{(I) R-CH=CH-CH}_2\text{OH} \xrightarrow{KMnO_4} \text{R-CH=CH-CHO} \xrightarrow{KMnO_4} \text{R-CH=CH-COOH}$$

$$\text{(II) } MnO_4^- + 4H^+ + 3e^- \rightleftharpoons MnO_2 + 2H_2O \text{ (weak alkaline solutions)}$$

**Fig. 11** Diagram of the major pathway for $KMnO_4$ stainings. $KMnO_4$ probably oxidises the functional groups of the side chains of the lignin from the alcohol to the carbonic acid (I). $MnO_2$ is finally deposited on the sections (II) (modified from Schmitt and Melcher 2004)

Subsequent poststaining with uranyl acetate and lead citrate leads to an excellent visualisation of cytoplasmic constituents and cell walls (Fig. 10).

However, analyses of wood cell walls are often performed with material which is directly embedded in Spurr's epoxy resin without the application of fixatives (e.g. Schmitt et al. 2006, Čufar et al. 2008). There is often no need to preserve living cells. Sections produced from such material are routinely stained with potassium permanganate ($KMnO_4$) which preferably reacts with lignin molecules and therefore indicates the lignin distribution in cross cell walls (see also Sect. 3.2). Schmitt and Melcher (2004) provided the chemical background of the staining process, which is based on the permanganate ion ($MnO_4^-$) acting as a strong oxidant in acidic and alkaline solutions and in high dilutions. $KMnO_4$ probably oxidises the functional groups of the alcohol of the lignin molecules to the aldehyde and carbonic acid (Fig. 11). Finally, the resulting water-insoluble reaction product manganese dioxide ($MnO_2$) is deposited on the ultrathin section visualising the reaction sites.

# 5 Application and Examples of Topochemical Detection of Lignin and Phenolic Extractives

## *5.1 UV-Scanning Profiles of Lignin Distribution in Wood Cell Walls*

Scanning UV microspectrophotometry enables direct imaging of the lignin distribution within individual cell wall layers and incorporates advances on the previous UV microscopy investigations of lignin topochemistry. Figure 12 shows typical two- and three-dimensional UV image profiles of lignin distribution. The localisation of phenolic extractives in parenchyma cells is also shown (Fig. 13). The grey and colour scales indicate different intensities of UV absorbance at $\lambda_{280\ nm}$ and $\lambda_{278\ nm.}$ The high resolution (0.25 µm × 0.25 µm per pixel) enables a high differentiation of the UV absorbance within individual cell wall layers.

The image profiles of spruce tracheids (Fig. 12a) are characterised by a high UV absorbance at the cell corners and compound middle lamellae ($abs_{280nm}$ 0.61–0.87) as compared to the adjacent S2 layers with a lower, slightly varying lignin distribution ($abs_{280nm}$ 0.35–0.54). As found by Lange (1954) and Fergus et al. (1969), the average lignin content in the compound middle lamella is about twice of that in the S2 of the tracheids. For detailed illustration, the scanning area is presented as a three-dimensional image as well as line image profile. In the three-dimensional profile, the compound middle lamella region stands out as a highly absorbing band which broadens towards a heavily lignified area at the cell corners. An example of a line image profile depicted by the marked cross line is shown in detail. The compound middle lamella of the scanned spruce tracheid is recorded as a pronounced peak while the S2 layers are characterised by a lower level on either side of the compound middle lamella. These graphical presentations allow a refined evaluation of the topochemical distribution of lignin within the cell walls. An unambiguous statistical presentation of the data is an additional asset.

The scanned beech fibre (Fig. 12b) shows a different absorbance level as compared to the softwood tracheids. In particular, the broad S2 layer displays a lower absorbance with values of $abs_{278nm}$ 0.16–0.29. The uniform level of absorbance in this wall layer corresponds to earlier results reported by Saka and Goring (1988), who predicted that lignin distribution across the width of the whole $S_2$ should be homogeneous. The compound middle lamella is distinguished by higher absorbance values as compared to those of the spruce tracheid.

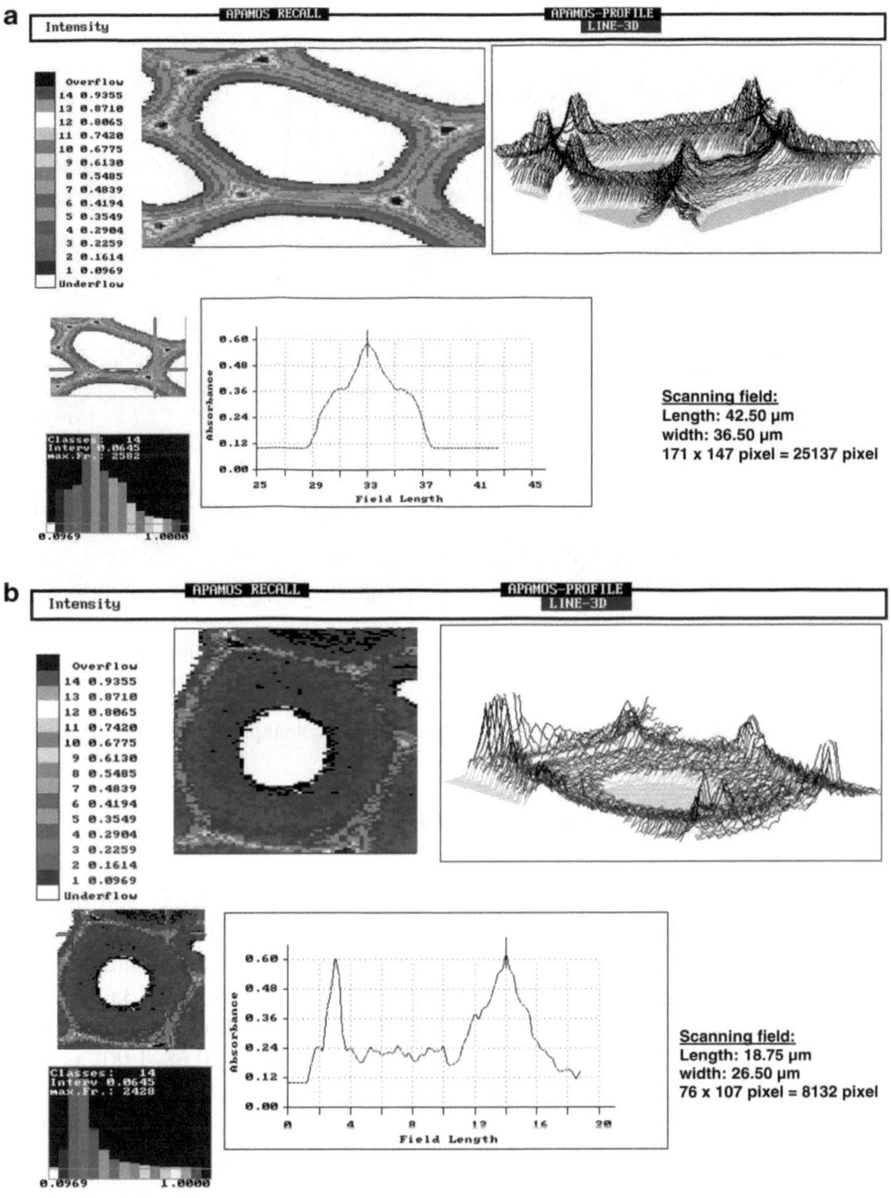


**Fig. 12** UV microscopic image profiles of an individual (**a**) tracheid of *Picea abies* and (**b**) fibre of *Fagus sylvatica* scanned with a geometrical resolution of 0.25 µm × 0.25 µm. The scales indicate the different UV absorbance values at a wavelength of 280 nm (absorbance maximum of softwood lignin) and 278 nm (absorbance maximum of hardwood lignin) (Koch and Grünwald 2004)

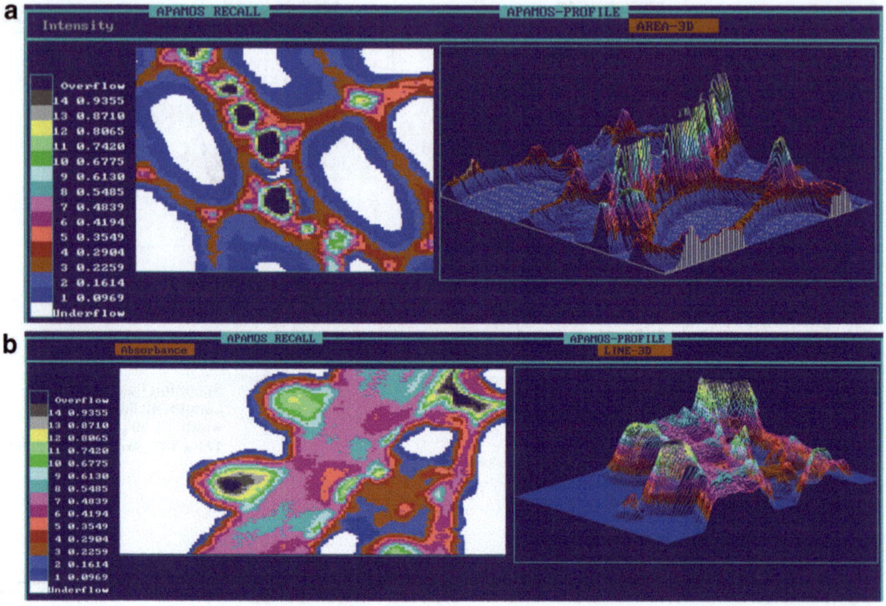


**Fig. 13** (**a**) UV microscopic image profiles of *Fagus sylvatica* tissue measured at $\lambda_{278nm}$ showing the deposition of phenolic extractives (arrows) in lumina of ray parenchyma cells (Koch et al. 2003), (**b**) detailed UV microscopic scanning profiles of deposited crystalline extractives on the vessel wall and within the pit membranes of *Intsia bijuga*. The colour pixels represent different UV absorbance values measured at $\lambda_{278nm}$ (Koch et al. 2006)

## 5.2 Topochemical Detection of Phenolic Extractives

Scanning UV microspectrophotometry can also be used to detect and quantify aromatic compounds associated with the woody tissue (Koch et al. 2003b). The presence of extractives can easily be visualised as spherical conglomerations of high absorbance as compared with the surrounding tissue. In Fig. 13a, the local deposition of extractives in the lumina of ray parenchyma cells of beech heartwood is emphasised by a significantly higher absorbance ($abs_{280nm}$ 0.68–1.00) as compared to the cell wall-associated lignins. The phenolic compounds are generally synthesised by parenchyma cells *in situ* and are highly condensed, making it impossible for them to penetrate into the interfibrillar spaces of the cell walls (Hillis 1987). The adjacent fibres do not seem to be impregnated, as evidenced by lower absorbance levels in these cells. The scanned fibres and parenchyma cells show the typical absorbance profile originating from the lignification of the different cell wall layers. In contrast to wood species with an obligatory heartwood formation, the cell walls of beech fibres are not impregnated, and the deposited phenolic extractives in the cell lumina do not contribute to an enhanced decay resistance (compare Koch and Kleist 2001; Kleist and Bauch 2001).

In addition, the topochemical distribution of phenolic deposits in the vessels of afzelia (*Afzelia* spp.) and merbau (*Intsia* spp.) heartwood was investigated by means of cellular UV microspectrophotometry (UMSP) in order to characterise the chemical composition and synthesis by pit membrane-associated enzymes (Koch et al. 2006). Figure 13b shows a detailed UV image profile of phenolic deposits, the vessel cell wall, and an associated parenchyma cell in merbau (*Intsia* spp.). The phenolic deposits as spherical conglomerations of high absorbance are silhouetted clearly against the surrounding vessel wall. Also here, the deposits are characterised by very high absorbances (0.7–1.0 overflow). Furthermore, on the outer part of the vessel S2, directly associated with the CML adjacent to an axial parenchyma cell, local high absorbance spots are detectable. These areas are localised in the pit canals and pit membranes. The UV intensities at 278 nm (0.7–0.9) of the pits are strong evidences for an impregnation of the membranes and pit canals with phenolic extractives.

## *5.3 UV-Absorbance Spectra of Lignin and Phenolic Extractives*

The lignification of individual cell wall layers and the deposition of phenolic extractives can also be studied by the evaluation of UV absorbance spectra in a wavelength range of 240–400 nm. In Fig. 14, typical UV absorbance spectra of individual cell wall layers in beech are presented. The UV spectra of the compound middle lamellae and S2 show the typical absorbance behaviour of a hardwood lignin with a distinct maximum at 278 nm and a local minimum at about 250 nm (Fergus and Goring 1970a; Takabe et al. 1992). The cell wall layers of vessels are generally characterised by higher absorbance values than that of fibres. This behaviour is based on the different chemical composition of lignin in both cell types. Fergus and Goring (1970a, b) and Terashima et al. (1986) proved that the lignin located in vessel walls consists predominantly of the strongly absorbing guaiacyl type units, while the fibre wall lignin contains more syringyl units showing a lower UV absorbance with increasing $OCH_3/C_9$ ratio (Musha and Goring 1975). In comparison, the UV spectra of spruce tracheids show generally higher absorbance values as compared to the hardwood fibres and vessels. A pronounced absorbance maximum at 280 nm usually indicates the presence of the strongly absorbing guaiacyl lignin (Musha and Goring 1975; Fujii et al. 1987).

The detected phenolic extractives in the ray parenchyma cells of beech heartwood have much higher absorbance values ($abs_{280nm}$ 0.75 and 0.95) than cell wall-associated lignins ($abs_{280nm}$ 0.20 and 0.38). Furthermore, their absorbance maxima display a bathochromic shift to a wavelength of 284 nm and a slight shoulder at a wavelength range of 320 nm. This spectral behaviour can be explained by the presence of chromophoric groups, e.g. conjugated double bonds. The higher degree of conjugation stabilises $\pi$–$\pi^*$ transitions resulting in absorbance bands shifted to higher wavelengths (Goldschmid 1971) which can be detected by UV

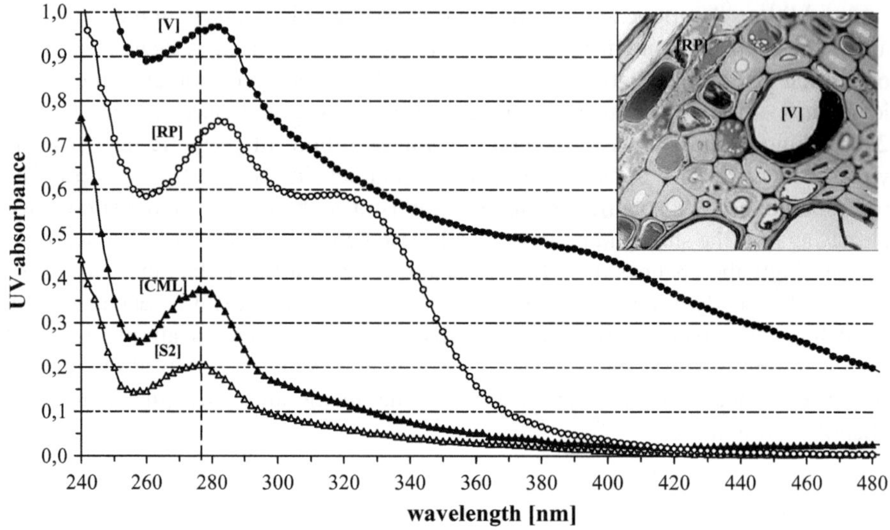


**Fig. 14** Representative UV absorbance spectra deposited phenolic compounds in the woody tissue of beechwood with red head (*open triangle*, low molecular phenolic compound in the secondary wall [S2], *open circle*, high condensed phenolic compound in the lumen of a ray parenchyma cell [RP], *filled circle*, high condensed phenolic compound in the lumen of a vessel [V], *filled triangle*, compound middle lamella of a fibre [CML]). UV photograph at 280 nm of discoloured beechwood tissue (Koch et al. 2003b)

microspectrophotometry. However, the technique does not allow the chemical identification of the condensed phenolic extractives (Koch et al. 2003a).

In the tropical wood species afzelia and merbau, the spectra of the phenolic deposits are characterised by a distinct absorbance maximum at 368 nm (Fig. 15). According to Hillis and Yazaki (1973), who have analysed the methanol–water extracts of the deposits in the cell lumen of merbau by high-performance liquid chromatography (HPLC), the deposits are composed of pure robinetin ($C_{15}H_{10}O_7$) and kaempferol ($C_{15}H_{10}O_2$). The UV spectrum of these both compounds has a distinct minimum at 280 nm. In this range, however, the lignified cell wall layers of vessels (Fig. 15 [3]) and fibres (Fig. 15 [4]) reveal a pronounced maximum due to the UV characteristics of the guaiacyl and syringyl units in lignin. Accordingly, lignin spectra can be basically distinguished from those of robinetin. The pit membranes and pit canals of associated vessel and parenchyma cells are impregnated by these compounds. These results verify the assumption that the synthesis of deposits in afzelia and merbau is regulated by means of pit membrane-associated enzymes.

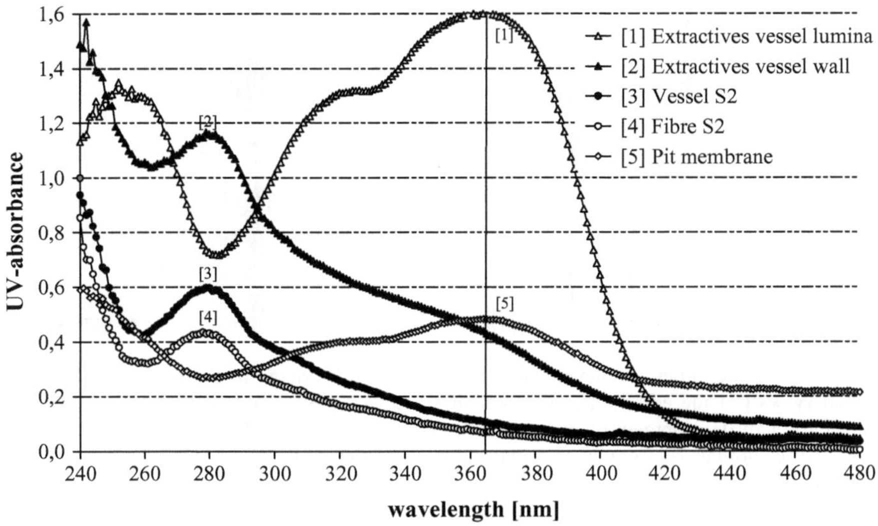


**Fig. 15** Representative UV absorbance spectra of individual cell wall layers (S2 of a vessel and a fibre), phenolic extractives deposited in vessel lumen, and extractives impregnating the vessel wall and pit membranes of *Intsia bijuga* (Koch et al. 2006)

## 5.4 *Topochemical Investigations on Cell Walls in Developing Xylem of Beech*

Cell wall thickening and lignification of beech xylem (*Fagus sylvatica* L.) were determined by means of light microscopy, cellular UV microspectrophotometry and transmission electron microscopy. It could be clearly demonstrated that the differentiation of first formed vessels in developing xylem was completed within 1 month after the onset of cambial divisions, whereas the differentiation of first formed fibres took about 2 months (Prislan et al. 2009). This finding is in good agreement with observations of Yoshinaga et al. (1997) on secondary wall thickening and lignification in oak xylem. The same authors also reported that cell wall thickening in oak fibres progressed in two phases, i.e. a faster beginning and a slow second phase. The observations for beech similarly showed that the rate of cell wall thickening was higher during the first weeks and lower during the last weeks. In general and independent of the position within a growth ring, walls of vessels and fibres closer to rays differentiated faster which was also found for conifers (Donaldson 2001).

Electron microscopy was used to follow cell wall formation after cell enlargement by the deposition of polysaccharidic material with subsequent lignification (Fig. 16). Lignin incorporation started in cell corner and radial intercorner middle lamella regions and proceeded centripetally towards the lumen. As the S1 started to lignify during S2 formation, the lignin content still increased in cell corner and intercorner middle lamella regions. The determination of lignin topochemistry by

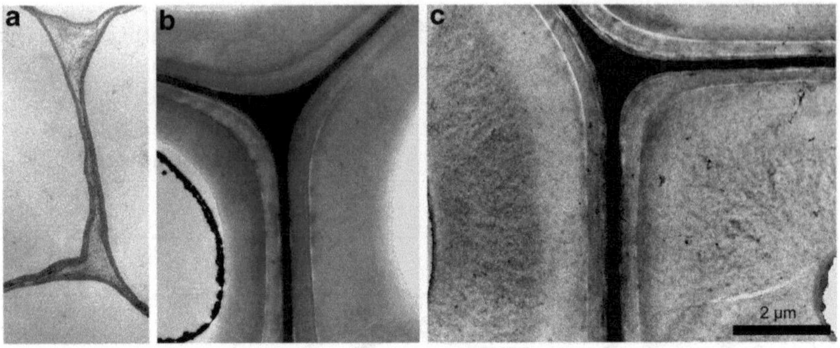


**Fig. 16** TEM micrographs of developing cell walls in beech xylem. (**a**) Young fibre wall without secondary wall at the end of the enlargement stage with well visible cell corner regions. (**b**) Secondary wall development of fibres, middle lamella regions and S1 layer completed and S2 layer still developing. (**c**) Completed wall development showing the typical wall layering of a fibre with broad S2 layer

means of UMSP revealed a homogeneous lignin distribution in the S2 layer of beech fibres as also described by Koch and Kleist (2001). These results were confirmed by TEM analyses of specimens after potassium permanganate staining (Fig. 16).

Point measurements with varying wavelengths were applied for analyses of the lignin composition. The spectra of vessel and fibre walls differed in the position of their maxima with the maxima for vessel walls at a wavelength of around 280 nm, whereas absorbance maxima of fibre walls were at around 278 nm (comp. Fig. 6). This indicates that the walls of these two cell types obviously have a different lignin composition with greater amount of strongly absorbing guaiacyl lignin in vessels and slightly more syringyl units in fibres. Several authors, e.g. Yoshinaga et al. (1997) and Terashima (2000), suggested that the reason for a different chemical composition of fibres and vessels is the successive incrustation of different monolignols (p-hydroxyphenyl, guaiacyl and syringyl) at different stages of differentiation. Yoshinaga et al. (1997) reported that vessel walls of oak showed maximal UV absorbance values at 280–285 nm during cell wall thickening with a shift down to about 273 nm at final stages of their differentiation. Point measurements also revealed that walls of ray and axial parenchyma cells as well as fibres have similarly positioned absorbance maxima at 278 nm.

When comparing the lignin contents of earlywood and latewood fibres of an individual growth ring in beech, latewood fibres displayed slightly higher absorbance values than earlywood fibres. Additionally, a shift of the maximum UV absorbance towards higher wavelengths in latewood fibres points to higher amounts of guaiacyl moieties in these walls as compared to walls of earlywood fibres. Takabe et al. (1992) who analysed the distribution of guaiacyl and syringyl lignin within growth rings of *Fagus crenata* also found that the lignin composition varied within earlywood and latewood. The observations also showed that the differentiation of the last formed fibres continued for approximately 1 month after cessation of cambial divisions. Similar observations were made for terminal latewood cells of different conifers like *Pinus radiata*, *Abies alba* and *Picea abies* (e.g. Donaldson 1992; Gričar et al. 2005; Gričar and Čufar 2008), but also for *Fagus sylvatica* (Čufar et al. 2008).

## 5.5 Topochemical Studies on Modified Lignin Distribution in the Xylem After Wounding

Wounding of trees induces various responses in woody tissues according to the compartmentalisation concept which is laid down in the so-called CODIT (*C*ompartmentalization *O*f *D*amage *I*n *T*rees) principle (e.g. Shigo and Marx 1977; Liese and Dujesiefken 1996). These responses are divided into passive reactions through structures already existing at the time of wounding (cell walls, cells, rays) and active reactions of living cells (mainly parenchyma cells) stimulated by the wounding (e.g. Schmitt and Liese 1993). Wound reactions are primarily aiming at the protection of inner xylem against air embolism to prevent dysfunction of large portions of the water-conducting system. Well-visible boundary layers consisting of several cell rows with very intensive reactions are finally formed around a wound as barriers against desiccation and the penetration of microorganisms as well as callus tissue which expands from the wound edge to the centre of the wound. Poplar was variously used to analyse wound reactions on a cellular and subcellular level (e.g. Frankenstein and Schmitt 2006). Among various reactions in xylem parenchyma cells, electron microscopy also revealed the formation of modified xylem at the wound edge along a transition zone between xylem formed prior to and after wounding. This zone regularly consists of a high number of unusually thick-walled fibres and could be clearly identified as xylem differentiating at the time of wounding. These modified poplar fibres deposited additional secondary wall (S2) material leading to different patterns of wall thickening. Some of these fibres developed an additional secondary wall layer clearly separated from the S2, whereas others showed a continuously thickened S2 layer (Fig. 17) (details in Frankenstein et al. 2005; Frankenstein and Schmitt 2006). Potassium permanganate staining of ultrathin sections resulted in a relatively strong staining of the secondary wall with heterogeneous lignin distribution within the S2, either with a distinctly darker outer S2 layer or in a patchy appearance of the entire S2 (Fig. 17a, b). These structural features point to an inhomogeneous lignin distribution also within the S2 layer.

The microdistribution of lignin in modified cell walls was also analysed by UV microspectrophotometry (Fig. 18a–c). Scanning analyses at a constant wavelength of 280 nm revealed a relatively low absorbance level for unmodified, thin-walled fibres of poplar and a homogeneous lignin distribution within their various wall layers (Fig. 18a). However, the spectra from modified, thick-walled fibres displayed increased lignin contents and an inhomogeneous lignin distribution. Elevated lignin

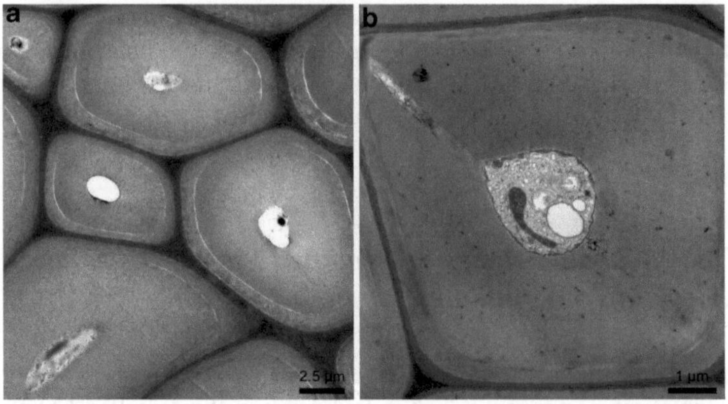


**Fig. 17** Electron micrographs from thick-walled poplar fibres in wound-associated xylem. Secondary wall S2 layers appear enormously thickened (**a** and **b**), whereby potassium permanganate staining resulted in an inhomogeneous staining indicating either a stronger lignification in the outer S2 layer (**a**) or a patchy lignin distribution within the S2 (**b**)

contents were recorded for fibre and vessel walls, whereby the lignin contents increased in the S2 layers of fibres from samples already taken after a few weeks of wound response. This feature appeared more pronounced in samples collected after some months of wound response. The lignin contents in compound middle lamella regions and cell corners mostly were significantly higher.

According to point measurements with varying wavelengths between 240 and 400 nm, absorbance maxima were determined (Fig. 19). At 270–272 nm, fibres from controls showed rather low lignin contents in their S2 layers, as compared with higher absorbance values in the compound middle lamella and cell corner regions. The position of the peak at around 272 is characteristic for a lignin type mainly composed of syringyl units. For modified fibres, the distinctly higher absorbance in their secondary walls and the slightly higher absorbance of the compound middle lamella regions confirmed increased lignin contents. Regarding lignin composition, within thickened secondary walls, the absorbance peak shifted slightly towards higher wavelength because of an increase in guaiacyl moieties and a reduced amount of associated p-hydroxy benzoic acid residues. The same effect was observed for middle lamella regions. The spectra obtained for cell corner regions showed the highest lignin concentrations of predominantly guaiacyl lignin, as deduced from the maximum at around 276–278 nm.

From these results obtained for poplar, it can be stated that in xylem fibres, differentiating at the time of wounding increased wall thicknesses and increased lignin contents and a slightly modified lignin composition may be induced. It is assumed that this wound response is part of the CODIT principle and adds a further mechanism contributing to an increased xylem resistance.

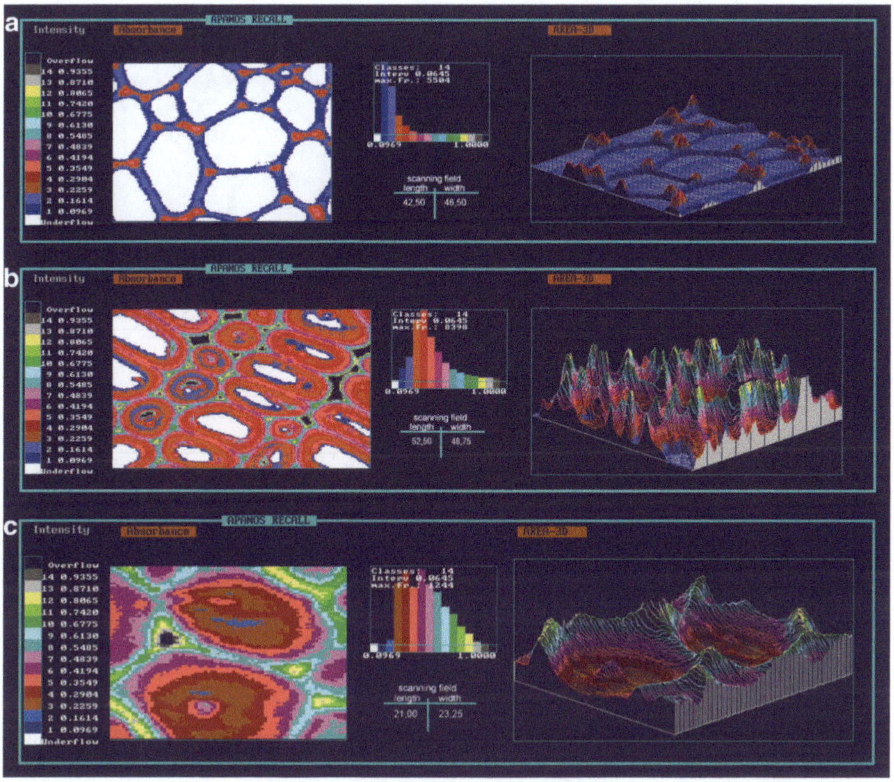


**Fig. 18** UV microscopic image profiles of xylem of unaffected poplar xylem (**a**) and wound-associated poplar xylem (**b** and **c**). In unaffected trees, xylem fibres exhibit thin walls with low absorbance values. Wound-associated poplar xylem at the edge of wounds develops thick-walled fibres with increasing absorbances indicating increased lignin contents in all wall layers (**b**) and an inhomogeneous lignin distribution (**c**) (Frankenstein and Schmitt 2006)

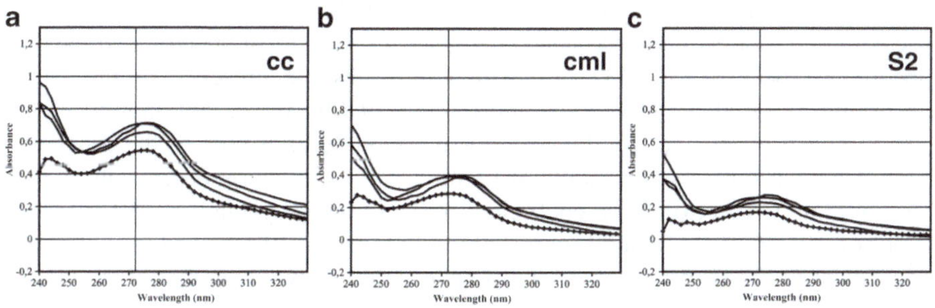


**Fig. 19** Point measurements with representative UV absorbance spectra of individual fibre wall layers (cc cell corner, cml compound middle lamella, S2 secondary wall S2 layer); wound-associated fibres show a generally higher absorbance than control fibres (*filled diamond*) (Frankenstein and Schmitt 2006)

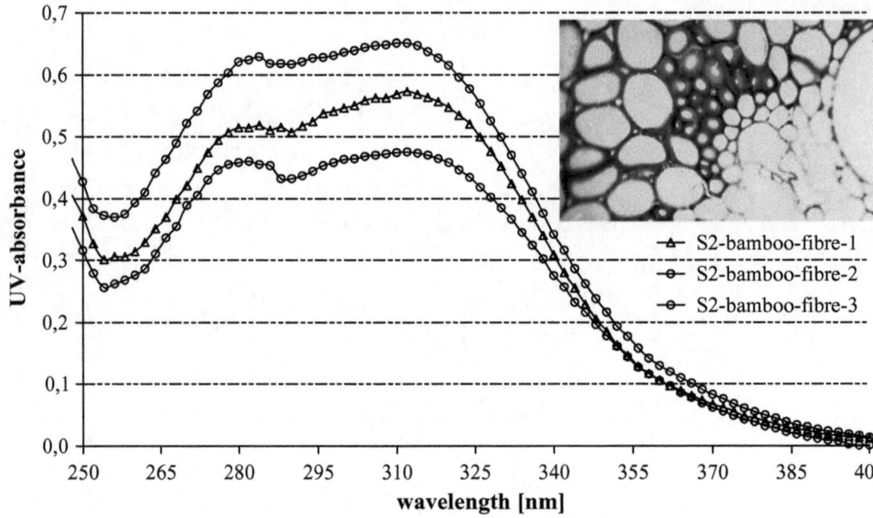


**Fig. 20** UV-absorption spectra of the $S_2$ cell walls of fibres from a vascular bundle at the middle culm wall

## *5.6 Lignin Distribution in Tropical Bamboo Species*

Bamboo lignin is considered to be composed of guaiacyl, syringyl and *p*-hydroxyphenylpropan units. As a unique feature, it also contains 5–10 % of *p*-coumaric acid ester (Higuchi 1987), located at the γ-positions of grass lignin, predominantly in syringyl units (Lu and Ralph 1999).

The distribution of lignin within the cell walls of the tropical bamboo *Gigantochloa levis* was studied topochemically by means of TEM and cellular UV microspectrophotometry (Lybeer and Koch 2005; Lybeer et al. 2006). All spectra curves of the epidermis, fibre, and parenchyma cell walls show a shoulder between 310 and 320 nm (Fig. 20), which can be linked to the presence of *p*-coumaroylation as demonstrated by Higuchi (1987). Most spectra also have an absorbance peak at 280–282 nm, which indicates the presence of the strong absorbing guaiacyl lignin (Fergus and Goring 1970b; Musha and Goring 1975). The UV absorbance of a specific anatomical region depends both on the concentration of the various structural units of lignin and the extinction coefficient of each structural unit. The extinction coefficient of the G (guaiacyl) unit at 280 nm is 3.5 times of that of the S (syringyl) unit (Fergus and Goring 1970a), and the extinction coefficient of the H (p-hydroxyphenylpropan) unit is lower than that of the G unit, but higher than that of the S unit (Faix and Schweers 1974). He and Terashima (1991) found a similar guaiacyl peak in the spectra from rice (*Oryza sativa* L.) and sugarcane (*Saccharum officinarum* L.).

In Fig. 21, the lignin distribution within a polylamellated fibre cell wall in mature bamboo is displayed, measured at $\lambda_{280\text{max}}$. The lamellation is generally

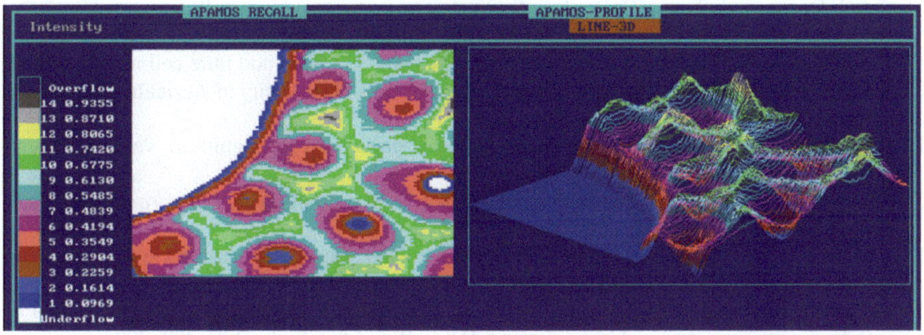


**Fig. 21** UV micrograph and 3D profiles of a bamboo fibre (*P. edulis*) measured at $\lambda_{280nm}$

described as alternating broad and narrow layers with different fibrillar orientation. The image profiles show a distinctly high absorbance of the cell corners and compound middle lamellae ($abs_{280nm}$ 0.81 to 0.94). The secondary cell wall is characterised by a uniform concentric structure, whereas the absorbance values decrease stepwise from the outer ($abs_{280nm}$ 0.74) towards the inner ($abs_{280nm}$ 0.29) parts of the wall.

## 6 Conclusions

The described methods and selected examples demonstrate that cellular UV microspectrophotometry and electron microscopy are ideally suited to study the topochemical distribution of lignin and phenolic extractives on a subcellular level. In particular, the application of the UV-scanning technique enables a direct imaging of lignin distribution and provides fundamental information on the topochemistry of lignin. Using these techniques, fine differences in the lignification of individual cell wall layers and the deposition of phenolic extractives can be analysed. The techniques can be used for a wide range of applications in wood biology and topochemistry. Currently, the topochemistry of genetically modified trees as well as thermally and chemically modified wood are studied by using cellular UV spectroscopy and electron microscopy.

## References

Boerjan W, Ralph J, Baucher M (2003) Lignin biosynthesis. Annu Rev Plant Biol 54:519–546

Brändström J (2001) Micro- and ultrastructural aspects of Norway spruce tracheids: a review. IAWA J 22:333–353

Ĉufar K, Gričar J, Zupančič M, Koch G, Schmitt U (2008) Anatomy, cell wall structure and topochemistry of water-logged archaeological wood aged 5,200 and 4,500 years. IAWA J 29:55–68

Daniel G, Duchesne I, Tokoh C, Bardage S (2004) The surface and intracellular nanostructure of wood fibres: electron microscope methods and applications. In: Schmitt U, Ander P, Barnett JR, Emons AMC, Jeronimidis G, Saranpää P, Tschegg S (eds) Wood fibre cell walls: methods to study their formation, structure and properties. Swedish University of Agricultural Sciences, Uppsala, pp 87–104

Dean JFD, Eriksson KE (1994) Laccase and the deposition of lignin in vascular plants. Holzforschung 48:21–33

Donaldson L (1992) Lignin distribution during latewood formation in *Pinus radiata* D. Don. IAWA Bull ns 13:381–387

Donaldson L (2001) Lignification and lignin topochemistry – an ultrastructural view. Phytochemistry 57:859–873

Donaldson L, Ryan KG (1987) A comparison of relative lignin concentration as determined by interference microscopy and bromination/EDXA. Wood Sci Technol 21:303–309

Eriksson I, Lidbrandt O, Westermark U (1988) Lignin distribution in birch (*Betula verrucosa*) as determined by mercurization with SEM- and TEM-EDXA. Wood Sci Technol 22:251–257

Faix O, Schweers W (1974) Vergleichende Untersuchungen an Polymermodellen des Lignins (DPH's) verschiedener Zusammensetzungen. 4. Mitteilung: UV-spektroskopische Untersuchungen. Holzforschung 28:94–98

Fengel D, Wegener G (1989) Wood-chemistry, ultrastructure, reactions. de Gruyter, Berlin

Fergus BJ, Goring DAI (1970a) The location of guaiacyl and syringyl lignins in birch xylem tissue. Holzforschung 24:113–117

Fergus BJ, Goring DAI (1970b) The distribution of lignin in birch wood as determined by ultraviolet microscopy. Holzforschung 24:118–124

Fergus BJ, Procter AR, Scott JAN, Goring DAI (1969) The distribution of lignin in sprucewood as determined by ultraviolet microscopy. Wood Sci Technol 3:117–138

Frankenstein C, Schmitt U (2006) Microscopic studies on modified wall structure and lignin topochemistry in xylem fibres of poplar after wounding. Maderas 8:93–106

Frankenstein C, Schmitt U, Waitkus C, Eckstein D (2005) Wound callus formation: a microscopic study on poplar (*Populus tremula* L. x *Populus tremuloides* Michx.). J Appl Bot Food Qual 79:44–51

Fromm J, Rockel B, Lautner S, Windeisen E, Wanner G (2003) Lignin distribution in wood cell walls determined by TEM and backscattered SEM techniques. J Struct Biol 143:77–84

Fujii T, Shimizu K, Yamaguchi A (1987) Enzymatic saccharification on ultrathin sections and ultraviolet spectra of Japanese hardwoods and softwoods. Mokuzai Gakkaishi 33:400–407

Fukazawa K (1992) Ultraviolet microscopy. In: Lin SY, Dence CW (eds) Methods in lignin chemistry. Springer, Berlin, pp 110–121

Glasser WG (1980) Lignin. In: Casey JP (ed) Pulp and paper. Chemistry and chemical technology. Wiley, New York, pp 39–111

Goldschmid O (1971) Ultraviolet spectra. In: Sarkanen KV, Ludwig CH (eds) Lignins, occurrence, formation, structure and reactions. Wiley, New York, pp 241–266

Gričar J, Čufar K (2008) Seasonal dynamics of phloem and xylem formation in silver fir and Norway spruce as affected by drought. Russ J Plant Physiol 55:538–543

Gričar J, Čufar K, Oven P, Schmitt U (2005) Differentiation of terminal latewood tracheids in silver fir tracheids during autumn. Ann Bot 95:959–965

Grosser D, Fengel D, Schmidt H (1974) Tamrit-Zypresse (*Cupressus dupreziana* A. CAMUS). Beitrag zur Ökologie, Anatomie und Chemie. Forstwiss Centralb 93:191–207

He L, Terashima N (1991) Formation and structure of lignin in monocotyledons. IV. Deposition process and structural diversity of the lignin in the cell wall of sugarcane and rice plant studied by ultraviolet microscopic spectroscopy. Holzforschung 45:191–198

Hepler PK, Newcomb EH (1963) The fine structure of young tracheary elements arising by redifferentiation of parenchyma in wounded *Coleus* stem. J Exp Bot 14:496–503

Hesse M, Meier H, Zeeh B (1991) Spektroskopische Methoden in der organischen Chemie. Thieme, Stuttgart

Higuchi T (1987) Chemistry and biochemistry of bamboo. Bamboo J 4:132–145
Higuchi T (1997) Biochemistry and molecular biology of wood. Springer, Berlin
Hillis WE (1987) Heartwood and tree exudates. Springer, New York
Hillis WE, Yazaki Y (1973) Polyphenols of Intsia Heartwoods. Phytochemistry 12:2491–2495
Jaffé HH, Orchin M (1962) Theory and application of ultraviolet spectroscopy. Wiley, New York
Joseleau JP, Ruel K (1997) Study of lignification by noninvasive techniques in growing maize internodes. An investigation by Fourier transform Infrared, CP/MAS $^{13}$C NMR spectroscopy and immunocytochemical transmission electron microscopy. Plant Physiol 114:1123–1133
Joseleau JP, Imai T, Kuroda K, Ruel K (2004) Detection in situ and characterization of lignin in the G-layer of tension wood fibres of *Populus deltoides*. Planta 219:338–345
Kim YS, Wi SG, Grünwald C, Schmitt U (2002) Immuno electron microscopic localization of peroxidases in the differentiating xylem of *Populus* spp. Holzforschung 56:355–359
Kindl H (1991) Biochemie der Pflanzen, 3rd edn. Springer, Berlin
Kleist G, Bauch J (2001) Cellular UV microspectrophotometric investigation of Sapelli heartwood (*Entandrophragma cylindricum* Sprague) from natural provenances in Africa. Holzforschung 55:117–122
Kleist G, Schmitt U (1999) Evidence of accessory components in vessel walls of Sapelli heartwood (*Entandrophragma cylindricum*) obtained by transmission electron microscopy. Holz Roh Werkstoff 57:93–95
Klemm D, Schmauder HP, Heinze T (2002) Cellulose. In: Vandamme E, De Baets S, Steinbüchel A (eds) Polysaccharide II: biopolymers: biology, chemistry, biotechnology, applications, vol 6. Wiley-VCH, Weinheim, pp 277–319
Koch G, Grünwald C (2004) Application of UV microspectrophotometry for the topochemical detection of lignin and phenolic extractives in wood fibre cell walls. In: Schmitt U, Ander P, Barnett JR, Emons AMC, Jeronimidis G, Saranpää P, Tschegg S (eds) Wood fibre cell walls: methods to study their formation, structure and properties. Swedish University of Agricultural Sciences, Uppsala, pp 119–130
Koch G, Kleist G (2001) Application of scanning UV microspectrophotometry to localise lignins and phenolic extractives in plant cell walls. Holzforschung 55:563–567
Koch G, Rose B, Patt R, Kordsachia O (2003a) Topochemical investigations on delignification of *Picea abies* [L.] Karst. during alkaline sulfite (ASA) and bisulfite pulping by scanning UV microspectrophotometry. Holzforschung 57:611–618
Koch G, Bauch J, Puls J (2003b) Topochemical characterisation of phenolic extractives in discoloured beechwood (*Fagus sylvatica* L.). Holzforschung 57:339–345
Koch G, Richter HG, Schmitt U (2006) Topochemical investigation on phenolic deposits in the vessels of afzelia (*Afzelia* spp.) and merbau (*Intsia* spp.) heartwood. Holzforschung 60:583–588
Kukkola EM, Koutaniemi S, Gustafsson M, Karhunen P, Ruel K, Lundell TK, Saranpää P, Brunow G, Teeri TH, Fagerstedt KV (2003) Localization of dibenzodioxocin substructures in lignifying Norway spruce xylem by transmission electron microscopy–immunogold labelling. Planta 217:229–237
Kukkola EM, Koutaniemi S, Pöllänen E, Gustafsson M, Karhunen P, Lundell TK, Saranpää P, Kilpeläinen I, Teeri TH, Fagerstedt KV (2004) The dibenzodioxocin lignin substructure is abundant in the inner part of the secondary wall in Norway spruce and silver birch xylem. Planta 218:497–500
Kukkola E, Saranpää P, Fagerstedt K (2008) Juvenile and compression wood cell wall layers differ in lignin structure in Norway spruce and Scots pine. IAWA J 29:47–54
Lange PW (1954) The distribution of the components in the plant cell wall. Svensk Papperstidn 57:563–567
Lehringer C, Gierlinger N, Koch G (2007) Topochemical investigation on tension wood fibres of *Acer* spp., *Fagus sylvatica* L. and *Quercus robur* L. Holzforschung 62:255–263
Lehringer C, Daniel G, Schmitt U (2009) TEM/FE-SEM studies on tension wood fibres of *Acer* spp., *Fagus sylvatica* L. and *Quercus robur* L. Wood Sci Technol 43:691–702

Liese W, Dujesiefken D (1996) Wound reactions of trees. In: Raychaudhuri SP, Maramorosch K (eds) Forest trees and palms-disease and control. Oxford/IBH, New Delhi, pp 20–42

Lu F, Ralph J (1999) Detection and determination of p-coumaroylated units in lignins. J Agric Food Chem 47:1988–1992

Lybeer B, Koch G (2005) Lignin distribution in the tropical bamboo species *Gigantochloa levis*. IAWA J 26:443–456

Lybeer B, Koch G, Van Acker J, Goetghebeur P (2006) Lignification and cell wall thickening in nodes of *Phyllostachys viridiglaucescens* and *Phyllostachys nigra*. Ann Botany 97:529–539

Maurer A, Fengel D (1991) Elektronenmikroskopische Darstellung von strukturellen Einzelheiten in Nadelholz-Zellwänden anhand sehr dünner Ultramikrotomschnitte. Holz Roh Werkstoff 49:53–56

Michalowicz G, Robert A (1990) The application of transmission electron microscopy for topochemical studies on aspen wood (*Populus tremula*): delignification during soda and soda/AQ pulping. Holzforschung 44:39–46

Musha Y, Goring DAI (1975) Distribution of syringyl and guaiacyl moieties in hardwoods as indicated by ultraviolet microscopy. Wood Sci Technol 9:45–58

Nimz HH (1981) Carbon-13 NMR spectra of lignins, 8 structural differences between lignins of hardwoods, softwoods, grasses and compression wood. Holzforschung 35:16–26

Novaes E, Kirst M, Chiang V, Winter-Sederoff H, Sederoff R (2010) Lignin and biomass: a negative correlation for wood formation and lignin content in trees. Plant Physiol 154:555–561

Parham RA (1974) Distribution of lignin in Kraft pulp as determined by electron microscopy. Wood Sci Technol 6:305–315

Paszner L, Behera NC (1989) Topochemistry of softwood delignification by alkali earth metal salt catalysed organosolv pulping. Holzforschung 43:159–168

Prislan P, Koch G, Schmitt U, Čufar K, Gričar J (2009) Topochemical investigation in developing wood of beech (*Fagus sylvatica*). Holzforschung 63:482–490

Röder T, Koch G, Sixta H (2004) Application of confocal Raman spectroscopy for the topochemical distribution of lignin and cellulose in plant cell walls of beech wood (*Fagus sylvatica* L.) compared to scanning UV microspectrophotometry. Holzforschung 58:480–482

Ruel K (2004) Immunolabelling of lignin sub-structures: a strategy for wood fibre wall topochemical analyses. In: Schmitt U, Ander P, Barnett JR, Emons AMC, Jeronimidis G, Saranpää P, Tschegg S (eds) Wood fibre cell walls: methods to study their formation, structure and properties. Swedish University of Agricultural Sciences, Uppsala, pp 131–140

Ruel K, Faix O, Joseleau JP (1994) New immunogold probes for studying the distribution of the different lignin types during plant cell wall biogenesis. J Trace Microprobe Techniques 12:247–265

Ruel K, Burlat V, Joseleau JP (1999) Relationship between ultrastructural topochemistry of lignin and wood properties. IAWA J 20:203–211

Saka S, Goring DAI (1988) Localization of lignins in wood cell walls. In: Higuchi T (ed) Biosynthesis and biodegradation of wood components. Academic, New York, pp 51–62

Saka S, Thomas RJ (1982) Evaluation of the quantitative assay of lignin distribution by SEM-EDXA-technique. Wood Sci Technol 16:1–8

Saka S, Whiting P, Fukazawa K, Goring DAI (1982) Comparative studies on lignin distribution by UV microscopy and bromination combined with EDXA. Wood Sci Technol 16:269–277

Sarkanen KV, Hergert HL (1971) Classification and distribution. In: Sarkanen KV, Ludwig CH (eds) Lignins: occurrence, formation, structure and reactions. Wiley, New York, pp 43–49

Schmitt U, Liese W (1993) Response of xylem parenchyma by suberization in some hardwoods after mechanical injury. Trees 8:23–30

Schmitt U, Melcher E (2004) Section staining with potassium permanganate for transmission electron microscopy: a useful tool for lignin localisation. In: Schmitt U, Ander P, Barnett JR, Emons AMC, Jeronimidis G, Saranpää P, Tschegg S (eds) Wood fibre cell walls: methods to study their formation, structure and properties. Swedish University of Agricultural Sciences, Uppsala, pp 105–117

Schmitt U, Singh AP, Frankenstein C, Mölle R (2006) Cell wall modifications in woody stems induced by mechanical stress. New Zealand J For Sci 36:72–86
Scott JAN, Procter AR, Fergus BJ, Goring DAI (1969) The application of ultraviolet microscopy to the distribution of lignin in wood: description and validity of the technique. Wood Sci Technol 3:73–92
Shigo AL, Marx HG (1977) Compartmentalization of decay in trees. USDA For Ser Agric Bull No 405, 74pp
Singh AP, Donaldson L (1999) Ultrastructure of tracheid cell walls in radiata pine (*Pinus radiata*) mild compression wood. Can J Bot 77:32–40
Smith CG, Rodgers MW, Zimmerlin A, Ferdinando D, Bolwell GP (1994) Tissue and subcellular immunolocalization of enzymes of lignin synthesis in differentiating and wounded hypocotyl of French bean (*Phaseolus vulgaris* L.). Planta 192:155–164
Spurr AR (1969) A low viscosity epoxy resin embedding medium for electron microscopy. J Ultrastruct Res 26:31–43
Takabe K (2002) Cell walls of woody plants: autoradiography and ultraviolet microscopy. In: Chaffey N (ed) Wood formation in trees. Taylor and Francis, London, pp 159–177
Takabe K, Fujita M, Harada H, Saiki H (1981) The deposition of cell wall components in differentiating tracheids of sugi. Mokuzai Gakkaishi 27:249–255
Takabe K, Miyauchi S, Tsunoda R, Fukazawa K (1992) Distribution of guaiacyl and syringyl lignins in Japanese beech (*Fagus crenata*): variation within annual ring. IAWA Bull 13:105–112
Terashima N (2000) Formation and ultrastructure of lignified plant cell walls. In: Kim YS (ed) New horizons in wood anatomy. Chonnam National University Press, Korea, pp 169–180
Terashima N, Fukushima K, Takabe K (1986) Heterogeneity in formation of lignin. VIII. An autoradiographic study on the formation of guaiacyl and syringyl lignin in Magnolia kobus DC. Holzforschung 40:101–105
Terashima N, Fukushima K, He L, Takabe K (1993) Comprehensive model of lignified plant cell wall. In: Jung HG, Buxton DR, Hatfield RD, Ralph J (eds) Forage cell wall structure and digestibility. Amer Soc Agriculture, Madison, WI, pp 247–270
Terashima N, Nakashima J, Takabe K (1998) Proposed structure for protolignin in plant cell walls, ACS Symposium Series 697. American Chemical Society, Washington, DC, pp 180–193
Timell TE (1969) The chemical composition of tension wood. Svensk Papperstidning 72:173–181
Timell TE (1986) Compression wood in gymnosperms. Springer, Berlin
Westermark U (1985) Bromination of different morphological parts of spruce (*Picea abies*). Wood Sci Technol 19:323–328
Westermark U, Lidbrandt O, Eriksson I (1988) Lignin distribution in spruce (*Picea abies*) determined by mercurization with SEM-EDXA technique. Wood Sci Technol 22:243–250
Yoshinaga A, Fujita M, Saiki H (1997) Secondary wall thickening and lignification of oak xylem components during latewood formation. Mokuzai Gakkaishi 43:377–383

# New Insights into Heartwood and Heartwood Formation

**Andreas Kampe and Elisabeth Magel**

**Abstract** The formation of true or obligate heartwood in the innermost living xylem tissues of many tree species is one of the most important ecological and economical secondary differentiation processes. Natural durability, biological, technological, and esthetic parameters of wood and its products depend on the presence, quality, and quantity of heartwood extractives. This chapter presents the actual knowledge about heartwood formation. The main focus is on the biochemical and molecular basis of heartwood formation as well as on new developments in using heartwood extractives as biocides or as pharmaceuticals. Biochemical studies, corroborated by gene expression studies, prove that (1) season of heartwood formation starts in summer, reaches a maximum in early fall, and ceases during dormancy; (2) the transient shift in metabolism towards an enhanced secondary metabolism is regulated by gene expression and protein de novo synthesis; and (3) programmed cell death (PCD) of parenchyma cells during heartwood formation shares some similarities with PCD during xylem formation. However, more work is needed (1) to find more similarities and/or differences between the two types of heartwood formation (*Robinia*-Type and *Juglans*-Type), (2) to elucidate the involvement of axial and ray parenchyma cells, and (3) to learn more about regulation of heartwood formation, such as the role of transcription factors or phytohormones.

---

A. Kampe • E. Magel (✉)
Wood Biology, Department of Wood Science, University of Hamburg, Leuschnerstrasse 91d, 21031 Hamburg, Germany
e-mail: elisabeth.magel@uni-hamburg.de

J. Fromm (ed.), *Cellular Aspects of Wood Formation*, Plant Cell Monographs 20,
DOI 10.1007/978-3-642-36491-4_3, © Springer-Verlag Berlin Heidelberg 2013

# 1 Introduction

The ecological and economic value of trees is basically determined by the formation of wood and the accumulation of biocides, such as phenolic substances, which determines its quality. During periods of cambial growth, the cambial initials divide, and depending on endogenous and exogenous factors, the cells which are formed in centrifugal and centripetal directions differentiate into phloem and xylem elements, respectively (Higuchi 1997). Within wood tissue, resulting from the ripening of inward-directed cambial derivatives, heartwood formation is the most prominent secondary differentiation process. Thus, in freshly felled logs of many trees, two different-colored wood zones can be observed: the pale-colored sapwood and the heartwood, which is usually much darker in color. By definition, heartwood is "the inner layers of wood, which, in the growing tree, have ceased to contain living cells, and in which the reserve materials (e.g., starch) have been removed or converted into heartwood substances" (IAWA 1964). Therefore, heartwood has no apparent physiological role because living cells and reserve materials are lacking (Hillis 1987). Sapwood is defined as "the portion of the wood that, in the living tree, contains living cells and reserve materials (e.g. starch)" (Anonymous 1957). Thus, it is the physiological active part of the tree. Its main function is conduction of water from roots to foliage (Taylor et al. 2002). In between the living, noncolored sapwood and the often dark-colored heartwood, the transition zone (tz) is located. In the transition zone, parenchyma cells are living. They are differently programmed genetically and the extractives accumulated in the heartwood are produced (Burtin et al. 1998; De Filippis and Magel 2012; Magel 2000; Spicer 2005; Taylor et al. 2002). The transition zone is defined as "the inner layers of the sapwood that are transitional between sapwood and heartwood in color and general character" (IAWA 1964).

This chapter reviews recent literature on heartwood, heartwood formation, and heartwood extractives which determine heartwood quality. Focus is exclusively on regular, obligate, or "true"-colored heartwood. Irregular, facultative heartwood or any other kind of discolorations is not included. Literature addressing these topics published between 2005 and 2012 is preferentially summarized. Literature prior to this period is covered by reviews on heartwood and heartwood formation given by Bamber and Fukazawa (1985), Hillis (1987), Magel et al. (1997), Magel (2002), Scheffer and Cowling (1966), Taylor et al. (2002), Spicer (2005), and Ziegler (1968). Topics are biochemical, molecular genetics, analytical, as well as pharmaceutical aspects of heartwood and characteristic heartwood components.

## 2 Proportion of Sapwood and Heartwood and Methods for Determination

Heartwood formation is recognized as a secondary differentiation process of xylem. It is an active program of tissue senescence, and thus, heartwood formation is an actively regulated stage of woody plant development (Spicer 2005). It is postulated that heartwood formation is a process to regulate the amount of sapwood in order to keep sapwood at an optimal level (Bamber 1976; Spicer 2005; Taylor et al. 2002) to meet the demand for structural and mechanical support, water transport, storage of reserve material, and the decrease in energy demands for maintaining living sapwood as a respiring tissue (Taylor et al. 2002). The amounts of sapwood and heartwood in a tree vary and are related to factors such as species, age of trees (or tree cambial age), social class of the tree, tree position in the stand, forest site type, soil properties, rate of growth, and thus height and dimensions of the tree (Bamber and Fukazawa 1985; Nawrot et al. 2008; Wang et al. 2010a, b). Heartwood and sapwood proportion in a mature tree is species specific, with number of sapwood rings ranging from 1 to 2 in *Catalpa speciosa* up to 80–100 in *Nyssa sylvatica* (Hillis 1987). Overall, heartwood proportion within species is under significant genetic control, but environmental influences seem to be equally important (Taylor et al. 2002).

Identification of sapwood and heartwood is attempted by several methods. "Macroscopic methods" such as visual assessment of light transmission, natural color changes, and differential staining of sapwood and heartwood by indicators or dyes are generally fast, uncomplicated, and suitable for low budgets (Forrester et al. 2010; Macfarlane et al. 2010; Pfautsch et al. 2011, 2012; Wang et al. 2010b). Using incident light microscopy enables the identification of the sapwood–heartwood border by the transition between open and blocked vessels and thus by a microscopic approach (Pfautsch et al. 2010). More expensive methods, successfully applied for the identification of sapwood and heartwood areas, are time-of-flight secondary ion mass spectrometry (TOF-SIMS, Saito et al. 2008a, b), computer tomography (CT, Wei et al. 2011), X-ray computed microtomography (micro-CT), electric resistivity tomography (ERT, Bieker and Rust 2010), direct-scanning X-ray densitometer, and heat-sensitive infrared imaging (Gartner 2002; Gjerdrum and Hoibo 2004). In *Eucalyptus* and *Corymbia* species, a near-infrared spectroscopy (NIR) was tested for the identification of sapwood area (Pfautsch et al. 2012). However, this method was in these species not yet competitive compared to macroscopic or microscopic methods in predicting the sapwood area of a stem. However, Sandberg and Sterley (2009) reported that the noncolored Norway spruce heartwood and sapwood in dried condition are correctly separated by using NIR spectroscopy and multivariate data analysis.

# 3 Color Characteristics of Heartwood

Heartwood is often darker in color than sapwood. The darker color is due to secondary metabolites, the heartwood extractives, which render heartwood its natural durability and esthetic characteristics. Thus commercial quality and value of wood are based predominantly on the quality and quantity of heartwood extractives (Ekeberg et al. 2006; Hillis 1987; Taylor et al. 2002). However, the natural durability and color not always correlates with the concentration of heartwood extractives (Hillis 1987; Taylor et al. 2002). A number of color determination techniques have been evaluated to collect quantitative parameters which show correlations to other wood properties (Vetter et al. 1990). In recent studies, the CIELab system (CIE, International Commission on Illumination; L*, a*, b* color space; L for lightness and a and b for the color-opponent dimensions, based on nonlinearly compressed CIE XYZ color space coordinates) was used to measure wood color parameters and correlate them with extractive content in important commercial tropical timber species such as *Acacia mangium*, *Vochysia guatemalensis* (Moya et al. 2012), or *Tectona grandis* (Lukmandaru et al. 2009; Moya and Marin 2011). The difference between heartwood and sapwood color for *Vochysia guatemalensis* (low color difference between sapwood and heartwood color) and *Acacia mangium* (color difference between heartwood and sapwood) is due to lower content of extractives in the sapwood of both species. These investigations confirmed that color parameters show species-specific correlations with, e.g., water or ethanol-toluene extractable contents of total extractives in *Acacia mangium*. In teak, color parameters are genetically determined and influenced by environmental factors such as site and growth conditions (Moya and Diego Marin 2011; Moya and Calvo-Alvarado 2012).

# 4 Sapwood and Heartwood: A Potent Biosorbent for Metal Ions

Besides many other features, sapwood and heartwood of tree species show a high potential as bioabsorbent for metal ions. For example, birch wood and spruce sapwood and heartwood exhibit a high potential of binding metal ions in the order $Fe^{3+} >> Pb^{2+} >> Cu^{2+} >> Fe^{2+} > Cd^{2+} > Zn^{2+} > Ni^{2+} > Mn^{2+} \geq Ca^{2+} \geq Sr^{2+} \geq Ba^{2+} >> Mg^{2+} >> K^{+} > Na^{+} = Li^{+}$ by acting as a weakly acid cation exchanger (Su et al. 2012). This activity is based on, e.g., carboxyl groups and phenolic hydroxyl groups. For heartwood powder of *Areca catechu*, promising results as an economic, time-saving and low-cost biosorbent for cadmium (II), lead (II), and copper (II) from aqueous synthetic wastewater are reported (Chakravarty et al. 2010a, b, 2012). Major metal binding groups are hydroxyl, carboxyl, amide, and amine groups for lead (II) and cadmium (II), whereas O–H, N–H, and C–O groups are the dominating binding sites for copper (II). Additionally, it seems very promising that woody biomass appears to be a potential agent in solving the global concern of depletion of terrestrial phosphorous resources by a concomitant increase of phosphorous load in water. By treatment of phosphorous-rich natural waters with iron-oxidizing bacteria and woody biomass, e.g., heartwood of conifers, the latter acts as a carrier for biogenic Fe-oxides and phosphorous. Besides adsorbing P from P-rich waters, it thus can be used further on as a source of P in plant cultivations (Takeda et al. 2010).

## 5 Characteristics of Ageing Sapwood and Heartwood (Moisture, Gas, Nutrient, Reserve Substances, Cell Death)

In most tree species, moisture content of the heartwood differs from sapwood. In hardwoods, it is species specific, while conifers show lower moisture content in the heartwood compared to sapwood (Taylor et al. 2002). Lowest concentrations are present in the transition zone (Nobuchi and Harada 1983). Kuroda et al. (2009) suggest that water distribution in the xylem plays a role in heartwood formation of *Cryptomeria japonica*. After investigating the tracheid-water content of both earlywood and latewood in sapwood, intermediate wood, and heartwood, they revealed that water content in the earlywood differed between sapwood (95–99 %), intermediate wood (7–12 %), and heartwood (4–100 %). Due to this, it was concluded that changes in water content of individual tracheids are closely related to heartwood formation. The factor contributing to decrease of moisture content in ageing sapwood tissues of hardwoods is the compartmentalization of vessels by tyloses, outgrowths developing from parenchyma cells via the pits, impeding water transport (Bamber and Fukazawa 1985; Déjardin et al. 2010; Hillis 1987; Nair et al. 1981; Spicer 2005; Taylor et al. 2002).

Decrease in moisture content with ageing of the sapwood tissue results in an increase of gas volume (Hillis 1987). This large volumetric proportion of gas in stem xylem has a substantial role for tree biomechanics by having an adaptive effect on surface stresses and lateral force a tree can withstand before breaking (Gartner et al. 2004). Moreover, it seems likely that stem gas may play important physiological roles related to many metabolic activities, e.g., water transport, respiration, storage, or heartwood formation (Gartner et al. 2004). Gas composition in woody tissues is low in oxygen and high in carbon dioxide. Reported and measured values range for $O_2$ from 1–1.5 % (Hillis 1987) up to 2–8 % during season and 15–20 % during dormancy (Spicer and Holbrook 2007b) and for $CO_2$ from 6–10 % (Spicer and Holbrook 2007b) up to 41–65 % (Hillis 1987). With increasing depth in the stem towards the sapwood-heartwood boundary, it was speculated that further oxygen depletion and carbon dioxide enrichment might cause cell death of the parenchyma cells and thus initiation of heartwood formation by limitation of respiration (Eklund and Klintborg 2000; Panshin and de Zeeuw 1980). In situ measurements of oxygen at different depths within stems of *Acer rubrum, Fraxinus americana*, *Quercus rubra*, and *Tsuga canadensis* confirmed depletion of stem oxygen levels with ageing sapwood tissues (Spicer and Holbrook 2005). However, it is unlikely that these oxygen contents lead to hypoxia of parenchyma cells at the sapwood–heartwood boundary and thus to cell death. These findings were manifested by measurement of sapwood respiration in *Acer rubrum*, *Fraxinus americana*, *Quercus rubra*, *Tsuga canadensis*, and *Pinus strobus*. Testing interactive effects of $CO_2$ and $O_2$ on sapwood respiration revealed that oxygen has a more pronounced inhibitory effect than $CO_2$. Application of $O_2$ and $CO_2$ levels similar to within-stem values resulted in moderate reduction of respiration. Thus it seems unlikely that alterations in $O_2$ and $CO_2$ levels with increasing age of the sapwood cause cell death during heartwood formation (Spicer and Holbrook 2007a, b).

A still open question during heartwood formation is whether trees recycle nutrients from senescing and dying tissues back into physiologically active xylem tissues. Several papers address this topic. In older literature, a generalized picture was drawn with higher contents of K, Ca, and Mg in the heartwood and resorption of P, K, and S (Hillis 1987; Saito et al. 2008b). This was refined by suggesting three types of mineral distribution within woody axes (Okado et al. 1993a, b in Taylor et al. 2002) and the grouping of nutrients into three categories: mobile (P, K), intermediately mobile (Mg, Zn), and immobile (Ca, Mn) (Myre and Camire 1994 in Taylor et al. 2002). In *Cryptomeria japonica*, radial transport of metals was investigated by injection of Rubidium (Rb) and Europium (Eu) in the sapwood, and a two-step process was found: active transport from the sapwood to outer heartwood via ray parenchyma cells and diffusion in the heartwood (Okada et al. 2011, 2012). Evaluating data of mineral nutrient concentrations in heartwood and sapwood of 22 species of gymnosperms and 71 species of angiosperms, it was found that P, N, and K contents are mostly lower in heartwood compared to sapwood. Concentration patterns of Ca and Mg showed great variations between wood species, in average with higher contents in the heartwood (Meerts 2002). Thus, Ca and Mg seem to be less mobile than P, N, and K with the latter (P, N, K) resorbed from senescing wood. At least for phosphorous, the translocation from senescing to younger sapwood tissues was recently shown for species such as bald cypress (*Taxodium distichum,* Galicki et al. 2008), balsam fir (*Abies balsamea* (L.) Mill.), white spruce (*Picea glauca* (Moench) Voss; Houle et al. 2008), *Quercus petraea* Liebl. (Vansteenkiste et al. 2007), and *Robinia pseudoacacia* L. (Passialis et al. 2008). Together with lowest heartwood/sapwood ratios of the P content ($<0.36$), it is assumed that P plays a crucial role in sapwood metabolism, e.g., in the form of adenine nucleotides, needed for heartwood formation (Magel 2000; Magel and Höll 1993). For other elements such as K, Ca, Mg, Mn, Fe, Al, or As, distribution pattern between sapwood and heartwood is more complex and depends in addition to the species under investigation (Meerts 2002), and environmental conditions (Galicki et al. 2008; Houle et al. 2008), highly on the method used (Augusto and Bert 2005). For carbon the by far dominating element concentrations are higher in heartwood than in bark or sapwood of *Pinus* species (De Aza et al. 2011).

Within the woody stem, the amount of living cells strongly decreases from the outer to the inner parts (Hauch and Magel 1998; Hillis 1987; Magel 2000; Spicer

2005; Taylor et al. 2002). At least for the longevity of ray parenchyma cells, there seem to be differences between gymnosperms and angiosperms. In most gymnosperms, a gradual loss of living cells across the sapwood up to the sapwood–heartwood transition zone is reported. However, in most angiosperms, ray parenchyma cells stay alive up to the innermost sapwood tissues and show an abrupt decrease at the transition zone (Spicer 2005, and references therein). Theories for the actual causes of cell death of both axial and ray parenchyma cells are numerous (formation of tylosis, gums, and/or extractives; decrease in moisture content; increase in gas volume, by an accelerated increase of $CO_2$; phytohormones); however, a final clue is missing (Spicer 2005 and references therein). Introducing modern biological methods such as molecular approaches and/or high-resolution imaging techniques or quantitative histochemistry (such as laser micro dissectioning techniques) should improve our understanding.

Parenchyma cells within the sapwood show profound morphological changes starting with differentiation within the cambial zone up to cell death at the sapwood–heartwood border and lack of protoplasm within the heartwood. Most obvious are, e.g., tylosis and thus compartmentalization of vessels; changes in amount, shape, and size of the cells; accumulation of extractives; and degeneration of the nucleus (Hillis 1987; Song et al. 2011; Spicer 2005; Taylor et al. 2002). The latter being an unmistakable sign of cell death. In *Bridelia retusa*, nuclei are present in axial and ray parenchyma cells of the sapwood and the transition zone, with a fusiform appearance within the transition zone (Nair et al. 1981). Within the ageing sapwood and the sapwood–heartwood transition zone of *Acacia auriculiformis*, nuclei accumulate arginine-rich histones and show different modes of desintegrity within ray and axial parenchyma cells contiguous and noncontiguous to vessels, pointing towards an increased activity of cells adjacent to vessels (Bhat and Patel 1980, 1982). In *Pinus*, differentiation and thus cell death of ray parenchyma cells during heartwood formation seems to be influenced by neighboring cell types. It is found that in ray parenchyma cells in contact with ray tracheids, cell death occurs early. Thus these parenchyma cells do not play a role in heartwood formation (Nakaba et al. 2008). The initiation of heartwood formation in branches of *Robinia pseudoacacia* starts within 4 years in pith regions. It is assumed that xylem ray parenchyma cells and small parenchyma cells in the premedullary zone are involved in the synthesis of heartwood extractives (Nakaba et al. 2012). Finally, the absence of protoplasm and of nuclei in ray parenchyma in middle and inner heartwood is a sign of ray parenchyma's cell death (Song et al. 2011). Heartwood formation is characterized by the death of parenchyma cells; however, it is assumed that their death is the result, not the cause, of heartwood formation (Bamber 1976).

Concomitant with cell death of parenchyma cells with ageing of the sapwood, quantity and quality of DNA and RNA decline and thus impede application of molecular techniques (Abe et al. 2011; Fukazawa and Higuchi 1966; De Filippis and Magel 1998; Magel et al. 2002; Rachmayanti et al. 2009).

Further proof for disintegration of parenchyma cells is analyses of lipids and lipolytic enzymes in *Robinia pseudoacacia*. The amount of phospholipids, a parameter of parenchyma cell integrity, decreases towards the transition zone, and in the heartwood, only trace amounts are found. Additionally, amounts of free sterols decline from sapwood to transition zone but reach a maximum in heartwood. Authors associated the decrease of both compounds to a first step of cell deterioration. The increase of free sterols in the heartwood may be due to a possible participation in defense mechanisms against pathogens (Hillinger et al. 1996).

Disintegration of parenchyma cells is accompanied by depletion of reserve substances, starch, nonstructural carbohydrates, and triacylglycerols. Starch grains abundant in the outer sapwood gradually decrease from middle sapwood towards the transition zone. Phosphatases hydrolyze starch in the middle sapwood to sugar phosphates and thus to precursors for the synthesis of either lipids or extractives. Small lipophilic droplets are found in the axial and ray parenchyma of the outer sapwood. These droplets increase remarkably in size in cells of the middle and inner sapwood as well as in the transition zone. It is assumed that within these tissues, lipids and carbohydrates provide carbon skeletons for the synthesis of heartwood extractives. The heartwood proper is free of storage material. Thus, starch and lipids are degraded during heartwood formation and play an important role as carbon skeletons for the biosynthesis of phenolic extractives and thus senescence of ray parenchyma cells (Bhat and Patel 1982; Baqui and Shah 1985; Hillis 1987; Lamlom and Savidge 2006; Magel 2000, 2002; Nair et al. 1981; Song et al. 2011; Ucar and Ucar 2011).

## 6 Biochemical Aspects of Heartwood Formation

Heartwood formation, a kind of programmed cell death in the oldest sapwood issues, is the final step in the life cycle of living xylem cells. In the transition zone between living sapwood and dead heartwood, parenchyma cells die genetically determined and extractives are accumulated. In the following, the current knowledge regarding the biochemical basis of these processes is summarized.

Based on macroscopic, microscopic, and biochemical findings, at least two types of obligate heartwood formation can be distinguished: Type I, *Robinia*-Type of heartwood formation, is characterized by the accumulation of phenolic extractives starting in the tissues transient between sapwood and heartwood with no phenolic precursors in the ageing sapwood (Bergström et al. 1999, 2003; Magel 2000; Magel et al. 1994; Nair et al. 1981; Roux and Paulus 1962). In Type II or *Juglans*-Type of heartwood formation, phenolic precursors gradually accumulate centripetally in the ageing sapwood tissues. In the transition zone, characteristic heartwood extractives are formed by primary (de novo biosynthesis) and secondary reactions (oxidation, hydrolysis of precursor substances; Burtin et al. 1998; Dellus et al. 1997a, b; Mayer et al. 2006).

At least for Type I of heartwood formation, the biochemical basis of this senescence process is thoroughly investigated and thus well understood in *Robinia pseudoacacia*. Heartwood formation in the sapwood-heartwood transition zone is characterized by a transient activation and shift of metabolic activities leading to in situ synthesis of heartwood extractives. It seems likely that cell death starts with the depletion of storage substances, both starch and lipids, due to an increase of the activity of hydrolyzing enzymes such as amylases, phosphatases, and lipases (Magel et al. 1997). This is facilitated by chemical and structural alterations of these reserve components with ageing of the sapwood. The tremendous imbalance between carbon skeletons available in the transition zone and carbon skeletons fixed as phenolics within the heartwood is overcome by a steady import of sucrose from outer sapwood tissues into the transition zone. Here sucrose is predominantly cleaved by sucrose synthase (SuSy), a strong marker for sink tissues in plants. High demands of energy and substrates during biosynthesis of the extractives require a concerted action of sucrose decomposition and metabolism via the "sucrose synthase pathway" (Xu et al. 1989). Due to the correlation of enhanced catalytic activities, enzyme protein levels, and SuSy-specific mRNA pools, regulation of this enzyme is obviously by gene expression. This concomitant with increased catalytic activities of a second sucrose cleaving enzyme, neutral invertase, assigns the transition zone as a strong sink tissue. In the beginning, products of sucrose breakdown, fructose, glucose, and UDP-glucose, are predominantly used for glycolytic and respiratory energy production (Höll and Lendzian 1973; Magel and Höll 1993), and to a minor extent for the synthesis of extractives. Synthesized phenolic heartwood extractives accumulate within the cytosol and might be strong inhibitors of mitochondrial electron flow. Thus mitochondria are regarded to be the first organelles ceasing activity (Hillis 1987; Ziegler 1968). Carbon skeletons accumulating due to nonfunctional mitochondria together with products of the accelerated oxidative pentose-phosphate pathway (Magel et al. 2001b) are increasingly used for the synthesis of phenolics via the general phenyl propanoid metabolism (Baqui and Shah 1985; Hauch and Magel 1998; Nobuchi and Harada 1985; Hillis 1987; Magel et al. 1991, 1994; Magel and Hübner 1997). Concomitantly, a transient expression and activation of key enzymes of the biosynthesis of secondary substances such as flavonoids, phenylalanine ammonia lyase (PAL), and chalcone synthase (CHS) is correlated with the accumulation of heartwood extractives (Magel and Hübner 1997). Like during senescence processes in herbaceous plants, this activation of enzymes enabling the in situ synthesis of heartwood phenolics seems to be related to ethylene production (Ke and Saltveit 1988; Magel 2000; Shain and Hillis 1973). The close correlation between primary (carbohydrate) and secondary (synthesis of heartwood phenols, flavonoids) metabolism during the synthesis of heartwood extractives is manifested by linear correlations of the catalytic activities of SuSy and PAL, and SuSy and CHS, with coefficients of correlation of 0.956 and 0.970, respectively (Magel 2000). From these results, we conclude that the type of programmed cell death that occurs during heartwood formation shares some biochemical features with, e.g., the type of programmed cell death that mature leaves undergo during senescence: Both processes require activation of hydrolytic enzymes, gene expression, and de novo protein synthesis. Using 2D SDS-PAGE (two-dimensional sodium dodecyl sulfate polyacrylamide gel electrophoresis) and electrospray ionization tandem mass spectrometry (ESI-MS/MS), the proteome of sapwood and transition zone of black locust was investigated (De Filippis and Magel 2012). Proteins strongly expressed in the transition zone are stress proteins such as heat-stress proteins and lectins, and enzymes involved in flavonoid biosynthesis. Moreover, changes in the biochemical and physiological state of the cells undergoing heartwood formation are profound and based on genomic changes at the posttranscriptional (DNA) level, as revealed by RAPD-PCR (random amplification of polymorphic DNA—polymerase chain reaction; De Filippis and Magel 1998).

For the biochemical basis of heartwood formation in the second type, Type II, or *Juglans*-Type, only few data are available. This type occurs in genera and species such as *Juglans* (Burtin et al. 1998), *Pseudotsuga* (Dellus et al. 1997a, b), *Platycarya* (Tanaka et al. 1998), *Prunus serotina* (Mayer et al. 2006), and *Eucalyptus globulus* (Morais and Pereira 2012). In these species, heartwood precursors, phenolic glycosides, and other low-molecular weight phenolic compounds augment in the ageing sapwood tissues (Burtin et al. 1998; Taylor et al. 2007). The presence of these substances seems to cause undesired discoloration within the sapwood during transport and storage (Koch 2004) or desired once during steaming (Burtin et al. 2000; Kollmann et al. 1951). Like in the *Robinia*-Type, ageing of the sapwood is characterized by a decrease of living parenchyma cells manifested by declining contents of protein and reserve substances (Burtin et al. 1998; Magel et al. 2001a). Heartwood formation and thus formation of heartwood extractives in the *Juglans*-Type is a two-step process: Moderate amounts of newly emerging flavanols are in situ synthesized by acceleration of metabolic pathways similar to Type I. Carbohydrates, most of all sucrose, are hydrolyzed by increased activities of acid invertases (Magel et al. 2001a) and cleaving products enable the in situ biosynthesis of flavanols via general phenyl propanoid metabolism and flavonoid biosynthetic pathways (Beritognolo et al. 2002; Magel et al. 2001a). Candidate enzymes involved (phenylalanine ammonia lyase, chalcone synthase, flavanone 3-hydroxylase, and dihydroflavonol 4-reductase) are regulated by gene expression and de novo protein synthesis (Beritognolo et al. 2002). However, the bulk of heartwood extractives is formed by secondary reactions such as hydrolysis of phenolic glycosides (Duroux et al. 1998), oxidation, interconversion, and polymerization of low-molecular weight phenols (Burtin et al. 1998; Dellus et al. 1997a, b). Whether oxidation of these phenolics is by peroxidases and/or nonenzymatic is still under discussion (Dehon et al. 2002). At least in *Juglans nigra*, glucose liberated during this deglycosylation is not further metabolized by the dying parenchyma cells and accumulates within the heartwood (Burtin et al. 1998; Magel et al. 2001a).

Independently of type of heartwood formation, histochemical, biochemical, and genetical (see below) results point to an initiation of processes of heartwood formation in late summer, with highest activities in autumn, and cessation during dormancy (Dehon et al. 2002; Hillis 1987; Magel 2000; Hauch and Magel 1998; Magel et al. 1997, 2001a; Nelson 1978; Nobuchi et al. 1984).

## 7 Heartwood Formation and Gene Expression

Despite of the economical and ecological importance of heartwood, the regulation of its formation remains poorly understood. This is due to limitations of molecular methods by few living cells within the transition zone and thus low amounts of DNA and RNA which in addition is of poor quality (Abe et al. 2011; De Filippis and Magel 1998; Fukazawa and Higuchi 1966; Magel et al. 2002; Rachmayanti et al. 2009). Accumulating phenolic components hamper biochemical and molecular approaches (Déjardin et al. 2010; Magel 2002; Magel et al. 2002). Overcoming these difficulties, quite a number of successfully conducted molecular investigations shed some more light on heartwood formation and underlying regulatory processes.

Expressed sequence tags (ESTs) from the transition zone of *Robinia* (Yang et al. 2003) revealed a total of 2,278 unigene sets. Genes representing secondary and hormone metabolism were more abundant in the transition zone compared to other tissues such as sapwood or bark/cambium (9.1 % versus 1.4 % and 0.8 %), whereas cell wall and structural metabolism genes were stronger represented in the sapwood (3.1 % versus 0.6 % and 0.9 %). Additionally, the number of ESTs of genes coding for chromatin and DNA metabolism, defense reaction, primary metabolism, protein synthesis and processing, and signal transduction is higher in transition zone in summer (harvested in July) compared with transition zone in fall (harvested end of November) or with sapwood. In general, highest expression of genes encoding sugar transport was present in sapwood, while structural genes coding for flavonoid biosynthesis were highest expressed in the transition zone. Here the total number of sequenced clones was high in summer (880) and strongly decreased until November (141). Functional classification of the ESTs in the transition zone of *Cryptomeria* (Yoshida et al. 2007) associated most of the clones to posttranslational modification, protein turnover, chaperones, followed by translation, ribosomal structure and biogenesis, intracellular trafficking, secretion, vesicular transport, signal transduction mechanisms, lipid transport and metabolism, transcription, RNA processing and modification, and carbohydrate transport and metabolism. Additionally, ESTs from drying and discoloring sapwood of *Cryptomeria* (Yoshida et al. 2012) revealed a preponderance of genes coding for secondary metabolism and defense-related proteins. ESTs studies were refined by employing cDNA microarray analyses of the transition zone of black locust (Yang et al. 2004) and walnut (Huang et al. 2010). Results of both studies corroborated the findings, manifesting the upregulation of genes coding for defense and cell rescue, gene expression and RNA metabolism, protein synthesis and processing, secondary metabolism, and signal transduction in the transition zone. Some of the genes, e.g., involved in signal transduction, were differentially regulated in summer and fall samples of the transition zone, pointing to a reaction of ray parenchyma cells to seasonal changes. These results showed that some of the genes differentially expressed in the transition zone overlap with those during xylem formation such as genes involved in transcriptional regulation, polyphenol and flavonoid biosynthesis, and programmed cell death (Huang et al. 2010; Moreau et al. 2005). In order to pinpoint candidate genes for heartwood formation, the expression of specific genes was evaluated by real-time PCR and semiquantitative RT-PCR approaches. Selection of the genes was by their abundance in the transition zone as found by EST or microarrays and involvement in metabolic pathways ascribed to heartwood formation such as glycolysis, carbohydrate metabolism, and synthesis of phenolic extractives. Following these methods, regulation of enzymes involved in sucrose cleavage such as sucrose synthase (Hauch and Magel 1998) and invertase, enzymes of glycolysis and lipid transfer (methionine adenosyltransferase and glutathione transferase; Yoshida et al. 2012), genes coding for enzymes involved in the syntheses of flavonoids (phenylalanine ammonia lyase, chalcone synthase, flavanone 3-hydroxylase, and dihydroflavonol 4-reductase, Beritognolo et al. 2002), and stilbene synthase (Preisig-Müller et al. 1999) were investigated. Results confirmed biochemical, spatial, and temporal data and elucidated that enzymes involved in the proposed shift in metabolism during heartwood formation are regulated by gene expression. In situ mRNA hybridization of, e.g., the dirigent protein involved in the biosynthesis of heartwood lignans in western red cedar (Patten et al. 2008) and of two lignin biosynthetic pathway enzymes, pinoresinol-lariciresinol reductase and phenylcoumaran benzylic ether reductase involved in the synthesis of heartwood lignans in *Pinus taeda* (Kwon et al. 2001), revealed tissue-specific expression and thus localization in axial and ray parenchyma cells with a preponderance in radial parenchyma cells. In *Juglans nigra,* the *JnCML*-like gene similar to grancal-cin-like calcium-binding EF-hand proteins in *Arabidopsis thaliana* and *Zea mays* containing EF-hand motifs is upregulated in the transition zone (Huang et al. 2011). As EF-hand motif proteins are associated with cell proliferation, differentiation, and programmed cell death, more light is shed on the involvement of programmed cell death during heartwood formation. First indications of the involvement of transcription factors in the regulation of heartwood formation are given by detection of JnKNAT3-like, a KNAT3-like homeobox gene from *Juglans nigra* L. which is highly expressed during summer in the sapwood-heartwood transition zone (Huang et al. 2009).

## 8 Heartwood Extractives, "Abnormal" Lignins, "Secondary" Lignification, and "Pseudolignification"

Independently of their synthesis either in Type I or Type II of heartwood formation, heartwood extractives are secondary metabolites, mostly phenolic compounds, and are responsible for color, biological (natural durability), technological, and commercial properties of heartwood and its products (Hillis 1987). Heartwood extractives show a large variety of different chemical structures and thus belong to different chemical classes such as tannins, terpenes, flavonoids, stilbenes, lignins, aromatic substances, and lectins (Hillis 1987). The quality and quantity of these extractives are species dependent, genetically determined, and under environmental control (Bito et al. 2011; Bush et al. 2011; Hillis 1987; Kwon et al. 2001; Taylor et al. 2002). Additionally, their amount depends on age of the tree, age of wood, and nature of the wood, with, e.g., lower amounts in juvenile compared to mature heartwood (Dünisch et al. 2010; Latorraca et al. 2011). As mentioned above, natural durability and color not always correlate with the concentration of heartwood extractives (Hillis 1987; Niamke et al. 2011; Taylor et al. 2002). Natural durability of heartwood extractives is due to both its biotoxic and antioxidant activity. In *Acacia catechu*, it was shown that heartwood extractives had higher antioxidant capacity than extracts from leaves or bark. Moreover it is discussed that heartwood extractives significantly protect plasmid DNA against strand scission induced by hydroxyl radicals (Guleria et al. 2011). Additionally, like in *Mangifera indica*, the antioxidant activity correlates with total phenol contents (Kawamura et al. 2011).

In the beginning of heartwood formation, these substances appear as lipophilic, apolar droplets in the vicinity of amyloplast. During disintegration of the protoplasma, these droplets stick to the pits, and further on, the content is released into the lumen of neighboring cells and cell walls (Magel et al. 1995; Déjardin et al. 2010). Thus, cell walls in heartwood are infiltrated with extractives, leading to reduced shrinkage and swelling capacity, and account for its durability. Using TOF-SIMS (time-of-flight secondary ion mass spectrometry), the discrimination of heartwood and sapwood at the tissue level is possible (Saito et al. 2008a, b). This method also enables the direct mapping of the morphological distribution of, e.g., heartwood extractives (Imai et al. 2005; Nakaba et al. 2012) and lignin components (Saito et al. 2012) within the cell wall. Using this method and UV microspectrophotometry (Dünisch et al. 2010; Imai et al. 2005; Nakaba et al. 2012), the localization of heartwood extractives in cell walls (and lumen of cells) was proven. Nevertheless, the chemical reactivity of heartwood extractives and preexisting cell wall components is poorly understood. It seems likely that infiltration of the cell walls involves a mechanism of an enzymatically initiated but chemically driven copolymerization between these phenolic derivatives and the preexisting cell wall macromolecular components including lignins and lignin-polysaccharide complexes (Monties 1987, 1989, 1991). Such a mechanism would allow to explain the occurrence of "abnormal" lignins (Gang et al. 1998) or the relative increase in lignin content of heartwood as reported for *Robinia* (Magel et al. 1995), *Quercus* (Jouin et al. 1988), and *Quebracho* (Fengel 1991) and would thus mitigate the concept of "secondary lignification" (Hergert 1977). Based on the occurrence of associations between phenolic extractives and cell wall components such as the lignin network, at least in angiosperm trees, a de novo synthesis of lignin during heartwood formation can be excluded. The term "secondary lignification" should therefore be changed into "pseudo lignification." In softwoods, in contrast, the lignification of, e.g., pit membranes of tracheids during heartwood formation (Hillis 1987; Bertaud and Holmbom 2004) justifies the term "secondary lignification."

## 9 Heartwood Extractives as Biocides

Heartwood's phenolic extractives render heartwood its specific appearance and manifest its natural durability by acting as potential bioprotectant towards a large variety of pathogens. "Natural" durability is mainly assigned to resistance against attack of wood-decaying fungi due to antifungal effects of the extractives. These effects have been studied for various species, e.g., *Myracrodruon urundeuva* (Sá et al. 2009), *Juniperus* (Clark et al. 1990), *Calocedrus macrolepis* (Yen et al. 2008), *Robinia pseudoacacia* (Pollet et al. 2008), *Cunninghamia lanceolata* (Wang et al. 2011a), and *Juglans regia* (Hashemi and Latibari 2011). Some extracts, like the hexane extract from the heartwood of *Dalbergia congestiflora* with medicarpin as the dominating antifungal compound, cause up to 100 % inhibition of fungal growth (Martinez-Sotres et al. 2012). However, biodeterioration of wood is also caused by insects such as termites, beetles, and marine borers. Heartwood of *Pinus thunbergii* is rich in alpha-terpineol and longifolene, components which inhibit growth of red tide plankton and is thus proposed as useful construction material (Tsuruta et al. 2011). In recent publications, emphasis was on heartwood and/or heartwood extractives exhibiting fungitoxic and/or termiticidal activities. This is of importance as termites cause great economic losses of timber preferentially in the tropics. In heartwood of *Pinus* sp., antioxidative properties are mainly due to the presence of morin, catechin, flavanone, and tannic acid. Its favorable effects for humans cause feeding deterrence and mortality in the termite *Reticulitermes flavipes* (Little et al. 2010). Dungani et al. (2012) reported that in heartwood of *Tectona grandis*, quinones seem to be the dominant toxic components for termites. Acetone–water extracts from oldest heartwood portions exhibited mortality rates of up to 100 % to the termite *Coptotermes curvignathus*. In wood of *Sextonia rubra*, rubrynolide was identified as the most effective protector substance against termites. Due to its potential to inhibit growth of tropical and invasive termites, it is discussed as a natural wood preservative to replace synthetic termiticides (Rodrigues et al. 2011). In *Cryptomeria japonica* heartwood, the resistance to termites correlates with the amount of specific extractives, which are localized preferentially in the cell wall of tracheids (Kijidani et al. 2011). Up to now, it is still an open question whether the durability of heartwood against termites is based exclusively on the quantity and quality of extractives or on morphological aspects like the heartwood's higher density (Kadir and Hale 2012). Besides phenolic components, lectins are toxic for termites. In the heartwood of *Myracrodruon urundeuva*, lectins with termiticidal activity against *Nasutitermes corniger* are present (Napoleao et al. 2011). It is assumed that termiticidal activity of lectins is complex and is based on various aspects such as the chitin-binding property, resistance to termite digestive enzymes, and the antibacterial effect on symbiotic bacteria of termites' gut. Besides being active repellents against termite attack, heartwood lectins exhibit also fungitoxic and bactericidic activities (Sa et al. 2009). The lectin from *Myracrodruon urundeuva* heartwood inhibits also growth of Gram-positive and Gram-negative bacteria and of phytopathogenic fungi. Thus, it is the first bioactive peptide found in heartwood. Additionally, active heartwood components which show both, resistance against termites and wood-decaying white-rot, brown-rot, and soft-rot fungi are found in the heartwood of softwood and hardwood species such as *Juniperus virginiana* L. (Eller et al. 2010; Mun and Prewitt 2011), *Dalbergia latifolia* L. (Sekine et al. 2009), *Caesalpinia echinata* (Da Silva et al. 2007), and *Prosopis kuntzei* (Scholz et al. 2012). Predominantly in the tropics, great losses of timber are due to combined attack of both wood destroying termites and fungi. Thus these components offer a chance to develop methods to find environmentally friendly biocides which are effective components against different vermins.

Concerning biocidic activity of heartwood and/or heartwood extractives, it is upcomingly discussed that highly durable wood seems to be protected against biodeterioration by a dual defensive function of the extractives: In addition to exhibiting some fungicidal and termiticidal activities based on their chemical structure (Kawamura et al. 2011), they also act as free radical scavengers/antioxidants and metal chelaters (Schultz and Nicholas 2000; Cheng et al. 2008). By this, the development of new commercial biocides or wood preservatives based on the synergistic effect of, e.g., commercial antioxidants and natural biocides is encouraged. Additionally, it is discussed that extractives could act as biocides by forming a mechanical barrier for decaying organisms. However, the relationship between the mechanical and the chemical aspects is poorly understood (Taylor et al. 2002). Moreover, it is not known how chemical ageing of heartwood extractives, which alters specific dynamic modulus and the damping coefficient of wood and heartwood (Zhang et al. 2011), may affect its biotoxic activity.

## 10 Heartwood Extractives as Pharmaceuticals

For long, heartwood extractives have served for a variety of mankind's needs, such as renewable resources, perfumes, dyes, varnishes and lacquers, or medicines (Hillis 1987). Ethanolic extracts from dried heartwood of various species were identified as herbal medicine and for therapeutic issues (Jeong et al. 2011). Bioassays with, e.g., the ethanol extract from heartwood of *Caesalpinia sappan* revealed promising results by attenuating collagen-induced arthritis in rats (Wang et al. 2011b), or heartwood extractives and lignans from *Streblus asper* had anti-Hepatitis B virus activities (Chen et al. 2012; Li et al. 2012b). During the last years, heartwood extractives were discussed as anticancer drugs. Wu et al. (2011a, b) identified bioactive compounds (benzofurans, neoflavonoids, neoflavone, phenanthrenedione, and chalcone) in heartwood extractives of *Pterocarpus santalinus*, which show inhibitory effects on elastase release from human neutrophils and anti-inflammatory effects and additionally exhibit potent cytotoxicity against cancer cell lines. An aqueous extract from the heartwood of *Acacia catechu* showed a dose-dependent growth inhibition of human carcinoma cells (Monga et al. 2011). It is assumed that this chemopreventive effect is based on the antioxidant activity of the components. Lignans and sesquiterpenoids from the heartwood of *Taiwania cryptomerioides* (Chang et al. 2003); flavonoids from *Dalbergia odorifera* (Choi et al. 2009); isoflavones, isoflavanones, flavanones, and isoflavans from the heartwood of *Dalbergia parviflora* (Umehara et al. 2009); and spirocyclic lignans from the heartwood of *Guaiacum* spp. (Chavez et al. 2011) are cytotoxic against human cancer lines, e.g., human breast cancer, induce apoptosis, and show antitumor properties. Additionally, a flavanone from *Dalbergia odorifera* heartwood modulates the regulation of antioxidative and anti-inflammatory activities by upregulation of specific cells (Li et al. 2012a). Thus, this flavanone is regarded as a powerful substance against neurodegenerative diseases which are induced by oxidative stress and neuroinflammation. Based on these findings, new cancer treatments and promising strategies are developed involving the chemopreventive and chemotherapeutic agents isolated from heartwood of different soft- and hardwood species (Songsiang et al. 2009; Shan et al. 2011). Besides being potential anticancer drugs, specific heartwood extractives are identified as antifungal agents, both against wood-decaying and human pathogenic fungi. Heartwood extractives of *Bagassa guianensis* Aubl. exhibit strong antifungal activity against the wood-destroyer *Pycnoporus sanguineus* and two human pathogenic fungi, *Candida glabrata* (yeast) and *Trichophyton rubrum* (Royer et al. 2012). Essential oils extracted from the heartwood of *Callitris neocaledonica* and *C. sulcata* (Cupressaceae) showed antifungal activity against clinical isolates of four human dermatophytic fungi and six yeasts (Waikedre et al. 2012). These promising results encourage the search for applications to counteract mortality due to fungal infections by elucidating wood extractives as antimycotic products. Besides these favorable characteristics of heartwood extractives, some of them show useful properties in diabetes (Achari et al. 2012).

## References

Abe H, Watanabe U, Yoshida K, Kuroda K, Zhang CH (2011) Changes in organelle and DNA quality, quantity, and distribution in the wood of *Cryptomeria japonica* over long-term storage. IAWA J 32:263–272

Achari B, Dutta PK, Roy SK, Prarthana Chakraborty P, Sengupta J, Bandyopadhyay D, Maity JK, Khan IA, Ding Y, Ferreira D (2012) Fluorescent pigment and phenol glucosides from the heartwood of *Pterocarpus marsupium*. J Nat Prod 75:655–660

Anonymous (1957) International glossary of terms used in wood anatomy (prepared by the International Association of Wood Anatomists). Trop Woods 107:1–36

Augusto L, Bert D (2005) Estimating stemwood nutrient concentration with an increment borer: a potential source of error. Forestry 78:451–455. doi:10.1093/forestry/cpi041

Bamber RK (1976) Heartwood, its function and formation. Wood Sci Technol 10:1–8. doi:10.1007/bf00376379

Bamber RK, Fukazawa K (1985) Sapwood and heartwood: a review. For Abstr 46:567–580

Baqui SA, Shah JJ (1985) Histoenzymatic studies in wood of *Acacia auriculiformis* Cunn. during heartwood formation. Holzforschung 39:311–320

Bergström B (2003) Chemical and structural changes during heartwood formation in *Pinus sylvestris*. Forestry 76:45–53. doi:10.1093/forestry/76.1.45

Bergström B, Gustafsson G, Gref R, Ericsson A (1999) Seasonal changes of pinosylvin distribution in the sapwood/heartwood boundary of *Pinus sylvestris*. Trees 14:65–71. doi:10.1007/pl00009754

Beritognolo I, Magel E, Abdel-Latif A, Charpentier JP, Jay-Allemand C, Breton C (2002) Expression of genes encoding chalcone synthase, flavanone 3-hydroxylase and dihydroflavonol 4-reductase correlates with flavanol accumulation during heartwood formation in *Juglans nigra*. Tree Physiol 22:291–300. doi:10.1093/treephys/22.5.291

Bertaud F, Holmbom B (2004) Chemical composition of earlywood and latewood in Norway spruce heartwood, sapwood and transition zone wood. Wood Sci Technol 38:245–256

Bhat K, Patel J (1980) Nuclear studies in relation to heartwood formation in *Ougeinia oojeinensis* Roxb. and *Garuga pinnata* Roxb. Caryologia 33:519–526

Bhat K, Patel J (1982) Nuclear behaviour during heartwood formation in *Acacia auriculiformis* A. Cann. Proc Indian Acad Sci B Part 3 Plant Sci 91:107–114. doi:10.1007/bf03167114

Bieker D, Rust S (2010) Non-Destructive estimation of sapwood and heartwood width in Scots pine (*Pinus sylvestris* L.). Silva Fennica 44:267–273

Bito N, Nakada R, Fukatsu E, Matsushita Y, Fukushima K, Imai T (2011) Clonal variation in heartwood norlignans of *Cryptomeria japonica*: evidence for independent control of agatharesinol and sequirin C biosynthesis. Ann For Sci 68:1049–1056. doi:10.1007/s13595-011-0118-7

Burtin P, Jay-Allemand C, Charpentier JP, Janin G (1998) Natural wood colouring in *Juglans* sp. (*J. nigra*, *J. regia* and hybrid *J. nigra* 23 × *J. regia*) depends on native phenolic compounds accumulated in the transition zone between sapwood and heartwood. Trees 12:258–264

Burtin P, Jay-Allemand C, Charpentier JP, Janin G (2000) Modifications of hybrid walnut (*Juglans nigra* 23 × *Juglans regia*) wood colour and phenolic composition under various steaming conditions. Holzforschung 54:33–38

Bush D, McCarthy K, Meder R (2011) Genetic variation of natural durability traits in *Eucalyptus cladocalyx* (sugar gum). Ann For Sci 68:1057–1066. doi:10.1007/s13595-011-0121-z

Chakravarty P, Sen Sarma N, Sarma HP (2010a) Biosorption of cadmium(II) from aqueous solution using heartwood powder of *Areca catechu*. Chem Eng J 162:949–955

Chakravarty P, Sen Sarma N, Sarma HP (2010b) Removal of lead(II) from aqueous solution using heartwood of *Areca catechu* powder. Desalination 256(1–3):16–21

Chakravarty P, Deka DC, Sarma NS, Sarma HP (2012) Removal of copper (II) from wastewater by heartwood powder of *Areca catechu*: kinetic and equilibrium studies. Desalin Water Treat 40:194–203. doi:10.5004/dwt.2012.2920

Chang ST, Wang SY, Kuo YH (2003) Resources and bioactive substances from *Taiwania* (*Taiwania cryptomerioides*). J Wood Sci 49:0001–0004. doi:10.1007/s100860300000

Chavez KJ, Feng XH, Fanders JA, Rodriguez E, Schroeder FC (2011) Spirocyclic lignans from *Guaiacum* (Zygophyllaceae) induce apoptosis in human breast cancer cell lines. J Nat Prod 74:1293–1297. doi:10.1021/np100891y

Chen H, Li J, Wu Q, Niu XT, Tang MT, Guan XL, Li J, Yang RY, Deng SP, Su XJ (2012) Anti-HBV activities of *Streblus asper* and constituents of its roots. Fitoterapia 83:643–649. doi:10.1016/j.fitote.2012.01.009

Cheng SS, Liu JY, Chang EH, Chang ST (2008) Antifungal activity of cinnamaldehyde and eugenol congeners against wood-rot fungi. Bioresour Technol 99:5145–5149. doi:10.1016/j.biortech.2007.09.013

Choi C, Choi Y, Cha MR, Kim Y, Yon G, Kim YK, Choi S, Kim Y, Ryu S (2009) Antitumor components isolated from the heartwood extract of *Dalbergia odorifera*. J Korean Soc Appl Biol Chem 52:375–379. doi:10.3839/jksabc.2009.067

Clark AM, McChesney JD, Adams RP (1990) Antimicrobial properties of heartwood, bark/sapwood and leaves of *Juniperus* species. Phytother Res 4:15–19

Da Silva CA, Bacellar Monteiro MB, Brazolin S, Carballeira Lopez GA, Richter A, Braga MR (2007) Biodeterioration of brazilwood *Caesalpinia echinata* Lam. (Leguminosae-Caesalpinioideae) by rot fungi and termites. Int Biodeterior Biodegradation 60:285–292

De Aza CH, Turrion MB, Pando V, Bravo F (2011) Carbon in heartwood, sapwood and bark along the stem profile in three mediterranean *Pinus* species. Ann For Sci 68:1067–1076. doi:10.1007/s13595-011-0122-y

De Filippis L, Magel E (1998) Differences in genomic DNA extracted from bark and from wood of different zones in *Robinia* trees using RAPD-PCR. Trees 12:377–384

De Filippis L, Magel E (2012) Identification of biochemical differences between the sapwood and transition zone in *Robinia pseudoacacia* L. by differential display of proteins. Holzforschung 66:543–549. doi:10.1515/hf.2011.178

Dehon L, Macheix JJ, Durand M (2002) Involvement of peroxidases in the formation of the brown coloration of heartwood in *Juglans nigra*. J Exp Bot 53:303–311. doi:10.1093/jexbot/53.367.303

Déjardin A, Laurans F, Arnaud D, Breton C, Pilate G, Leplé JC (2010) Wood formation in angiosperms. C R Biol 333:325–334. doi:10.1016/j.crvi.2010.01.010

Dellus V, Mila I, Scalbert A, Menard C, Michon V, Herve du Penhoat CLM (1997a) Douglas-fir polyphenols and heartwood formation. Phytochemistry 45:1573–1578. doi:10.1016/s0031-9422(97)00245-8

Dellus V, Scalbert A, Janin G (1997b) Polyphenols and colour of Douglas fir heartwood. Holzforschung 51:291–295

Dungani R, Ul Haq Bhat I, Abdul Khalil HP, Naif A, Hermawan D (2012) Evaluation of antitermitic activity of different extracts obtained from Indonesian teakwood (*Tectona grandis* L.f.). BioResources 7:1452–1461

Dünisch O, Richter HG, Koch G (2010) Wood properties of juvenile and mature heartwood in *Robinia pseudoacacia* L. Wood Sci Technol 44:301–313. doi:10.1007/s00226-009-0275-0

Duroux L, Delmotte F, Lancelin JM, Keravis G, Jay-Allemand C (1998) Insight into naphthoquinone metabolism: ß-glucosidase-catalysed hydrolysis of hydrojuglone ß-D-glucopyranoside. Biochem J 333:275–283

Ekeberg D, Flæte PO, Eikenes M, Fongen M, Naess-Andresen CF (2006) Qualitative and quantitative determination of extractives in heartwood of Scots pine (*Pinus sylvestris* L.) by gas chromatography. J Chromatogr A 1109:267–272. doi:10.1016/j.chroma.2006.01.027

Eklund L, Klintborg A (2000) Ethylene, oxygen and carbon dioxide in woody stems during growth and dormancy. In: Savidge R, Barnett J, Napier R (eds) Cell and molecular biology of wood formation. BIOS Scientific Publishers, Oxford

Eller FJ, Clausen CA, Green F, Taylor SL (2010) Critical fluid extraction of *Juniperus virginiana* L. and bioactivity of extracts against subterranean termites and wood-rot fungi. Ind Crops Prod 32:481–485. doi:10.1016/j.indcrop. 2010.06.018

Fengel D (1991) Chemical studies on the wood of *Quebracho colorado* (part 1). Holzforschung 45:259–263

Forrester DI, Collopy JJ, Morris JD (2010) Transpiration along an age series of *Eucalyptus globulus* plantations in southeastern Australia. For Ecol Manag 259:1754–1760. doi:10.1016/j.foreco.2009.04.023

Fukazawa K, Higuchi T (1966) Studies on the mechanism of heartwood formation. IV. RNA content in the ray parenchyma cells. J Jpn Wood Res Soc 12:221–226

Galicki SJ, Davidson GR, Threlkeld ST (2008) Element mobility in bald cypress xylem. Tree Ring Res 64:39–46. doi:10.3959/2007-7.1PD

Gang DR, Fujita M, Davin LB, Lewis NG (1998) The 'abnormal lignins': mapping heartwood formation through the lignan biosynthetic pathway. In: Lewis NG, Sarkanen S (eds) Lignin and lignan biosynthesis, vol 697. ACS Symposium Series, Washington, DC

Gartner BL (2002) Sapwood and inner bark quantities in relation to leaf area and wood density in Douglas-fir. IAWA J 23:267–285

Gartner B, Moore JR, Gardiner BA (2004) Gas in stems: abundance and potential consequences for tree biomechanics. Tree Physiol 24:1239–1250

Gjerdrum P, Hoibo O (2004) Heartwood detection in Scots pine by means of heat-sensitive infrared images. Holz Roh- und Werkst 62:131–136. doi:10.1007/s00107-004-0467-0

Guleria S, Tiku AK, Singh G, Vyas D, Bhardwaj A (2011) Antioxidant activity and protective effect against plasmid DNA strand scission of leaf, bark, and heartwood extracts from *Acacia catechu*. J Food Sci 76:C959–C964. doi:10.1111/j.1750-3841.2011.02284.x

Hashemi SKH, Latibari AJ (2011) Evaluation and identification of walnut heartwood extractives for protection of poplar wood. BioResources 6:59–69

Hauch S, Magel E (1998) Extractable activities and protein content of sucrose-phosphate synthase, sucrose synthase and neutral invertase in trunk tissues of *Robinia pseudoacacia* L. are related to cambial wood production and heartwood formation. Planta 207:266–274. doi:10.1007/s004250050482

Hergert HL (1977) Secondary lignification in conifer trees. In: Cellulose Chemistry and Technology, vol 48. ACS Symposium Series. AMERICAN CHEMICAL SOCIETY, pp 227–243. doi:10.1021/bk-1977-0048.ch016

Higuchi T (1997) Biochemistry and molecular biology of wood. Springer, Berlin

Hillinger C, Holl W, Ziegler H (1996) Lipids and lipolytic enzymes in the trunkwood of *Robinia pseudoacacia* L. during heartwood formation. 1. Radial distribution of lipid classes. Trees 10:366–375. doi:10.1007/bf02185640

Hillis WE (1987) Heartwood and tree exudates. Springer, Berlin

Höll W, Lendzian K (1973) Respiration in sapwood and heartwood of *Robinia pseudoacacia*. Phytochemistry 12:975–977. doi:10.1016/0031-9422(73)85002-2

Houle D, LaFleche MR, Duchesne L (2008) Sequential extractions of elements in tree rings of balsam fir and white spruce. Commun Soil Sci Plant Anal 39:1138–1146. doi:10.1080/00103620801925836

Huang Z, Meilan R, Woeste K (2009) A *KNAT3*-like homeobox gene from *Juglans nigra* L., *JnKNAT3*-like, highly expressed during heartwood formation. Plant Cell Rep 28:1717–1724. doi:10.1007/s00299-009-0771-6

Huang Z, Tsai CJ, Harding S, Meilan R, Woeste K (2010) A cross-species transcriptional profile analysis of heartwood formation in black walnut. Plant Mol Biol Rep 28:222–230. doi:10.1007/s11105-009-0144-x

Huang Z, Surana P, Kihara D, Meilan R, Woeste K (2011) *JnCML-like*, an EF-hand motif-containing gene seasonally upregulated in the transition zone of black walnut (*Juglans nigra* L.). Am J Mol Biol 1:140–155

IAWA (1964) Multilingual glossary of terms used in wood anatomy. IAWA J 40:186

Imai T, Tanabe K, Kato T, Fukushima K (2005) Localization of ferruginol, a diterpene phenol, in *Cryptomeria japonica* heartwood by time-of-flight secondary ion mass spectrometry. Planta 221:549–556. doi:10.1007/s00425-004-1476-2

Jeong SJ, Koh W, Kim B, Kim SH (2011) Are there new therapeutic options for treating lung cancer based on herbal medicines and their metabolites? J Ethnopharmacol 138:652–661. doi:10.1016/j.jep. 2011.10.018

Jouin D, Tollier MT, Monties B (1988) Lignification of oak wood. I. Lignin determinations in sapwood and heartwood. Cellul Chem Technol 22:231–243

Kadir R, Hale MD (2012) Comparative termite resistance of 12 Malaysian timber species in laboratory tests. Holzforschung 66:127–130. doi:10.1515/hf.2011.116

Kawamura F, Ramle SFM, Sulaiman O, Hashim R, Ohara S (2011) Antioxidant and antifungal activities of extracts from 15 selected hardwood species of Malaysian timber. Eur J Wood Wood Prod 69:207–212. doi:10.1007/s00107-010-0413-2

Ke D, Saltveit E Jr (1988) Plant hormone interaction and phenolic metabolism in the regulation of russet spotting in iceberg lettuce. Plant Physiol 88:1136–1140

Kijidani Y, Toyoumi A, Kitahara R, Fujimoto H (2011) Termite resistance and extractives distribution of sapwood and heartwood of Obi-Sugi cultivars (*Cryptomeria japonica*). Mokuzai Gakkaishi 57:195–202

Koch G (2004) Biologische und chemische Untersuchungen über Inhaltsstoffe im Holzgewebe von Buche (*Fagus sylvatica* L.) und Kirschbaum (*Prunus serotina* Borkh.) und deren Bedeutung für Holzverfärbungen. Mitteilungen der Bundesforschungsanstalt fürr Forst- und Holzwirtschaft. Nr. 216.

Kollmann F, Keylwerth R, Kübler H (1951) Verfärbungen des Vollholzes und der Furniere bei der künstlichen Holztrocknung. Eur J Wood Wood Prod 9:383–390

Kuroda K, Yamashita K, Fujiwara T (2009) Cellular level observation of water loss and the refilling of tracheids in the xylem of *Cryptomeria japonica* during heartwood formation. Trees 23:1163–1172. doi:10.1007/s00468-009-0356-6

Kwon M, Davin LB, Lewis NG (2001) In situ hybridization and immunolocalization of lignan reductases in woody tissues: implications for heartwood formation and other forms of vascular tissue preservation. Phytochemistry 57:899–914. doi:10.1016/s0031-9422(01)00108-x

Lamlom SH, Savidge RA (2006) Carbon content variation in boles of mature sugar maple and giant sequoia. Tree Physiol 26:459–468

Latorraca JVF, Dünisch O, Koch G (2011) Chemical composition and natural durability of juvenile and mature heartwood of *Robinia pseudoacacia* L. An Acad Bras Cienc 83:1059–1068

Li B, Lee DS, Jeong GS, Kim YC (2012a) Involvement of heme oxygenase-1 induction in the cytoprotective and immunomodulatory activities of 6,4′-dihydroxy-7-methoxyflavanone in murine hippocampal and microglia cells. Eur J Pharmacol 674:153–162. doi:10.1016/j.ejphar.2011.11.016

Li LQ, Li J, Huang Y, Wu Q, Deng SP, Su XJ, Yang RY, Huang JG, Chen ZZ, Li S (2012b) Lignans from the heartwood of *Streblus asper* and their inhibiting activities to Hepatitis B virus. Fitoterapia 83:303–309. doi:10.1016/j.fitote.2011.11.008

Little NS, Schultz TP, Nicholas DD (2010) Termite-resistant heartwood. Effect of antioxidants on termite feeding deterrence and mortality. Holzforschung 64:395–398. doi:10.1515/hf.2010.053

Lukmandaru G, Ashitani T, Takahashi K (2009) Color and chemical characterization of partially black-streaked heart-wood in teak (*Tectona grandis*). J For Res 20:377–380

Macfarlane C, Bond C, White DA, Grigg AH, Ogden GN, Silberstein R (2010) Transpiration and hydraulic traits of old and regrowth eucalypt forest in southwestern. Aust For Ecol Manage 260:96–105. doi:10.1016/j.foreco.2010.04.005

Magel E (2000) Biochemistry and physiology of heartwood formation. In: Savidge R, Barnett J, Napier R (eds) Cell and molecular biology of wood formation. BIOS Scientific Publishers, Oxford

Magel E (2002) Biochemistry and quantitative histochemistry of wood. In: Chaffey NJ (ed) Wood formation in trees: cell and molecular biology techniques. Taylor & Francis, London. doi:10.1201/9780203166444.ch14

Magel E, Höll W (1993) Storage carbohydrates and adenine nucleotides in trunks of *Fagus sylvatica* L. in relation to discolored wood. Holzforschung 47:19–24. doi:10.1515/hfsg.1993.47.1.19

Magel E, Hübner B (1997) Distribution of phenylalanine ammonia lyase and chalcone synthase within trunks of *Robinia pseudoacacia* L. Bot Acta 110:314–322

Magel E, Drouet A, Claudot A, Ziegler H (1991) Formation of heartwood substances in the stem of *Robinia pseudoacacia* L. I. Distribution of phenylalanine ammonium lyase and chalcone synthase across the trunk. Trees 5:203–207. doi:10.1007/bf00227526

Magel E, Jay-Allemand C, Ziegler H (1994) Formation of heartwood substances in the stemwood of *Robinia pseudoacacia* L. II. Distribution of nonstructural carbohydrates and wood extractives across the trunk. Trees 8:165–171. doi:10.1007/bf00196843

Magel E, Monties B, Drouet A, Jay-Allemand C, Ziegler H (1995) Heartwood formation: biosynthesis of heartwood extractives and 'secondary' lignification. Eurosilva: contribution

to forest tree physiology. Results from Eurosilva projects, presented at Dourdan, France, 7–10 Nov 1994
Magel E, Hillinger C, Hoell W, Ziegler H (1997) Biochemistry and physiology of heartwood formation: role of reserve substances. In: Rennenberg H, Eschrich W, Ziegler H (eds) Trees: contributions to modern tree physiology. Backhuys Publishers, Leiden
Magel E, Abdel-Latif A, Hampp R (2001a) Non-structural carbohydrates and catalytic activities of sucrose metabolizing enzymes in trunks of two *Juglans* species and their role in heartwood formation. Holzforschung 55:135–145. doi:10.1515/hf.2001.022
Magel E, Hillinger C, Wagner T, Höll W (2001b) Oxidative pentose phosphate pathway and pyridine nucleotides in relation to heartwood formation in *Robinia pseudoacacia* L. Phytochemistry 57:1061–1068. doi:10.1016/s0031-9422(01)00091-7
Magel E, Hauch S, De Filippis L (2002) Random amplification of polymorphic DNA and reverse transcription polymerase chain reaction of RNA in studies of sapwood and heartwood. In: Chaffey NJ (ed) Wood formation in trees: cell and molecular biology techniques. Taylor & Francis, London, New York
Martinez-Sotres C, Lopez-Albarran P, Cruz-de-Leon J, Garcia-Moreno T, Rutiaga-Quinones JG, Vazquez-Marrufo G, Tamariz-Mascarua J, Herrera-Bucio R (2012) Medicarpin, an antifungal compound identified in hexane extract of *Dalbergia congestiflora* Pittier heartwood. Int Biodeterior Biodegrad 69:38–40. doi:10.1016/j.ibiod.2011.11.016
Mayer I, Koch G, Puls J (2006) Topochemical investigations of wood extractives and their influence on colour changes in American black cherry (*Prunus serotina* Borkh.). Holzforschung 60:589–594
Meerts P (2002) Mineral nutrient concentrations in sapwood and heartwood: a literature review. Ann For Sci 59:713–722
Monga J, Chauhan CS, Sharma M (2011) Human epithelial carcinoma cytotoxicity and inhibition of DMBA/TPA induced squamous cell carcinoma in Balb/c mice by *Acacia catechu* heartwood. J Pharm Pharmacol 63:1470–1482. doi:10.1111/j.2042-7158.2011.01354.x
Monties B (1987) Composition chimique des bois de chenes. relations avec la qualite des eaux de vie. Connaissance Vigne Vin 21:169–190
Monties B (1989) Lignins. Methods in plant biochemistry. In: Dey PM, Harborne JB (eds) Plant phenolics, vol 1. Academic, San Diego
Monties B (1991) Plant cell wall as lignocellulosic composite: relations with lignin structures and functions. Anim Feed Sci Technol 32:159–175
Morais MC, Pereira H (2012) Variation of extractives content in heartwood and sapwood of *Eucalyptus globulus* trees. Wood Sci Technol 46:709–719. doi:10.1007/s00226-011-0438-7
Moreau C, Aksenov N, Lorenzo M, Segerman B, Funk C, Nilsson P, Jansson S, Tuominen H (2005) A genomic approach to investigate developmental cell death in woody tissues of *Populus* trees. Genome Biol 6:R34
Moya R, Calvo-Alvarado J (2012) Variation of wood color parameters of *Tectona grandis* and its relationship with physical environmental factors. Ann For Sci 69:947–959. doi:10.1007/s13595-012-0217-0
Moya R, Marin JD (2011) Grouping of *Tectona grandis* (L.f.) clones using wood color and stiffness. New For 42:329–345. doi:10.1007/s11056-011-9255-y
Moya R, Fallas RS, Bonilla PJ, Tenorio C (2012) Relationship between wood color parameters measured by the CIELab system and extractive and phenol content in *Acacia mangium* and *Vochysia guatemalensis* from fast-growth plantations. Molecules 17:3639–3652. doi:10.3390/molecules17043639
Mun SP, Prewitt L (2011) Antifungal activity of organic extracts from *Juniperus virginiana* heartwood against wood decay fungi. For Prod J 61:443–449
Nair MNB, Shah JJ, Pandalai RC (1981) Wood anatomy and histochemical-changes of sapwood during heartwood formation in *Bridelia retusa* Spreng. Proc Ind Acad Sci Plant Sci 90:425–433

Nakaba S, Kubo T, Funada R (2008) Differences in patterns of cell death between ray parenchyma cells and ray tracheids in the conifers *Pinus densiflora* and *Pinus rigida*. Trees 22:623–630. doi:10.1007/s00468-008-0220-0

Nakaba S, Yamagishi Y, Sano Y, Funada R (2012) Temporally and spatially controlled death of parenchyma cells is involved in heartwood formation in pith regions of branches of *Robinia pseudoacacia* var. *inermis*. J Wood Sci 58:69–76. doi:10.1007/s10086-011-1221-y

Napoleao TH, Gomes FS, Lima TA, Santos NDL, Sa RA, Albuquerque AC, Coelho LCBB, Paiva PMG (2011) Termiticidal activity of lectins from *Myracrodruon urundeuva* against *Nasutitermes corniger* and its mechanisms. Int Biodeterior Biodegrad 65:52–59. doi:10.1016/j.ibiod.2010.05.01

Nawrot M, Pazdrowski W, Szymanski M (2008) Dynamics of heartwood formation and axial and radial distribution of sapwood and heartwood in stems of European larch (*Larix decidua* Mill.). J For Sci 9:409–417

Nelson ND (1978) Xylem ethylene, phenol oxidizing enzymes, and nitrogen and heartwood formation in walnut and cherry. Can J Bot 56:626–634

Niamke FB, Amusant N, Charpentier JP, Chaix G, Baissac Y, Boutahar N, Adima AA, Kati-Coulibaly S, Jay-Allemand C (2011) Relationships between biochemical attributes (non-structural carbohydrates and phenolics) and natural durability against fungi in dry teak wood (*Tectona grandis* L. f.). Ann For Sci 68:201–211. doi:10.1007/s13595-011-0021-2

Nobuchi T, Harada H (1983) Physiological features of the "white zone" of Sugi (*Cryptomeria japonica* D. Don) cytological structure and moisture content. Mokuzai Gakkaishi 29:824–832

Nobuchi T, Harada H (1985) Ultrastructural changes in parenchyma cells of Sugi (*Cryptomeria japonica* D. Don) associated with heartwood formation. Mokuzai Gakkaishi 31:965–973

Nobuchi T, Sato R, Iwata R, Harada H (1984) Season of heartwood formation and the related cytological structure of ray parenchyma cells in *Robinia pseudoacacia* L. Mokuzai Gakkaishi 30:628–636

Okada N, Hirakawa Y, Katayama Y (2011) Application of activable tracers to investigate radial movement of minerals in the stem of Japanese cedar (*Cryptomeria japonica*). J Wood Sci 57:421–428. doi:10.1007/s10086-011-1188-8

Okada N, Hirakawa Y, Katayama Y (2012) Radial movement of sapwood-injected rubidium into heartwood of Japanese cedar (*Cryptomeria japonica*) in the growing period. J Wood Sci 58:1–8. doi:10.1007/s10086-011-1218-6

Panshin AJ, de Zeeuw C (1980) Textbook of wood technology. Part 1. Formation, anatomy, and properties of wood. McGraw-Hill, New York

Passialis C, Voulgaridis E, Adamopoulos S, Matsouka M (2008) Extractives, acidity, buffering capacity, ash and inorganic elements of black locust wood and bark of different clones and origin. Eur J Wood Wood Prod 66:395–400. doi:10.1007/s00107-008-0254-4

Patten AM, Davin LB, Lewis NG (2008) Relationship of dirigent protein and 18s RNA transcript localization to heartwood formation in western red cedar. Phytochemistry 69:3023–3037. doi:10.1016/j.phytochem.2008.06.020

Pfautsch S, Bleby TM, Rennenberg H, Adams MA (2010) Sap flow measurements reveal influence of temperature and stand structure on water use of *Eucalyptus regnans* forests. For Ecol Manag 259:1190–1199. doi:10.1016/j.foreco.2010.01.006

Pfautsch S, Keitel C, Turnbull TL, Braimbridge MJ, Wright TE, Simpson RR, O'Brien JA, Adams MA (2011) Diurnal patterns of water use in *Eucalyptus victrix* indicate pronounced desiccation-rehydration cycles despite unlimited water supply. Tree Physiol 31:1041–1051. doi:10.1093/treephys/tpr082

Pfautsch S, Macfarlane C, Ebdon N, Meder R (2012) Assessing sapwood depth and wood properties in *Eucalyptus* and *Corymbia* spp. using visual methods and near infrared spectroscopy (NIR). Trees 26:963–974. doi:10.1007/s00468-011-0674-3

Pollet C, Jourez B, Hébert J (2008) Natural durability of black locust (*Robinia pseudoacacia* L.) wood grown in Wallonia, Belgium. Can J For Res 38:1366–1372. doi:10.1139/x07-244

Preisig-Müller R, Schwekendiek A, Brehm I, Reif HJ, Kindl H (1999) Characterization of a pine multigene family containing elicitor-responsive stilbene synthase genes. Plant Mol Biol 39:221–229. doi:10.1023/a:1006163030646

Rachmayanti Y, Leinemann L, Gailing O, Finkeldey R (2009) DNA from processed and unprocessed wood: factors influencing the isolation success. Forensic Sci Int Genet 3:185–192. doi:10.1016/j.fsigen.2009.01.002

Rodrigues AMS, Amusant N, Beauchene J, Eparvier V, Lemenager N, Baudasse C, Espindola LS, Stien D (2011) The termiticidal activity of *Sextonia rubra* (Mez) van der Werff (Lauraceae) extract and its active constituent rubrynolide. Pest Manage Sci 67:1420–1423. doi:10.1002/ps.2167

Roux DG, Paulus E (1962) Condensed tannins.13. interrelationships of flavonoid components from heartwood of *Robinia pseudoacacia*. Biochem J 82:324–330

Royer M, Rodrigues AMS, Herbette G, Beauchene J, Chevalier M, Herault B, Thibaut B, Stien D (2012) Efficacy of *Bagassa guianensis* Aubl. extract against wood decay and human pathogenic fungi. Int Biodeterior Biodegrad 70:55–59

Sá R, Gomes F, Napoleão T, Santos N, Melo C, Gusmão N, Coelho L, Paiva P, Bieber L (2009) Antibacterial and antifungal activities of *Myracrodruon urundeuva* heartwood. Wood Sci Technol 43:85–95. doi:10.1007/s00226-008-0220-7

Saito K, Mitsutani T, Imai T, Matsushita Y, Fukushima K (2008a) Discriminating the indistinguishable sapwood from heartwood in discolored ancient wood by direct molecular mapping of specific extractives using time-of-flight secondary ion mass spectrometry. Anal Chem 80:1552–1557. doi:10.1021/ac7021162

Saito K, Mitsutani T, Imai T, Matsushita Y, Yamamoto A, Fukushima K (2008b) Chemical differences between sapwood and heartwood of *Chamaecyparis obtusa* detected by ToF-SIMS. Appl Surf Sci 255:1088–1091

Saito K, Watanabe Y, Shirakawa M, Matsushita Y, Imai T, Koike T, Sano Y, Funada R, Fukazawa K, Fukushima K (2012) Direct mapping of morphological distribution of syringyl and guaiacyl lignin in the xylem of maple by time-of-flight secondary ion mass spectrometry. Plant J 69:542–552

Sandberg K, Sterley M (2009) Separating Norway spruce heartwood and sapwood in dried condition with near-infrared spectroscopy and multivariate data analysis. Eur J For Res 128:475–481. doi:10.1007/s10342-009-0296-0

Scheffer TC, Cowling EB (1966) Natural resistance of wood to microbial deterioration. Annu Rev Phytopathol 4:147–170. doi:10.1146/annurev.py.04.090166.001051

Scholz G, Pfeffer A, Ibiza-Palacios MS, Oliver-Villanueva JV (2012) Preliminary study on the durability of *Prosopis kuntzei* Harms. Eur J Wood Wood Prod 70:361–364. doi:10.1007/s00107-011-0533-3

Schultz TP, Nicholas DD (2000) Naturally durable heartwood: evidence for a proposed dual defensive function of the extractives. Phytochemistry 54:47–52. doi:10.1016/S0031-9422(99)00622-6

Sekine N, Ashitani T, Murayama T, Ogiyama K, Takahashi K (2009) The bioactive extracts of heartwood of *Dalbergia latifolia*. Mokuzai Gakkaishi 55:29–36

Shain L, Hillis WE (1973) Ethylene production in xylem of *Pinus radiata* in relation to heartwood formation. Can J Bot 51:1331–1335

Shan T, Ma Q, Guo K, Liu J, Li W, Wang F, Wu E (2011) Xanthones from mangosteen extracts as natural chemopreventive agents: potential anticancer drugs. Curr Mol Med 11:666–677

Song K, Liu B, Jiang X, Yin Y (2011) Cellular changes of tracheids and ray parenchyma cells from cambium to heartwood in *Cunninghamia lanceolata*. J Trop For Sci 23:478–487

Songsiang U, Wanich S, Pitchuanchom S, Netsopa S, Uanporn K, Yenjai C (2009) Bioactive constituents from the stems of *Dalbergia parviflora*. Fitoterapia 80:427–431. doi:10.1016/j.fitote.2009.06.002

Spicer R (2005) Senescence in secondary xylem: heartwood formation as an active developmental program. In: Holbrook NM, Zwieniecki MA (eds) Vascular transport in plants. Elsevier Academic, Amsterdam. doi:10.1016/B978-012088457-5/50024-1

Spicer R, Holbrook NM (2005) Within-stem oxygen concentration and sap flow in four temperate tree species: does long-lived xylem parenchyma experience hypoxia? Plant Cell Environ 28:192–201

Spicer R, Holbrook NM (2007a) Parenchyma cell respiration and survival in secondary xylem: does metabolic activity decline with cell age? Plant Cell Environ 30:934–943. doi:10.1111/j.1365-3040.2007.01677.x

Spicer R, Holbrook NM (2007b) Effects of carbon dioxide and oxygen on sapwood respiration in five temperate tree species. J Exp Bot 58:1313–1320. doi:10.1093/jxb/erl296

Su PP, Granholm K, Pranovich A, Harju L, Holmbom B, Ivaska A (2012) Metal ion sorption to birch and spruce wood. BioResources 7:2141–2155

Takeda I, Somura H, Mori Y (2010) Recovery of phosphorus from natural water bodies using iron-oxidizing bacteria and woody biomass. Ecol Eng 36:1064–1069

Tanaka T, Jiang ZH, Kouno I (1998) Distribution of ellagic acid derivatives and a diarylheptanoid in wood of *Platycarya strobilacea*. Phytochemistry 47:851–854

Taylor A, Gartner B, Morrell J (2002) Heartwood formation and natural durability – a review. Wood Fiber Sci 34:587–611

Taylor AM, Brooks JR, Lachenbruch B, Morrell JJ (2007) Radial patterns of carbon isotopes in the xylem extractives and cellulose of Douglas-fir. Tree Physiol 27:921–927

Tsuruta K, Yoshida Y, Kusumoto N, Sekine N, Ashitani T, Takahashi K (2011) Inhibition activity of essential oils obtained from Japanese trees against *Skeletonema costatum*. J Wood Sci 57:520–525. doi:10.1007/s10086-011-1209-7

Ucar MB, Ucar G (2011) Characterization of methanol extracts from *Quercus hartwissiana* wood and bark. Chem Nat Compd 47:697–703

Umehara K, Nemoto K, Matsushita A, Terada E, Monthakantirat O, De-Eknamkul W, Miyase T, Warashina T, Degawa M, Noguchi H (2009) Flavonoids from the heartwood of the Thai medicinal plant *Dalbergia parviflora* and their effects on estrogenic-responsive human breast cancer cells. J Nat Prod 72:2163–2168. doi:10.1021/np900676y

Vansteenkiste D, Van Acker J, Stevens M, Le Thiec D, Nepveu G (2007) Composition, distribution and supposed origin of mineral inclusions in sessile oak wood – consequences for microdensitometrical analysis. Ann For Sci 64:11–19. doi:10.1051/forest:2006083

Vetter RE, Coradin VR, Martinol EC, Camargos JAA (1990) Wood colour – a comparison between determination methods. IAWA J 11:429–439

Waikedre J, Vitturo CI, Molina A, Theodoro PNET, do Rosario Rodrigues Silva M, Espindola LS, Maciuk A, Fournet A (2012) Antifungal activity of the essential oils of *Callitris neocaledonica* and *C. sulcata* heartwood (Cupressaceae). Chem Biodivers 9:644–653. doi:10.1002/cbdv.201100229

Wang XC, Wang CK, Zhang QZ, Quan XK (2010a) Heartwood and sapwood allometry of seven Chinese temperate tree species. Ann For Sci 67:410. doi:10.1051/forest/2009131

Wang YL, Liu GB, Kume T, Otsuki K, Yamanaka N, Du S (2010b) Estimating water use of a black locust plantation by the thermal dissipation probe method in the semiarid region of Loess Plateau, China. J For Res 15:241–251. doi:10.1007/s10310-010-0184-y

Wang J, Li JA, Li SJ, Freitag C, Morrell JJ (2011a) Antifungal activities of *Cunninghamia lanceolata* heartwood extractives. BioResources 6:606–614

Wang YZ, Sun SQ, Zhou YB (2011b) Extract of the dried heartwood of *Caesalpinia sappan* L. attenuates collagen-induced arthritis. J Ethnopharmacol 136:271–278. doi:10.1016/j.jep.2011.04.061

Wei Q, Leblon B, La Rocque A (2011) On the use of X-ray computed tomography for determining wood properties: a review. Can J For Res 41:2120–2140

Wu SF, Chang FR, Wang SY, Hwang TL, Lee CL, Chen SL, Wu CC, Wu YC (2011a) Anti-inflammatory and cytotoxic neoflavonoids and benzofurans from *Pterocarpus santalinus*. J Nat Prod 74:989–996. doi:10.1021/np100871g

Wu SF, Hwang TL, Chen SL, Wu CC, Ohkoshi E, Lee KH, Chang FR, Wu YC (2011b) Bioactive components from the heartwood of *Pterocarpus santalinus*. Bioorg Med Chem Lett 21:5630–5632

Xu DP, Sung SJS, Loboda T, Kormanik PP, Black CC (1989) Characterization of sucrolysis *via* the uridine diphosphate and pyrophosphate dependent sucrose synthase pathway. Plant Physiol 90:635–642. doi:10.1104/pp. 90.2.635

Yang J, Park S, Kamdem DP, Keathley DE, Retzel E, Paule C, Kapur V, Han KH (2003) Novel gene expression profiles define the metabolic and physiological processes characteristic of wood and its extractive formation in a hardwood tree species, *Robinia pseudoacacia*. Plant Mol Biol 52:935–956. doi:10.1023/a:1025445427284

Yang J, Kamdem DP, Keathley DE, Han KH (2004) Seasonal changes in gene expression at the sapwood – heartwood transition zone of black locust (*Robinia pseudoacacia*) revealed by cDNA microarray analysis. Tree Physiol 24:461–474. doi:10.1093/treephys/24.4.461

Yen TB, Chang HT, Hsieh CC, Chang ST (2008) Antifungal properties of ethanolic extract and its active compounds from *Calocedrus macrolepis* var. *formosana* (Florin) heartwood. Bioresour Technol 99:4871–4877. doi:10.1016/j.biortech.2007.09.037

Yoshida K, Nishiguchi M, Futamura N, Nanjo T (2007) Expressed sequence tags from *Cryptomeria japonica* sapwood during the drying process. Tree Physiol 27:1–9

Yoshida K, Futamura N, Nishiguchi M (2012) Collection of expressed genes from the transition zone of *Cryptomeria japonica* in the dormant season. J Wood Sci 58:89–103. doi:10.1007/s10086-011-1234-6

Zhang T, Bai SL, Bardet S, Almeras T, Thibaut B, Beauchene J (2011) Radial variations of vibrational properties of three tropical woods. J Wood Sci 57:377–386. doi:10.1007/s10086-011-1189-7

Ziegler H (1968) Biologische Aspekte der Kernholzbildung. Eur J Wood Wood Prod 26:61–68. doi:10.1007/bf02615811

# Part II
# Control of Wood Formation by Endogenous and Exogenous Factors

# The Role of Hormones in Controlling Vascular Differentiation

**Roni Aloni**

**Abstract** The vascular system in plants is induced and controlled by streams of inductive hormonal signals. Auxin produced in young leaves is the primary controlling hormone in vascular differentiation. Its downward transport pathways, major controlling mechanisms, and sensitivity of cells to auxin are clarified. Cytokinin, from the root cap moves upward, increases the sensitivity to auxin and stimulates cambial cell divisions. Gibberellin produced in mature leaves moves non-polarly and promotes elongation, regulates cambium activity, and induces long fibers and long tracheids. Transgenic plants with elevated bioactive gibberellin concentrations grow rapidly and yield numerous longer fibers and longer tracheids. Centrifugal movement of ethylene from maturing vessels induces the radial vascular rays. In conifer trees, jasmonate, which promotes defense response, is mediated by ethylene and induces traumatic resin ducts. In addition the role of the hormonal signals in regulating gradients of cell size and density, in controlling the type of differentiating vascular element and how they have shaped wood evolution are elucidated.

## 1 Introduction

The aim of this chapter is to integrate molecular and organismal findings in a holistic approach for elucidating the hormonal mechanisms that control vascular differentiation in the whole plant and their evolutionary aspects. This knowledge provides the hormonal understanding required for improving wood formation in trees which has enormous economic importance.

---

R. Aloni (✉)
Department of Molecular Biology and Ecology of Plants, Tel Aviv University, Tel Aviv 69978, Israel
e-mail: alonir@post.tau.ac.il

J. Fromm (ed.), *Cellular Aspects of Wood Formation*, Plant Cell Monographs 20,
DOI 10.1007/978-3-642-36491-4_4, 

Research on vascular differentiation has expanded at an impressive rate focusing mainly on molecular genetic aspects of xylem cell and tissue differentiation, and the topic has been intensively reviewed (e.g., Turner and Sieburth 2003; Fukuda 2004; Sieburth and Deyholos 2006; Baucher et al. 2007; Jung and Park 2007; Turner et al. 2007; Caño-Delgado et al. 2010; Spicer and Groover 2010; Scarpella and Helariutta 2010; Nieminen et al. 2012; Ursache et al. 2012). The *Arabidopsis* plant has become a useful model system for studying wood formation (Chaffey et al. 2002; Zhang et al. 2011). We also witness a dramatic progress in the analysis of the genes that regulate xylem cell differentiation mainly by using the *Zinnia elegans* liquid cell culture (Yamaguchi et al. 2010a, b, 2011; Oda and Fukuda 2012; Bollhöner et al. 2012). Furthermore, numerous transcription factors which regulate various aspects of vascular differentiation and secondary growth were uncovered (Groover 2005; Groover and Robischon 2006; Demura and Fukuda 2007; Bollhöner et al. 2012).

I provide new evidence on auxin transport pathways (Figs. 1, 2, and 3), elucidate the nature of the mobile gibberellin signal (Dayan et al. 2012), and discuss the role of the new hormone strigolactone (Agusti et al. 2011), the importance of ethylene production in maturing vessel elements (Pesquet and Tuominen 2011), and how the duration of cell differentiation influences the gradient in tracheid size along the stem (Anfodillo et al. 2012).

The chapter presents updated information and concepts on the development of the secondary xylem, namely, the wood, in trees and herbaceous model plants. In particular, I focus on the hormonal signals and how they regulate the differentiation of various types of vascular cells in the axial and radial directions, including the induction of traumatic resin ducts, how tree size influences cambial activity and wood formation, and the control of cell size and density along the plant axis. The relationships between phloem and xylem and vascular differentiation in tumorous tissues are also discussed. Finally, I clarify how the xylem cells and the cambium have been specialized during evolution from tracheids to vessels and fibers and from diffuse-porous to ring-porous wood.

## 2 Control of Vascularization by Moving Hormonal Signals

The plant vascular system is composed of xylem (water-conducting tissue) and phloem (food-conducting tissue) which are complex, being composed of several types of cells, which are induced by a few hormonal signals (Roberts et al. 1988; Evert 2006). Plant hormones are organic molecules that at low concentrations influence plant growth and development. The hormones or their precursors are synthesized in various locations and move through specific transport pathways to sites where they regulate growth and differentiation.

Along the plant axis, the vascular cells are induced and controlled by longitudinal streams of inductive signals (Sachs 1981; Aloni 1987, 2001), while the vascular

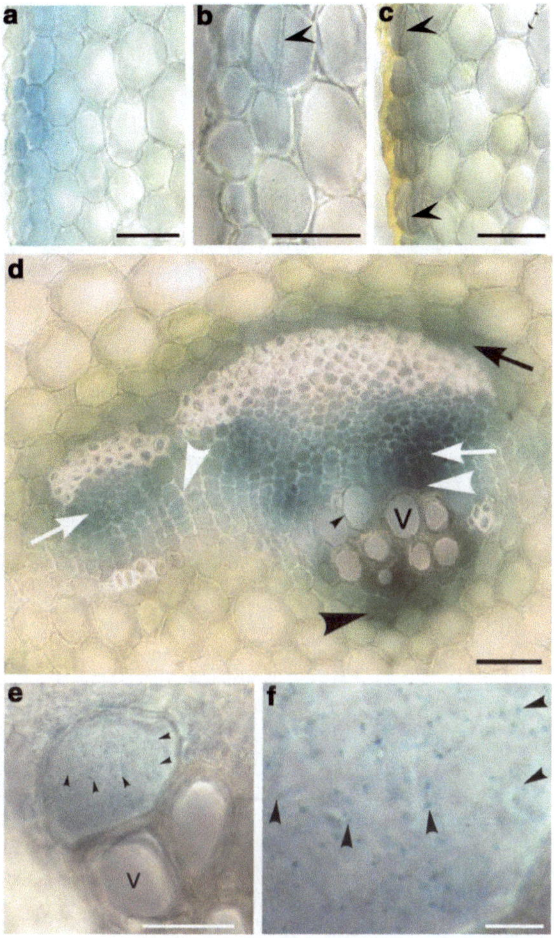


**Fig. 1** Cross sections in stems of the *GH3* transformant of white clover, *Trifolium repens* (L.) cv. Haifa, with the naturally occurring 749-bp auxin-responsive promoter region of the soybean gene *GH3* (Hagen et al. 1991), contained at least three auxin-responsive elements (AuxRE) fused to the *GUS* gene, transformed according to Larkin et al. (1996), and grown as described by Schwalm et al. (2003). The histochemical blue staining for GUS activity in the sections shows free auxin in the epidermis (**a**, **b**) and developing phellogen (**c**), auxin pathways in vascular bundles (**d**) and in a differentiating vessel (**e**, **f**). *Arrowheads* mark early cell divisions inducing the phellogen (**b**, **c**). Auxin transport through the vascular bundles occurs in the cambium (*white arrowheads*), in the sieve tubes (*white arrows*), in a differentiating vessel (*small black arrowhead*), in the bundle sheath (*black arrow*), and in xylem parenchyma cells (*large black arrowhead*). Blue-stained vesicles (*arrowheads*) are observed along actin filaments in a differentiating vessel (**e**) which is enlarged in (**f**). V, vessel. Bar = 10 μm in (**f**), 50 μm in (**d**, **e**), 100 μm in (**a–c**) (Roni Aloni and Cornelia I. Ullrich, unpublished)

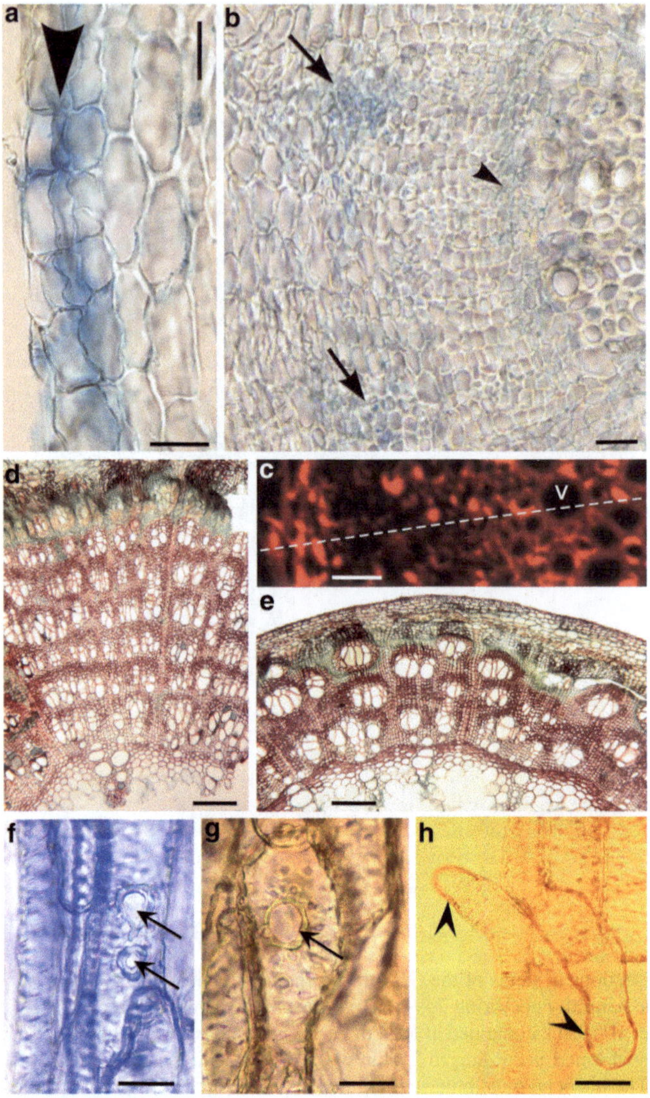


**Fig. 2** Cross sections in the *DR5::GUS* transformant (Ulmasov et al. 1997) of *Arabidopsis thaliana* (L.) Heynh. stem (**a**–**c**). The blue GUS staining (**a**, **b**) shows free auxin in the developing phellogen (*arrowhead* in **a**), in the sieve tubes (*arrows* in **b**), and cambium (*arrowhead* in **b**). Cryomicrotome prepared cross section (12 μm thick) incubated overnight with mouse monoclonal hybridoma antibodies, raised against IAA (Mertens et al. 1985), and then labeled with red fluorescent Alexa conjugate as the secondary antibody (**c**). The broken line (in **c**) shows the radial line used to measure the total auxin concentrations with a confocal laser scanning microscopy (the results are shown in Fig. 3). (Markus Langhans, Cornelia I. Ullrich and Roni Aloni, unpublished). Two transverse sections made along the same internode of *Phaseolus vulgaris* L. taken below a site of auxin (0.1 % NAA (w/w)) application, at 5 mm (**d**), and 40 mm below the applied auxin (**e**), showing an increase in vessel diameter and decrease in vessel density with increasing distance

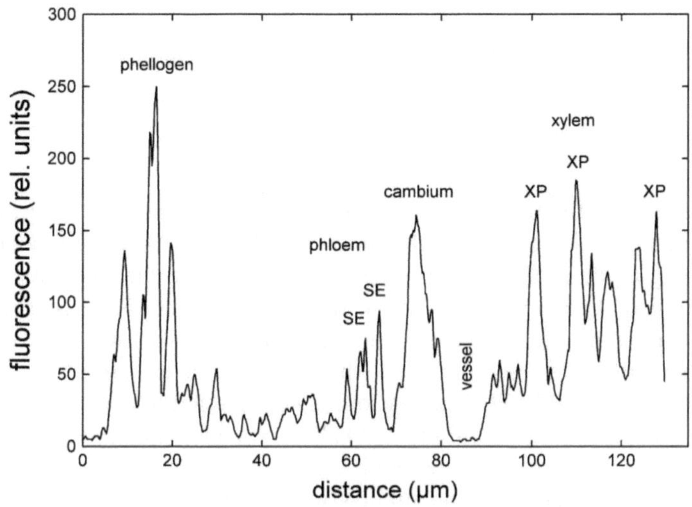


**Fig. 3** Pattern of fluorescence originating from the *Arabidopsis thaliana* cells along the marked radial broken line shown in Fig. 2c revealing high total auxin levels in the phellogen, sieve tubes (SE), cambium, and xylem parenchyma (XP). (Markus Langhans, Cornelia I. Ullrich, and Roni Aloni, unpublished)

rays are induced and regulated by radial hormonal flows (Lev-Yadun and Aloni 1995; Aloni et al. 2000).

## 2.1 *Auxin (IAA) Is the Young-Leaf Signal*

The auxin hormone, namely, indole-3-acetic acid (IAA), is the most common naturally occurring auxin. IAA is the major shoot signal which regulates all aspects of vascular differentiation in plants (Aloni 2001). The pioneering study of Jacobs (1952) demonstrated that the auxin produced in young leaves is the limiting and controlling signal of xylem regeneration around a wound. The polar transport of IAA from young shoot organs (Aloni 2010) downward via the procambium and cambium to the root tips (Aloni 1987; Aloni et al. 2006a) induces and controls vascular differentiation and wood formation. The continuity of the vascular tissues

**Fig. 2** (continued) from the auxin source, as well as changes of vessel pattern from layers (**d**) to bundles (**e**) (Aloni and Zimmermann 1983). Longitudinal views of tracheids with perforations (**f**, **g**) induced by high concentration of auxin (0.5 % NAA in lanolin (w/w)) in the hypocotyls of young *Pinus pinea* (L.) seedlings with one or two openings (*arrows*) on side walls (**f**, **g**). Intrusively growing tracheid tips (**h** *arrowheads*) in a young hypocotyls of *P. pinea* induced by high concentration of gibberellin at low auxin background (1 % $GA_3$ and 0.1 % NAA in lanolin (w/w)). V, vessel. Bars = 50 μm (**a**, **b**, **f–h**), 100 μm (**c–e**)

along the plant axis is a result of the steady polar flow of IAA from leaves to roots (Sachs 1981; Aloni 1987; Berleth et al. 2000; Scarpella and Helariutta 2010).

### 2.1.1 Free and Conjugated Auxin

The percentage of the bioactive IAA hormone, namely, the free auxin, is usually very low in a given tissue and might range from 1 % to 5 % of the total amount of auxin (Ljung et al. 2002) and up to 25 %, depending on the tissue and the plant species studied (Ludwig-Müller 2011). Most of the auxin in cells is covalently bound to other molecules, and these conjugated molecules are inactive as hormonal signals and serve as a reservoir from which the free auxin can be released. Free auxin (detected by *DR5::GUS* expression) is not detected in the apical bud and the youngest leaf primordia (Aloni et al. 2003) or youngest flower primordia (Aloni et al. 2006b), which are all loaded with conjugated auxin, detected by antibodies (Aloni et al. 2003, 2006b). Bound auxin is accumulated in the shoot apex, youngest leaves, and flowers likely due to the local upward polar auxin flow from the free-auxin-producing young leaves (Reinhardt et al. 2003; Benková et al. 2003; Scheres and Xu 2006). The youngest leaf primordia likely start as sinks for IAA and become sources of free auxin during leaf development (Aloni et al. 2003; Aloni 2010; Teale et al. 2006).

This developmental auxin pattern stimulates me to put forward the following new concept, proposing that due to the importance and requirement for continuous IAA supply from the shoot organs to the rest of the plant, the shoot establishes a large reservoir of conjugated auxin molecules at the shoot apex and youngest shoot organs, from which the IAA will later be hydrolyzed continuously in young growing leaves. Thus, the massive bound auxin pool in the young shoot organs guarantees the constant supply of free auxin, which induces and continuously regulates wood formation and phloem differentiation along the tree axis (Aloni 2001).

### 2.1.2 The Canalization Hypothesis

The orderly pattern of vascular tissues from leaves to roots was explained by the *canalization hypothesis* (Sachs 1981). According to this hypothesis, IAA flow which starts by diffusion induces a polar IAA transport system which promotes IAA movement and leads to canalization of the IAA flow along a narrow file of cells. The continuous polar transport of IAA through these cells induces a further complex sequence of events which terminates in the formation of a vessel (Sachs 1981, 2000). Molecular evidence supports the *canalization hypothesis* demonstrating that rearrangement of polar IAA flow changes tissue polarity through modification of the site of a PIN-FORMED (PIN) protein (an essential component involved in IAA efflux) on the plasma membrane (Sauer et al. 2006; Teale et al. 2006; Friml 2010; Wabnik et al. 2010; Kleine-Vehn et al. 2011).

Repeating some of Sachs's (1981) basic experiments, such as local auxin application, wounding, or auxin accumulation (during de novo organ formation), and analyzing them with molecular tools reveal that these treatments lead to rearrangements in the subcellular polar localization of the PIN auxin transport components. This auxin effect on PIN polarity is cell-specific, does not depend on PIN transcription, and involves the indole-3-acetic acid-auxin response factor (Aux/IAA-ARF) signaling pathway. The experiments show that IAA acts as polarizing signal, which links individual cell polarity with tissue and organ polarity through control of PIN polar targeting (Sauer et al. 2006).

### 2.1.3 Sensitivity to Auxin

A novel putative auxin transport facilitator family called PIN-LIKES (PILS) was recently uncovered (Barbez et al. 2012). The PILS proteins are required for auxin-dependent regulation of plant growth by determining the cellular sensitivity to auxin. PILS proteins regulate intracellular auxin accumulation at the endoplasmic reticulum and thus the free-IAA levels available for nuclear auxin signaling. The analysis of the PILS proteins suggests that intracellular auxin transport and, therefore, the auxin compartmentalization might be evolutionarily older than the directional cell-to-cell PIN-dependent auxin transport mechanism. The identification of a novel protein family for the regulation of intracellular auxin homeostasis highlights the evolutionary and developmental importance of intracellular auxin transport (Barbez et al. 2012).

During the evolution of temperate deciduous hardwood trees, the ring-porous trees have developed from diffuse-porous trees. The analysis of their developmental biology and anatomy suggests that the cambium in ring-porous trees has become very sensitive to extremely low IAA stimulation (Aloni 1991, 2001), which will be clarified below regarding wood evolution.

### 2.1.4 The IAA Transport Pathways

The polar movement of free auxin from leaves to roots occurs via specific transport pathways (Aloni 2010), which are discussed below, with new experimental evidence (Figs. 1, 2, and 3). Most of the studies on auxin transport focused on the molecular mechanisms that control the transport of auxin (e.g., Gälweiler et al. 1998; Geldner et al. 2003; Teale et al. 2006; Kleine-Vehn et al. 2011; Runions and Friml 2011), and the complex routes of IAA transport (Uggla et al. 1996, 1998; Aloni 2010) were poorly investigated. Free auxin produced in young leaves moves polarly mainly in the vascular tissues (Sachs 1981; Roberts et al. 1988; Aloni 2001), specifically in the cambium (Uggla et al. 1996; Sundberg et al. 2000), and through xylem parenchyma (Gälweiler et al. 1998; Palme and Gälweiler 1999; Booker et al. 2003), starch sheath, and the root pericycle (Aloni et al. 2006a). In addition, there is

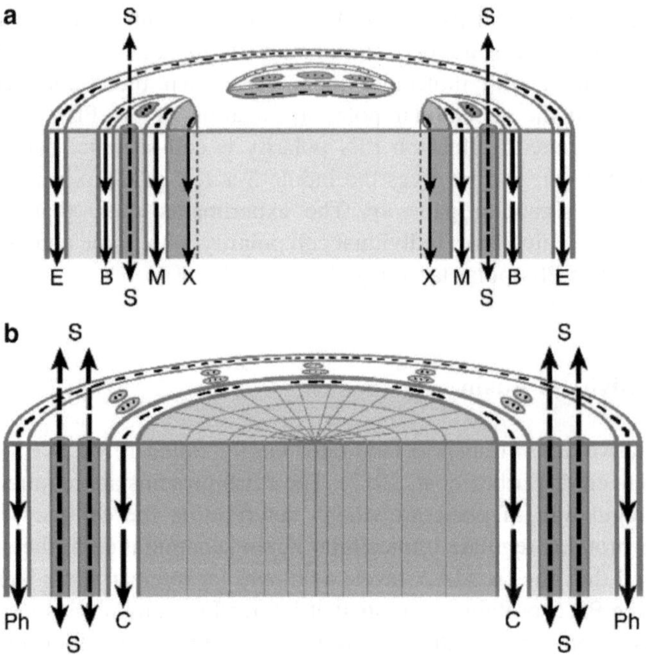


**Fig. 4** Schematic diagrams illustrating the free-IAA transport pathways in the primary shoot (**a**) and the secondary body (**b**). In the *internal route*, IAA moves polarly through the vascular meristem (M) (in the primary body), which becomes the cambium (C) in the secondary body, through the (X) xylem parenchyma cells and the (B) bundle sheath (in primary shoot). In the *peripheral route*, IAA moves polarly through the epidermis (E) and the phellogen (Ph). In the *nonpolar route*, IAA moves up and down in the (S) sieve tubes

evidence for rapid nonpolar movement of the hormone through sieve tubes (Morris et al. 1973; Goldsmith et al. 1974). This nonpolar auxin flow in the phloem conduits originates in mature leaves (Morris et al. 1973). Furthermore, there are indications that auxin flows in the epidermis (Barker-Bridgers et al. 1998; Swarup et al. 2001; Friml and Palme 2002). These fragments of important information have been incorporated into a general concept model for explaining where IAA moves in the plant body (Aloni 2010). Here, I add new experimental data supporting the *auxin pathway model* (Fig. 4).

All living cells in the plant body are capable of transporting IAA, but only those through which free auxin is canalized become specialized to transport the hormone rapidly, resulting in canalized files of cells (Sachs 1981). During plant development, initial auxin flows are canalized into three main routes of IAA transport (Aloni 2010). These flow patterns start during embryogenesis. The first two pathways which originate in young leaves induce and control the vascular and protective tissues, while the third pathway, which is activated later during leaf maturation, controls the activity of the phloem conduits. The auxin transport model describes the following three main routes:

1. *The internal route*—is a complex pathway that can be subdivided into the following longitudinal components (Fig. 4): primary shoot (a), primary root (is not shown in Fig. 4), and secondary body (b) which is produced between the primary parts in woody dicotyledonous and gymnosperm species. Each of these components has its unique anatomy and physiology as follows:

*In the primary shoot* (Fig. 4a), IAA from young leaves moves polarly downward through the vascular bundles in three distinct streams: (1) via the vascular meristem (M in Fig. 4), namely, the procambium or early stages of developing cambium, (2) in the surrounding (B in Fig. 4) bundle sheath at the phloem pole (black arrow in Fig. 1d), and (3) through parenchyma cells at the (X in Fig. 4) xylem pole (large black arrowhead in Fig. 1d). This auxin flow through the xylem parenchyma cells inhibits lateral bud development in *Arabidopsis* (Booker et al. 2003) and likely in other species.

*In the secondary body* (Fig. 4b), the internal polar IAA streams descending from the primary shoot (Fig. 4a) merge into one pathway, which occurs in the cambium (white arrowheads in Fig. 1d, arrowhead in Fig. 2b; C in Fig. 4b; cambium in Fig. 3) and differentiating vessel elements (small black arrowhead in Fig. 1d, Fig. 1e, f).

2. *The peripheral route*—courses through the protective and adjunct tissues. The auxin in this pathway originates in the epidermal cells of young leaves, in auxin-producing trichomes and stomata (Aloni et al. 2003), and moves polarly toward the root tips through the epidermis (Fig. 1a, b; E in Fig. 4a) and subepidermal cell layers (Fig. 1a, b) and through the phellogen (Figs. 1c and 2a; phellogen in Fig. 3; Ph in Fig. 4b) which produces the cork.

3. *The nonpolar route*—which originates in mature leaves courses in the phloem conduits, where auxin moves rapidly (Morris et al. 1973; Goldsmith et al. 1974) up and down via the sieve tubes (white arrows in Fig. 1, arrows in Fig. 2b; SE in Fig. 3; S in Fig. 4). This fast auxin flow is considered a housekeeping signal that reduces callose levels in the sieve tubes (Aloni 1995). The nonpolar IAA flow can also remove the dormancy callose and promotes the resumption of phloem activity in spring (Aloni et al. 1991; Aloni and Peterson 1997), [whereas the nonpolar cytokinin flow in the sieve tubes (Kudo et al. 2010) increases callose levels on the sieve plates and can plug the sieve tubes for winter dormancy (Aloni et al. 1990a)].

A radial distribution pattern of endogenous IAA was detected across the cambium region in *Pinus sylvestris* showing a peak in the cambial zone where cell division takes place, steeply decreasing toward the mature xylem and phloem. The IAA content was measured, by gas chromatography–mass spectrometry technique, in serial 30-μm thick longitudinal tangential sections obtained across the cambial region (Uggla et al. 1996). A technical sampling problem might have caused the absence of a clear IAA peak in the phloem. Because the circumference of a tree is rounded, it is possible that the tangential sections of the cambium included also the thin phloem tissue, and therefore, a clear peak of IAA in the phloem could not be detected in the pine trees, as might have been expected from studies of angiosperm species (Morris et al. 1973; Goldsmith et al. 1974). It is also possible that the quantity of IAA in the phloem conduits of *P. sylvestris* is extremely low and therefore could not be detected.

By studying free-auxin patterns (by *DR5::GUS* expression) in cross sections of *Trifolium* and *Arabidopsis*, we confirm the presence of free auxin in the cambium (white arrowhead in Fig. 1d, arrowhead in Fig. 2b) and clearly detect also IAA in the sieve tubes (white arrows in Fig. 1d, arrows in Fig. 2b). Additionally, we detect IAA in the phellogen (Fig. 1c, Fig. 2a). Furthermore, by analyzing the fluorescence pattern (Fig. 3) originating from cells analyzed with auxin antibodies (detecting bound and free auxin) along a radial line (shown in Fig. 2c) in a cross section, we found clear high auxin concentration peaks in the phellogen, sieve tubes, cambium, and xylem parenchyma cells (Markus Langhans, Cornelia I. Ullrich, and Roni Aloni, unpublished). Thus, by using cross-section analysis of both *DR5::GUS* expression and auxin immunolocalization, we show the anatomy of each specific transporting cell and demonstrate the three IAA pathways in the (1) cambium, (2) sieve tubes, and (3) phellogen.

### 2.1.5 Hormonal Cross Talks and Interactions

It has long been known that hormones influence each other's biosynthesis and activity (Taiz and Zeiger 2006; Dettmer et al. 2009). Effects induced by one hormone may be mediated by another (see below). Auxin interactions with other hormones are common. The auxin–cytokinin cross talk regulates various aspects of vascular differentiation and development (Moubayidin et al. 2009; Bishopp et al. 2011a); so are the interactions between auxin and ethylene (Muday et al. 2012); likewise, is the influence of jasmonate which may be mediated by ethylene (Hudgins and Franceschi 2004).

## 2.2 *Cytokinins Are the Root Cap Signals*

Cytokinins (CKs) are adenine derivatives, and the most common CK is zeatin. CKs produced in the root cap are major hormonal signals of the root (Aloni et al. 2004, 2005, 2006a; Miyawaki et al. 2004). The CKs exist as free and bound molecules. The free forms induce cell divisions in meristematic tissues and promote vascular differentiation along the plant (Saks et al. 1984). Cytokinins applied to the basal side of *Coleus* internodes (and IAA on their apical side) induced both sieve tube (Aloni et al. 1990a) and vessel (Baum et al. 1991) regeneration around a wound.

Cytokinin signaling regulates cambial activity. An expression peak for genes encoding CK receptors was detected in the dividing cambial cells of both *Populus* and *Betula* trees (Nieminen et al. 2008). The quadruple *Arabidopsis* mutant *atipt1;3;5;7*, in which 4 genes encoding cytokinin biosynthetic isopentenyl-transferases are disrupted by T-DNA insertion, was unable to form cambium and showed reduced thickening of the root and stem (Matsumoto-Kitano et al. 2008).

CKs increase the sensitivity of the cambium to the auxin signal, which has promoted the development of ring-porous wood during the evolution of trees (Aloni 1991, 2001).

CKs regulate many developmental activities in plants; they promote leaf development and delay leaf senescence, and they promote breaking of bud dormancy in deciduous trees during spring and shoot development. CKs stimulate lateral bud growth, thus breaking shoot apical dominance (Taiz and Zeiger 2006). In shoot apical dominance, the free auxin represses local biosynthesis of CK in the stem nodes, but after decapitation, CKs, which are thought to be derived mainly from the root caps, are also locally biosynthesized in the nodes along the stem (Tanaka et al. 2006). The CK synthesized in the cap of the primary active root induces root apical dominance, by inhibiting the development of lateral root primordia adjacent to the active root cap, which gives priority to the primary root in competition with its own lateral roots (Aloni et al. 2006a).

CKs produced in the root cap are transported upward from root to shoot in the xylem, via vessels and tracheids (Aloni et al. 2005; Sakakibara et al. 2006; Kuroha and Satoh 2007). The upward CK transport is regulated by the transpiration stream, where the hormone moves mainly to developing shoot organs with high transpiration rates (Aloni et al. 2005). However, CKs move also non-polarly in the phloem (Kudo et al. 2010). The phloem-transported cytokinin toward the root apex can regulate polar auxin transport and maintain vascular patterning at the root tip (Bishopp et al. 2011b).

$NO_3^-$, the nitrate (but not $NH_4^+$) supply to nitrogen-depleted roots, causes a rapid upregulation of *IPT* genes resulting in an increase of CK content in the root, which is transported via the xylem upward into the shoot (Gessler et al. 2004; Miyawaki et al. 2004; Rahayu et al. 2005; Sakakibara et al. 2006; Ruffel et al. 2011). CKs rapidly downregulate the expression of *IPT1,3,5,7* (Miyawaki et al. 2004), which may emphasize the regulatory role of root-to-shoot CK mass transport on shoot CK synthesis. This suggests that synthesis of CK in the shoot could guarantee the CK availability in an emergency under conditions of insufficient CK supply from the root, e.g., under nitrogen deficiency (Miyawaki et al. 2004; Takei et al. 2004) and in young shoot organs of large trees (Gessler et al. 2004) which are further away from the roots. Under conditions of insufficient CK supply from the root, nitrate-responsive *IPT3* is expressed in the phloem (Takei et al. 2004; Miyawaki et al. 2004) and produces CK in the shoot, which could regulate cell divisions in the shoot's cambium.

## *2.3 Gibberellins Are the Mature Leaf Signals*

Gibberellins (GA) are a large family of more than 125 tetracyclic diterpenes. Some of them are essential endogenous regulators which promote cell and stem elongation and many other developmental functions in plants, while most of the GAs are inactive and may serve as precursors of the bioactive $GA_1$ and $GA_4$, which promotes stem elongation (Taiz and Zeiger 2006).

Gibberellins are the mature leave signals, which stimulate cell divisions in the stem's cambium and induce fiber differentiation along the stem (Hess and Sachs 1972; Aloni 1979, 1985, 2001; Dayan et al. 2012). The GA signal which triggers enhanced secondary xylem differentiation in *Arabidopsis* is graft transmissible, suggesting that the GA is a mobile signal (Ragni et al. 2011).

Interestingly, the precursor of the gibberellin hormone ($GA_1$), namely, $GA_{20}$, produced in mature leaves of tobacco can flow non-polarly via the phloem, from the mature leaves to sink organs, namely, to both stem and root [$GA_{19}$ (the precursor of $GA_{20}$) is not mobile]. When the mature leaf-induced $GA_{20}$ precursor arrives to the cambium, it is converted, by local cambial activity of the GA20-oxidase, to the bioactive gibberellin form ($GA_1$), which activates the cambium. The bioactive gibberellin hormone produced in mature leaves is also mobile and moves in the phloem. Therefore, the removal of mature leaves substantially depletes the endogenous GA concentrations in the stem, which impairs cambial activity, fiber differentiation, and shoot elongation (Dayan et al. 2012). It is suggested that the bioactive $GA_4$ in the cambium is produced locally from its precursor, namely, $GA_9$, which might also be mobile. The levels of $GA_4$ and $GA_9$ were very low in tobacco (compare to $GA_1$ and $GA_{20}$) and therefore have not been studied (Dayan et al. 2012).

The bioactive gibberellins ($GA_1$ and $GA_4$) were predominantly found in the expansion zone of differentiating xylem cells in *Populus* suggesting their main role in early stages of wood formation, including cell elongation (Israelsson et al. 2005). Stems of transgenic plants with elevated GA concentrations grow rapidly, produce longer fibers (Eriksson et al. 2000; Biemelt et al. 2004; Dayan et al. 2010), and enhance wood production (Dayan et al. 2010). Furthermore, gibberellin also regulates lignin biosynthesis (Aloni et al. 1990b; Tokunaga et al. 2006) and is involved in reaction wood formation (Jiang et al. 2008; Nugroho et al. 2012).

Gibberellins can regulate the transition from juvenile to adult phases; thus in conifers, exogenous GA application can induce the reproductive phase and cone production in young trees. Conversely, in woody angiosperms including many fruit trees, gibberellins promote vegetative growth by the inhibition of flowering (Goldschmidt and Samach 2004), and when bioactive GAs are applied to mature woody plants, they may induce rejuvenation (Frydman and Wareing 1973, 1974; Taiz and Zeiger 2006).

## 2.4 Ethylene ($C_2H_4$)

The gas ethylene ($C_2H_4$) is a plant hormone which is synthesized in many tissues in response to stress. Wounding, flooding, wind, bending, high auxin levels, elevated cytokinin concentrations, and methyl jasmonate promote ethylene synthesis in trees. Elevated $C_2H_4$ concentrations inhibit stem elongation and may promote leaf and fruit abscission. Ethylene stimulates defense responses to injury or disease and reduces vessel width (Aloni et al. 1998; Hudgins and Franceschi 2004; Taiz and Zeiger 2006).

Interestingly, ethylene stimulates tracheary element (TE) differentiation in *Zinnia elegans* cell culture (Pesquet and Tuominen 2011). The $C_2H_4$ peaks at the time of TE maturation correlating with the activity of the ethylene biosynthetic 1-aminocyclopropane-1-carboxylic acid (ACC) oxidase and the maturing *Zinnia* TEs accumulate ethylene (Pesquet and Tuominen 2011). Blocking ethylene signaling by using silver thiosulfate (STS) appears to block TE maturation (see Bollhöner et al. 2012).

In wood, the ethylene produced in the differentiating tracheary elements diffuses in the centrifugal direction, and this radial ethylene flow through the cambium initials induces the vascular ray (by promoting cell divisions in the cambial fusiform initials) and the enlargement of existing rays (Lev-Yadun and Aloni 1995; Aloni et al. 2000). In poplar trees, ethylene and its 1-aminocyclopropane-1-carboxylate (ACC) synthase promote cambial cell division and wood formation (Love et al. 2009). When the centrifugal transport of ethylene outward to the plant environment is blocked by flooding water, the ethylene accumulates in the cortex or the bark, and the resulting local high $C_2H_4$ concentrations can induce aerenchyma (Li et al. 2006), which enables aeration of flooded stems and roots. High ethylene concentrations also promote lateral and adventitious root formation (Aloni et al. 2006a; Kuroha and Satoh 2007) by local interruptions to the polar auxin flow, causing local sites of high IAA concentrations which induce the adventitious root tips.

Asymmetric patterns of elevated ethylene concentrations may be induced in stems and branches in response to gravistimulation, bending, and wind, which could probably be promoted by asymmetric distribution of elevated IAA concentrations. However, no clear modifications in IAA distribution pattern could be detected in poplar and pine trees following gravistimulation (Hellgren et al. 2004), which might indicate a technical difficulty in tissue sampling. Asymmetric ethylene distributions can promote reaction wood formation: the formation of compression wood in conifers and tension wood in hardwoods (Timell 1986; Zobel and van Buijtenen 1989). Thus, endogenous ethylene produced in response to leaning of a stem stimulates cell division in the cambium giving rise to an eccentric tension wood formation in poplar trees (Love et al. 2009).

In conifers, the ethylene hormone induces chemical defenses against insects and pathogens. Wounding and ethylene can promote traumatic resin-duct formation in conifer woods (Hudgins and Franceschi 2004; Hudgins et al. 2006). The resin-duct epithelial cells produce oleoresin terpenoids, which protect the tree from insects and their associated pathogens (Keeling and Bohlmann 2006; Ralph et al. 2007; Schmidt et al. 2011).

## 2.5 *Abscisic Acid from the Root Meristems*

Abscisic acid (ABA) is the long-distance stress signal produced in the root's meristematic cells (Koiwai et al. 2004) as soil is drying. ABA is transported upward through the xylem from the roots to the shoot for regulating the closure of stomata under stress and retarding vascular meristematic activity (Taiz and Zeiger 2006). However, ABA can also be produced in the phloem of the shoot and even in the stomata themselves (Koiwai et al. 2004). The flow rate of water through tracheids and vessels is crucially affected by stomata opening and closing; ABA closes the stomata and could retard shoot development, while cytokinin from the root caps which is transported upward through the xylem (Aloni et al. 2005) has a positive effect on stomata opening (Dodd 2003), on cambium activity (Nieminen et al. 2008; Matsumoto-Kitano et al. 2008), and on shoot development (Taiz and Zeiger 2006). ABA, which is the universal stress hormone of higher plants, has a central role in plant developmental plasticity; it is likely involved in slowing down and stopping cell division in the cambium and wood differentiation in trees toward their winter dormancy by retarding and ending their cambium activity. Evidently, in intact *Eucommia ulmoides* trees, the expression of the auxin-binding protein1 (ABP1), one of the putative receptors of auxin, is promoted by auxin during the period of cambium activity and is inhibited by ABA during dormancy (Hou et al. 2006). In addition, ABA plays a crucial role in promoting plant tolerance to cold (Galiba et al. 2009), which may be used to improve plant sustainable survival (Xue-Xuan et al. 2010).

## 2.6 *Jasmonates*

The phytohormone jasmonic acid (JA) and its volatile methyl ester (MeJA) are fatty acid-derived cyclopentanones. A positive effect of jasmonate application on cambium activity indicated a stimulatory role of JA in secondary growth, suggesting that JA signaling can promote cambial cell divisions (Sehr et al. 2010).

Jasmonates (JAs) activate defense-related genes against insects and pathogens (Howe 2004). The MeJA moves in both the phloem and xylem pathways. The MeJA enters into the phloem and moves in the sieve tube sap along with photoassimilates. MeJA promotes its own transport; whole plant experiments suggest that enhanced transport of both MeJA and sugar may be due to MeJA enhancing the energy of the plasma membrane (Thorpe et al. 2007). In Pinaceae, application of methyl jasmonate (MeJA) can induce traumatic resin-duct formation (Hudgins et al. 2003; Hudgins and Franceschi 2004; Huber et al. 2005). This MeJA-induced defense response is mediated by ethylene, which promotes the formation of traumatic resin-secreting epithelial cells (Hudgins and Franceschi 2004). In *Picea abies*, treatment with methyl jasmonate induces the expression of isoprenyl diphosphate synthase genes during the formation of traumatic resin ducts, promoting the terpenoid-based oleoresin accumulation in the ducts, which protect against herbivores and pathogens. The trunk response to the MeJA treatment was detected up to 60 cm above the site of application (Schmidt et al. 2011).

## 2.7 *Brassinosteroids (BRs)*

A group of naturally occurring plant polyhydroxysteroids with wide range of biological activities interact with other hormones including auxin, gibberellins, and ethylene (Zhang et al. 2009; Clouse 2011). Nanomolar levels of Brassinosteroids (BRs) stimulate tracheary element formation in isolated mesophyll cells of *Zinnia elegans* (Iwasaki and Shibaoka 1991) and regulate expression of genes associated with xylem formation (Fukuda 1997, 2004). BR levels increased dramatically prior to the morphogenesis of tracheary elements in cultured *Zinnia* cells, showing that BRs are necessary for the initiation of the final stage of tracheary element differentiation (Yamamoto et al. 1997, 2001). BRs from the epidermis might influence the differentiation of vascular tissues (Savaldi-Goldstein et al. 2007) and possibly control tissue-type specificity of vascular cell proliferation (Carlsbecker and Helariutta 2005).

## 2.8 *Strigolactones Are Novel Root Hormones*

Strigolactones (SLs), a group of terpenoid lactones derived from carotenoids, are a novel class of plant hormones recently described to be involved in the repression of shoot branching (Klee 2008; Gomez-Roldan et al. 2008; Umehara et al. 2008; Domagalska and Leyser 2011). The SLs are produced in the root and move upward via the xylem (Kohlen et al. 2011) to the stem, where they inhibit lateral bud development. Strigolactones positively regulate cambial activity and vascular differentiation (Agusti et al. 2011). The SL signaling in the vascular cambium itself is sufficient for cambium stimulation, and it interacts with the auxin signaling pathway (Agusti et al. 2011).

The strigolactones were known as root signals that stimulate parasitic plants' germination (Cook et al. 1972). These SLs were recently found to attract arbuscular mycorrhizal fungi facilitating the uptake of soil nutrients (Akiyama et al. 2005). From evolutionary point of view, I assume that plant roots started to release strigolactones to the environment as a primary signal for attracting arbuscular mycorrhizal fungi. Much later during evolution, the parasitic plants have emerged, eventually adapting themselves to use this root signal for their germination, thus ensuring the presence of a nearby host root.

Interestingly, strigolactone concentration increases as a response to inorganic phosphate deficiency (López-Ráez et al. 2008; Kohlen et al. 2011). This fact stimulates to suggest that in response to phosphate stress the strigolactones promote main stem elongation on the expenses of branch development, thus giving the plant an advantage in its competition for light against neighboring plants. As SLs promote vascular development in the main stem (Agusti et al. 2011), they also improve nutrient supply to the actively growing apical bud. In contrast, the cytokinin produced in the root cap (Aloni et al. 2004, 2005, 2006a) is a general promoting signal for buds and leaf development; its concentration increases following an increase in nitrate ($NO_3^-$) supply (Takei et al. 2001, 2004; Sakakibara et al. 2006; Ruffel et al. 2011), whereas the strigolactone promotes only main stem elongation by inhibiting lateral bud development. Thus, root-specific signaling shapes shoot developmental architecture and vascular differentiation in response to phosphate and nitrate levels in the soil.

# 3 The Vascular System

The plant vascular system connects the shoot organs with the roots and enables efficient long-distance transport between them. The development of the vascular system in a plant is an open type of differentiation, continuing as long as the plant grows from apical and lateral meristems. The continuous development of new vascular tissues enables regeneration and adaptation to changes in the environment (Aloni 2001; Scarpella and Helariutta 2010).

## *3.1 Vascular Meristems*

In the plant body, the vascular tissues are produced from embryonic tissues, called vascular meristems, whose stem cells retain the ability to divide and continually multiply. This ensures flexibility and adaptation of the vascular system to constant changes inside the plant and in its environment. Two meristematic stages are distinguished: the procambium and the cambium.

The procambium is the apical meristem that produces primary phloem and primary xylem in the embryo and in young parts of the shoots and roots. The cambium is a lateral meristem found in gymnosperms and dicotyledons. It develops in the older parts of the plant axis, where it produces secondary phloem and the secondary xylem (Larson 1994; Evert 2006; Matte Risopatron et al. 2010).

## *3.2 Venation Pattern Formation in Leaves*

Vascular differentiation in a leaf is limited to early stages of primordium development. The gradual pattern of IAA production during leaf development was explained by the *leaf venation hypothesis* (Aloni 2001). This hypothesis was confirmed experimentally (Aloni et al. 2003) showing that the primary sites of IAA production are the developing hydathodes at the leaf margin (when hydathodes mature, they passively exudate water through their epidermal pore in response to high root pressure). During leaf-primordium development, there are gradual shifts in the sites and concentrations of IAA production, progressing from the hydathode of the elongating tip, continuing downward along the expanding blade margins, and ending in the central regions of the lamina (Aloni 2001, 2010; Aloni et al. 2003). This pattern of sites of IAA concentrations was also confirmed by auxin analysis by GC-MS/MS techniques (Müller et al. 2002). Likewise, flowers, fruits, and seeds produce free auxin in a gradual shifting pattern during their development (Aloni et al. 2006b).

In leaves of *Arabidopsis* plants treated with auxin transport inhibitors, the vascular tissues became progressively confined toward the leaf margin. When the concentration of auxin transport inhibitor was increased, the vascular elements were more restricted to the margin, indicating that the leaf vascular system depends on inductive signals from the leaf margin (Mattsson et al. 1999; Sieburth 1999). By impairing the auxin flow, with an auxin transport inhibitor, the secondary and tertiary veins did not or were poorly developed (Mattsson et al. 1999; Sieburth 1999), likely because the continuous high auxin concentrations near the leaf margin inhibited the formation of local sites of low IAA production in the central regions of the lamina.

Scarpella et al. (2006) elegantly demonstrated in *Arabidopsis* leaf primordia that the auxin-efflux-associated protein, PIN1, is polarly expressed in the cell membranes prior to pre-procambial formation, clearly showing the IAA flow directions and pathways in the primordium prior to procambium formation. Integrated polarities in all emerging veins indicate auxin drainage toward preexisting veins, but veins could display divergent polarities until they become connected at both ends (Scarpella et al. 2006).

## 3.3 Relationships Between Phloem and Xylem

Plant vascular systems are usually composed of phloem and xylem. In plant tissue cultures, grown on a solid medium, low auxin levels induced sieve elements with no tracheary elements, while high auxin concentrations resulted in the differentiation of both phloem and xylem (Aloni 1980, 1987), but even on the high concentrations, at the surface further away from the auxin-containing medium, only phloem with no xylem developed (Aloni 1980). High auxin concentrations applied to decapitated *Luffa* stems induced xylem differentiation in its phloem anastomoses (Aloni 1995) indicating the need for high hormonal stimulation for xylem differentiation. Along a wild-type plant axis, xylem does not differentiate in the absence of phloem, though strands of phloem (with no xylem) and phloem anastomoses are common in stems of many plant species (Aloni and Sachs 1973; Roberts et al. 1988; Aloni and Peterson 1990; Aloni and Barnett 1996). It has been suggested that the differentiation of phloem strands and phloem anastomoses between the strands is controlled by streams of low auxin levels (Aloni 1987, 1995).

Conversely, in leaves, the proximity between the sites of auxin synthesis (Aloni et al. 2003) and the site of differentiating vascular cells probably results in relatively high local IAA concentrations at the differentiating sites. This may explain why xylem can differentiate in the absence of phloem at the freely ending veinlets (Lersten 1990; Horner et al. 1994) and the hydathodes (Raven et al. 2005) which are sites of free-auxin production (Aloni et al. 2003). In *Oxalis stricta*, there are virtually no sieve tubes in any terminal vein, while *Polygonum convolvulus*, at the other extreme, has sieve tubes extending to the tips of most terminal veins (Horner et al. 1994). In most of the studied species, freely ending veinlets in leaves may display different relations between phloem and xylem in the same plant (Lersten 1990; Lersten and Curtis 1993; Horner et al. 1994).

## *3.4 Vascular Differentiation in Roots*

The hormonal control of vascular differentiation and regeneration in both stem and root follows similar general principles (Aloni et al. 2006a). The vascular system is continuous along the entire plant as it is induced and controlled by the continuous polar IAA movement along the plant body (Sachs 1981; Aloni 1987; Berleth et al. 2000), from the hydathodes of young leaves downward to the root tips. Primary vascular differentiation in the root vascular cylinder is characterized by a radial pattern of alternating strands of xylem and phloem. It has been suggested that the radial pattern of the root protoxylem vs. protophloem strands is induced by alternating longitudinal polar streams of high IAA vs. low IAA concentrations, respectively (Aloni et al. 2006a). The downward transport of cytokinin via the phloem might influence vascular patterning in the root meristem by regulating the downward IAA transport (Bishopp et al. 2011b).

The secondary xylem along the plant axis, in both the stem (Aloni and Zimmermann 1983; Aloni 2001; Leitch 2001; Olson and Rosell 2013) and the root (Fahn 1964), shows a gradual and continuous increase in vessel width and decrease in vessel density with increasing distance from the young leaves. These gradual vascular patterns are probably induced by a gradient of decreasing IAA concentrations from the young leaves toward the root tips (Aloni and Zimmermann 1983). Both cytokinin and IAA regulate root vascular differentiation (Aloni et al. 2006a) and root gravitropism (Aloni et al. 2004); these two hormones, together with ethylene produced in maturing vessels (Pesquet and Tuominen 2011; Bollhöner et al. 2012), regulate lateral root initiation (Aloni et al. 2006a; Hirota et al. 2007).

## *3.5 Regulation of Vascular Tissues in Plant Tumors*

The advantage of studying vascular differentiation in tumorous tissues is that they clarify the importance of polarity, hormonal concentration, and gradients in normal organized tissues. The most studied tumorous plant tissues are the crown galls induced by *Agrobacterium tumefaciens* on many plant species (Aloni and Ullrich 2007). The tumor is an unorganized tissue resulting from an unbalanced hormonal production characterized by a few local random sites of extremely high auxin and high cytokinin concentrations, which, therefore, result in impaired tissue polarity. High ethylene production in these galls is likely a result of the high IAA and high CK production by the *A. tumefaciens*-transformed plant cells (Aloni et al. 1998).

Plant tumors induced by *A. tumefaciens* were considered unorganized or partly organized masses (Sachs 1991 and references therein). However, a three-dimensional pattern analysis of the phloem and xylem in the *A. tumefaciens*-induced crown galls unveiled a sophisticated vascular network of continuous vascular strands extending from the host plant up to the tumor surface. The development of these strands indicates synthesis of auxin by the *A. tumefaciens*-transformed plant cells located immediately beneath the surface of the fast growing tumor (Aloni et al. 1995). The strands in a tumor contain both xylem and phloem and are interconnected by phloem anastomoses, consisting of sieve tubes but not vessels. In the adjacent stem tissues, the tumor induces the development of pathologic xylem characterized by narrow vessels, absence of fibers, and giant rays. These anatomical features have triggered us to propose the *gall-constriction hypothesis* (Aloni et al. 1995) which explains the mechanism that gives priority in water supply to the growing gall over the host shoot. The hypothesis proposes that a growing gall retards the development of its host shoot by decreasing vessel diameter in the shoot tissues adjacent to the tumor, which substantially reduces water supply to the upper parts of the shoot. It was further postulated that the controlling signal that induces the narrow vessels in the host is the hormone ethylene (Aloni et al. 1995), which is known to reduce vessel diameter (Yamamoto et al. 1987). The *gall-constriction hypothesis* was experimentally confirmed by showing that tumor-induced ethylene is a limiting and controlling factor of crown gall morphogenesis; very high ethylene levels were produced continuously by a growing crown gall during a few weeks, up to 140 times more ethylene than in wounded but not infected control stems, reaching a maximum at five weeks after infection (Aloni et al. 1998; Wächter et al. 1999). Tumor-induced ethylene diminished vessel diameter in the host stem and enlarged the surface (through which high transpiration occurs) of the tumor (Aloni et al. 1998), thus giving priority in water supply to the growing gall over the host shoot. Comparison between the development of plant and animal tumors has shown an analogous requirement for neovascularization in both, presaging possible strategies for prevention and therapy (Ullrich and Aloni 2000). The discovery that plant tumors produce ethylene required for gall development (Aloni et al. 1998; Wächter et al. 1999) has promoted to develop ethylene insensitive fruit trees, which are tumor free.

# 4 Hormonal Regulation of Xylem Cell Differentiation

In the xylem, the conducting cells are the tracheary elements. They function in long-distance water transport, as nonliving cells after autolysis of their cytoplasm. Tracheary elements are characterized by secondary wall thickenings (Oda and Fukuda 2012) which enable them to retain their shape when dead, despite the pressure of the surrounding cells. The two fundamental types of xylem conduits are the tracheid (typical to gymnosperms) and the vessel (of angiosperms) which is built of vessel elements. Among the vessels are the fibers, which are the supporting cells. Fibers may function as living cells which retain their protoplasts for relatively long periods (Fahn and Leshem 1963), or support the plant body as dead cells (Fahn 1990). In conifers, resin ducts and glands are common for protecting the tree from insects. Secondary vascular tissues are characterized by vascular rays, oriented in the radial direction, which are usually built of parenchyma cells, but may also contain radial vascular elements (Fahn 1990; Evert 2006).

## *4.1 Tracheid Differentiation*

A tracheid is a non-perforated long cell with bordered pits. Tracheids are both the conducting and supporting cells that build the “softwood” of gymnosperms. Auxin movement through the cambium of pine trees induces the differentiation of tracheids from cambium initials (Larson 1969; Uggla et al. 1996; Sundberg et al. 2000). In young pine seedlings, tracheids (Fig. 2f–h) can also redifferentiate from parenchyma cells by application of both auxin and gibberellin (Kalev and Aloni 1998; Aloni et al. 2000). When only auxin is applied, it induces very short tracheids, while gibberellin, in the presence of auxin, promotes tracheid elongation by stimulating intrusive growth of both the upper and lower ends of the differentiating tracheids (Fig. 2h). Tracheid differentiation could also be promoted by ethylene (Aloni et al. 2000).

## *4.2 Vessel Differentiation*

A vessel is a long continuous tube made up of a few or numerous vessel elements connected end to end by perforation plates and limited in length by imperforated walls at both extremities. Vessels, and not vessel elements, are the operating units in the “hardwood” of angiosperms. Their dimensions are important parameters for understanding long-distance water transport, wood quality, xylem pathology, wood adaptation, and evolution. Vessels do not end randomly in the stem of young trees; e.g., in *Populus* and *Olea*, vessel endings are significantly higher at the nodes (Salleo et al. 1984). Therefore, the nodes are considered “safety zones,” because gaseous emboli and fungal spores fail to pass through the endings. In diffuse-porous trees, the longest vessels are about 1 m long, whereas in ring-porous trees the largest earlywood vessels are extremely long and can reach the length of the tree itself (Zimmermann and Jeje 1981). An increase in vessel diameter markedly increases the efficiency of water transport, owing to decrease in resistance to flow, whereas increase in both diameter and length decreases the safety of water conduction, in terms of cavitation (Tyree and Zimmermann 2002; Hacke et al. 2006; Fu et al. 2012). The vessels are induced by the polar flow of auxin originating in young leaves (Jacobs 1952; Sachs 1981; Aloni 2010; Scarpella and Helariutta 2010). High IAA concentrations stimulate rapid cell differentiation resulting in narrow vessels, while low IAA levels induce slow differentiation, which permits more time for cell expansion until secondary wall deposition, therefore resulting in wide vessels (Aloni and Zimmermann 1983). Along the plant axis, the IAA induces gradual gradients of increasing vessel diameter and decreasing vessel density from leaves to roots (Fig. 2d, e). The hormonal mechanism that regulates these vessel gradients (Aloni and Zimmermann 1983) as well as the mechanism that regulates the formation of large earlywood vessels in ring-porous trees (Aloni 1991) will be clarified below.

## 4.3 *Fiber Differentiation*

Fibers are long and narrow cells with thick secondary walls that are usually heavily lignified. Differentiation of fibers in the "hardwood" of angiosperms and in the phloem is induced by gibberellin in the presence of auxin, and the GA which induces fibers originates in mature leaves (Hess and Sachs 1972; Aloni 1979, 1987; Dayan et al. 2012). The numerous fibers in the latewood of ring-porous trees are induced during the summer and autumn by mature leaves that produce GAs (Aloni 1991, 2001; Aloni et al. 1997), while the IAA from the mature leaves is transported into the sieve tubes (Morris et al. 1973; Goldsmith et al. 1974). When both gibberellin and auxin were applied exogenously to stems from which all the leaves and buds had been excised (to remove endogenous hormone production from leaves and buds), high IAA concentrations stimulated rapid differentiation of short fibers with thick secondary walls, while high levels of GA resulted in long fibers with thin secondary walls (Aloni 1979; Roberts et al. 1988). Both gibberellin and auxin also regulate lignin biosynthesis (Aloni et al. 1990b; Tokunaga et al. 2006). Applications of gibberellin and auxin to various industrial plants increase long fiber production (Aloni 1985). In addition, cytokinin promotes fiber differentiation (Aloni 1982; Saks et al. 1984), whereas ethylene can retard fiber formation (Aloni et al. 1998). Understanding the gibberellin biosynthesis pathway (e.g., Taiz and Zeiger 2006) enables molecular manipulation of GA production. Overexpression of GA 20-oxidase, a gene encoding the enzyme responsible for the rate-limiting step involved in GA synthesis, enhances fiber yield (Eriksson et al. 2000). The transgenic tobacco plants and poplar trees showed higher levels of GAs in their shoots and an increase in fiber length (Eriksson et al. 2000; Eriksson and Moritz 2002). This genetic manipulation also increased the GA inactivation, due to GA 2-oxidase catalysis. We have therefore used another approach to elevate gibberellin concentrations by silencing the GA 2-oxidase (i.e., preventing deactivation of the bioactive gibberellin), which elevates the bioactive GA concentrations in our transgenic model plants (tobacco) and promoted rapid shoot elongation, increased fiber production, increased fiber size, and decreased fiber lignification (Dayan et al. 2010). Low lignification can reduce the cost of paper and other fibrous materials. Endogenous bioactive gibberellin concentrations could be boosted up by inducing both the overexpression of GA 20-oxidase and silencing the GA 2-oxidase genes (Dayan et al. 2010), which could result in synergistic effects. These molecular manipulations could also modify lignin metabolism and change lignin structure and content.

## 4.4 Ray Differentiation

The radial component of the secondary vascular tissues is the vascular rays. Usually, the appearance of rays indicates the transition from procambium to cambium. The rays serve as radial transport pathways between the xylem and phloem and vice versa. The rays are induced and controlled by radial moving signals, and they are shaped (their longitudinally elongated structure) by axial signal flows (Lev-Yadun and Aloni 1995). The ethylene, which was recently shown to be synthesized in maturing tracheary elements (Pesquet and Tuominen 2011; Bollhöner et al. 2012), moves centrifugally through the cambium (from the differentiating tracheary elements toward the phloem) and is the major hormonal signal which promotes the initiation of rays and their regulation (size and spacing) in the cambium (Lev-Yadun and Aloni 1995; Aloni et al. 2000). In young trees, there is a natural gradual increase in ray size with increasing distance from the pith and with growing distance from the young leaves. These developmental patterns might result from a gradual decrease of IAA concentrations and a gradual increase in ethylene synthesis with increasing distance from the young leaves. In conifers, the ray initials occupy about 10 % of the cambium surface, whereas in woody angiosperms, the ratios are more variable ranging from 0 % rays (in rayless wood) to about 25 % ray volume. A substantial increase in ray dimensions occurs in response to wounding. This wound effect on ray size can be experimentally induced by ethylene application. The proliferation of ray parenchyma cells on the cut surface following wounding enables the recovery of the injury and regeneration of wood after wounding (Lev-Yadun and Aloni 1995; Aloni et al. 2000).

## *4.5 Resin-Duct Differentiation*

Resin ducts lined with resin-secreting epithelial cells are a common feature in conifers (Evert 2006). They protect the tree from insects and their associated pathogens (Keeling and Bohlmann 2006; Ralph et al. 2007; Schmidt et al. 2011). In the Pinaceae, there are genera (e.g., *Pinus, Picea, Larix*, and *Pseudotsuga*) which produce resin ducts as a natural feature and in response to injury will also produce traumatic resin ducts, other genera (e.g., *Abies, Tsuga, Cedrus*, and *Pseudolarix*) which produce only traumatic resin ducts in response to wounding, and there are genera (e.g., *Cupressus* and *Juniperus*) which never produce any resin ducts (Fahn 1990; Evert 2006). The naturally occurring resin-duct system is built of longitudinal and radial (occurring inside large vascular rays) ducts. The largest number of resin ducts is produced when the cambium of an injured branch is intensively active (Fahn and Zamski 1970; Fahn 1990). In *Pinus halepensis*, auxin which enhances radial growth of wood also promoted resin-duct formation (Fahn and Zamski 1970); however, this resin-duct formation is not a direct auxin effect because the resin ducts developed only about one month after the auxin application. The application of the ethylene-releasing agent ethrel (2-chloroethylphosphonic acid) to *P. halepensis* seedlings promoted the production of longitudinal resin ducts in their wood (Yamamoto and Kozlowski 1987). Jasmonate, more specifically, methyl jasmonate (MeJA), which activates defense-related genes, seems to be the primary signal which induces traumatic resin-duct formation in conifers (Hudgins et al. 2003; Hudgins and Franceschi 2004; Huber et al. 2005; Schmidt et al. 2011), and this jasmonate-induced defense response is mediated by ethylene (Hudgins and Franceschi 2004). This was evident in studies on *Pseudotsuga menziesii* which showed that the MeJA-induced ethylene production earlier and 77-fold higher than wounding. Pretreatment of *P. menziesii* stems with an ethylene response inhibitor (1-methylcyclopropene) inhibited the MeJA and wound responses (Hudgins and Franceschi 2004). Severe stress and wounding might induce large resin cavities which damage the wood for technological use. This damage could likely be prevented in trees with low ethylene sensitivity. Similarly, we demonstrated that crown gall tumor development which is also regulated by ethylene was inhibited on the ethylene insensitive tomato, the *Never ripe* mutant (Aloni et al. 1998). In vitro production of pine trees from shoots is well established (Hargreaves et al. 2005). Regularly, to increase survival of adventitious originated plants, the culture jars are aerated to reduce the buildup of ethylene concentrations during the culture process. Therefore, I propose that conifer trees with lowered sensitivity to ethylene would be selected during the tissue-culture process by keeping the jars closed. Thus, only tissues of ethylene insensitive trees will survive. Various lines with different sensitivities to ethylene can be selected and will be analyzed for traumatic resin-duct formation and afterward for their resistance under field conditions. It is expected that selected tree lines with lowered ethylene sensitivity will show decreased response to wounding and consequently limited traumatic resin-duct formation. Lowering ethylene sensitivity in conifers may reduce their insect resistance. This issue can be solved genetically by introducing into the selected trees toxin genes against insects (Gordon et al. 2007; Gurevitz et al. 2007). For example, transgenic *Pinus radiata* trees containing a *Bacillus thuringiensis* toxic gene displayed variable levels of resistance to insect damage, with one transgenic line being highly resistant to feeding damage (Grace et al. 2005). Anti-attractant treatments against insects may also be used (Erbilgin et al. 2007).

## 5 Vascular Differentiation in Branch Junction

The *segmentation hypothesis* (Zimmermann 1983) suggests that the xylem at a branch junction consists of narrow vessel elements or tracheids that form a bottle-neck for water transport. The hydraulic segmentation of branches from the main stem gives priority of water supply to the main stem over the branches (Zimmermann 1983; Aloni et al. 1997). Spiral vascular tissues and circular vessels are induced by the movement of auxin in circular patterns (Sachs and Cohen 1982). Circular vessels are often produced in the upper side of branch junctions, and their size and frequency increase continuously with age and branch width (Lev-Yadun and Aloni 1990). The circular vessels do not function in water transport, and they actually interrupt and reduce water flow in the upper side of the branch junction. Therefore, the long-distance water transport into branches occurs preferably through the branch sides and its lower region (Aloni et al. 1997). The young-leaf biomass, which produces IAA, regulates the development of a branch and its success in competition with other branches. Greater IAA production induces more longitudinal vessels, or tracheids, parallel to the wood grain which gives the branch improved access to the water resources of the tree. Thus, IAA export from a branch regulates branch vigor (Kramer and Borkowski 2004; Kramer 2006). Pruning young trees improves the growth of their main stem and their wood quality. Pruning branch-stem junctions stops their circular pattern formation, resulting in the production of uniform trunk wood without knots in the post-pruning annual rings and optimization of mechanical wood properties at the junction sites, which is important for wood industry.

## 6 Wood Gradients

### 6.1 *Regulation of Vessel Size and Density Along the Tree Axis*

The *six-point hypothesis* (Aloni and Zimmermann 1983) suggests that the auxin hormone descending from young leaves to root tips acts as a morphogenetic signal which forms polar concentration gradients along the plant axis. Such longitudinal IAA gradients in the vascular cambium provide directional and location information to differentiating cells along the morphogenetic fields. The decreasing gradient of IAA concentrations along the tree axis from leaves to roots results in a general and gradual increase in tracheid dimensions, or vessel diameter, which is associated with a decrease in vessel density, with increasing distance from the young leaves (Aloni and Zimmermann 1983; Zimmermann 1983; Aloni 1987; Leitch 2001). Accordingly, the narrow tracheids or vessels differentiate near the young leaves where the highest auxin concentrations are expected, while the widest tracheids and vessels are formed in the base of the trunk and in the root at the greatest distance from the auxin sources. The gradual increase in vessel diameter from leaves to roots is associated with a gradual decrease in vessel density. Hence, vessel density is generally greater in branches, where the vessels are narrow, than in roots, where they are wide (Aloni and Zimmermann 1983; Leitch 2001). We proposed that the general increase in vessel size and decrease in vessel density along the plant axis is regulated by a gradient of decreasing IAA concentrations from leaves to roots (Aloni and Zimmermann 1983), which is based upon the assumption that the steady polar flow of IAA from leaves to roots controls these polar changes in the vascular system. IAA concentration controls the rate of cell differentiation and cell expansion before secondary wall deposition. High auxin concentrations near the young leaves induce narrow vessels (Fig. 2d) because of their rapid differentiation, allowing only limited time for cell growth. Conversely, low IAA concentrations further down result in slow differentiation, which permits more cell expansion before secondary wall deposition and thereby results in wide vessels (Fig. 2e).

Recently, an important support to the *six-point hypothesis* (Aloni and Zimmermann 1983) was documented by studying the duration of cell differentiation along the stem of a *Picea abies* tree (Anfodillo et al. 2012). This study demonstrates that the duration of the expansion phase is positively correlated with the lumen area of the tracheids and that the lumen area of the conduits from the top of the tree (11.5 m in height) to its base was linearly dependent on the time during which the differentiating tracheids remained in the expansion phase (Anfodillo et al. 2012). Their results show that at the top of the tree's trunk (9 m from the ground), the tracheid expansion time was 7 days, at 6 m aboveground the cells expended for 14 days, and at 3 m for 19 days (Anfodillo et al. 2012). Therefore, the tracheids at the base of the tree, which have the longest period of cell expansion before secondary wall deposition, become the widest conduits.

Vessel density is controlled by auxin concentration; accordingly, high auxin concentrations (near the sites of IAA production) induce greater density (Fig. 2d), while low concentrations (further down, toward the roots) diminish density (Fig. 2e). Consequently, vessel density decreases from leaves to roots (Aloni and Zimmermann 1983; Leitch 2001). The hypothesis was experimentally confirmed by showing that various auxin concentrations applied to decapitated stems induce substantial gradients of increasing vessel diameter and decreasing vessel density from the auxin source (Fig. 2d, e) toward the roots (Aloni and Zimmermann 1983). High auxin concentration yielded numerous vessels that remained narrow because of their rapid differentiation; low auxin concentration resulted in slow differentiation and therefore in fewer and wider vessels.

Studies on transgenic plants with altered levels of IAA confirmed the general relations between IAA concentration and vessel size and density. Thus, auxin overproducing plants (i.e., ones overexpressing the *iaaM* gene) contained many more vessel elements than did control plants, and their vessels were narrow (Klee et al. 1987); conversely, plants with lowered IAA levels (i.e., expressing the *iaaL* gene as an anti-auxin gene) contained fewer vessels of generally larger size (Romano et al. 1991).

### 6.2 Cambial Activity Reflects the Social Status of a Forest Tree

An important study on cambium dynamics and wood formation in a 40-year-old *Abies alba* plantation near Nancy, France, has shown that the timings, duration, and rate of tracheid production change according to the social status (relative size and vitality) of a tree in the forest (Rathgeber et al. 2011). The study demonstrates clear gradients of cambial activity related to the crown area and the height of the trees. Cambial activity started earlier, stopped later, and therefore lasted longer in dominant trees than in intermediate and suppressed ones. Cambial activity was more intense in dominant trees than in the smaller trees. It was estimated that about 75 % of tree-ring width variability was attributable to the rate of cell production and only 25 % to extend cambial duration. Interestingly, growth duration was correlated to tree height, while growth rate was correlated to crown area (Rathgeber et al. 2011).

Vigorous crowns produce more auxin in their young leaves and more bioactive gibberellins in their mature leaves. The synergistic effects of these two hormones upgrade cambial activity and enhance tracheid production. Together with the expected elevated hormonal production, a larger crown also provides higher sugar contents, as was found in the outer wood of the most productive poplar clones (Deslauriers et al. 2009). These results suggest that gradients in cambial activity and of wood formation are strongly related to tree size and vigor. It is likely that the dominant trees are genetically superior and therefore their seeds should be collected for future plantations.

## 7 Hormonal Wood Evolution

### 7.1 From Tracheids to Vessels and Fibers

Vessel elements are more efficient conductors of water than tracheids, since water flows through vessel elements occurring via openings, namely, perforations, rather than diffusion through the cell walls of tracheids (Tyree and Zimmermann 2002).

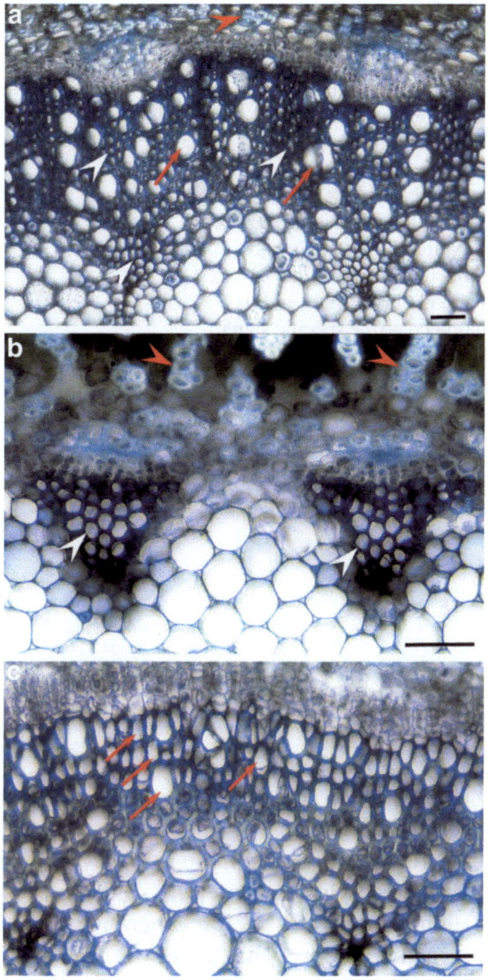

**Fig. 5** Cross sections of *Ephedra campylopoda* C.A. Mey stems, showing a relatively primitive (built of tracheids [*white arrowheads*], vessels [*arrows*], and fibers [*red arrowheads*]) vascular system in an intact stem (**a**), a stem treated for 1 month with gibberellin (1 % $GA_3$ in lanolin (w/w)) (**b**) or with auxin (0.2 % NAA in lanolin (w/w)) (**c**). The sections show the typical xylem built of tracheids and vessels (**a**), gibberellin-induced xylem differentiation of only tracheids (with no vessels) and many fibers in the phloem (**b**), auxin-induced continuous layers of mainly vessels (with no fibers) (**c**). Bars = 100 μm (**a–c**) (Pua Feigenbaum and R Aloni, unpublished)

Tracheids appeared in ancient land plants (Gerrienne et al. 2011) about 430 million years ago, while vessel elements were recorded much later, about 140 million years ago, and became dominant in angiosperms (Raven et al. 2005). Vessel elements have evolved independently from tracheids in several diverse groups of plants, making them an excellent example of parallel evolution (Bailey 1944). The naturally occurring perforated tracheids are very rare in conifer trees (Bannan 1958). The suggestion that these perforations have been induced by polar auxin movement was supported experimentally by the formation of perforations in tracheids following the application of high auxin concentration to hypocotyls of young pine seedlings (Aloni et al. 2000). Therefore, these tracheids with perforations (Fig. 2f, g) support the general view about the evolutionary origin of vessels from tracheids. I would like to suggest that it is likely that during xylem evolution, the transporting tissues have become more sensitive to the auxin stimulation (Barbez et al. 2012) rather than an increase in auxin stimulation occurred.

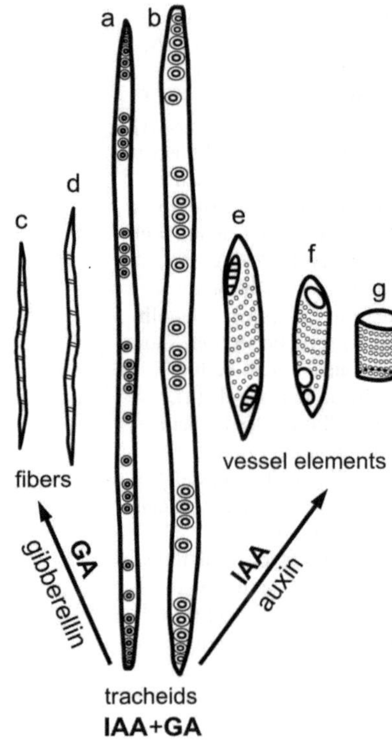


**Fig. 6** Schematic diagrams illustrating the role of auxin (IAA) and gibberellin (GA) in shaping the evolution of vessel elements and fibers from the long tracheids of primitive woods. The tracheids characterized by bordered pits are induced by a mixture of both auxin and gibberellin (**a**, **b**). During plant evolution, gibberellin has become the specific signal for fibers with simple pits (**c**, **d**) and IAA the signal for short vessel elements with perforation plates (**e**–**g**)

Fibers, like vessels, have originated from tracheids of more primitive plants. IAA movement through the cambium of conifer trees induces the differentiation of tracheids from the cambium initials (Savidge 1996; Uggla et al. 1996). In isolated half hypocotyls of young pine seedlings, replacing the cotyledons by auxin application induced very short regenerative tracheids, which originated from very young parenchyma cells, while there was need for both auxin and gibberellin for inducing the differentiation of long tracheids (Kalev and Aloni 1998; Aloni et al. 2000) (Fig. 2h). Furthermore, Pua Feigenbaum and Roni Aloni (unpublished) have shown that in young stems of *Ephedra campylopoda* which regularly produces tracheids, vessels, and fibers (Fig. 5a), a gibberellin application promoted fiber formation (Fig. 5), while auxin application induced vessel differentiation (Fig. 5c) with no fibers.

All these results suggest that during vascular evolution, the original hormonal mechanism that induced the differentiation of tracheids in primitive plants has become more specific in higher plants. Thus, from the ancient inducing mechanism for typically elongated tracheids (a combination of both auxin and gibberellin), each vascular element in higher plants is mainly induced and regulated by one specific hormone: *Auxin* by itself induces short vessel elements (Jacobs 1952; Sachs 1981; Aloni 2010), whereas *gibberellin*, in the presence of auxin, has become the specific signal which induces long fibers (Aloni 1979, 1987; Dayan et al. 2012).

This means that the well-known evolutionary transition from tracheids to fibers and vessel elements reflects the hormonal specialization which has occurred during plant evolution (Fig. 6).

## 7.2 From Diffuse-Porous to Ring-Porous Wood

The vascular cambium is the lateral meristem that actively divides and produces the secondary xylem and phloem (Larson 1994). The polar transport of IAA from young leaves (Aloni 2010) to roots through the vascular cambium (Fig. 4), in the presence of cytokinin (Matsumoto-Kitano et al. 2008; Nieminen et al. 2008) from the root cap [possibly also CK from the sieve tubes (Sakakibara et al. 2006)], with gibberellin originating in mature leaves (Dayan et al. 2012), and ethylene from maturing vessel elements (Pesquet and Tuominen 2011; Bollhöner et al. 2012), keeps the cambium active during the growing season. Deciduous trees lose their leaves during periods of extreme environmental conditions, which is promoted by ABA (Hou et al. 2006), and then their cambium becomes dormant. In temperate deciduous hardwood trees, the size differences of vessels in the earlywood and latewood are quite marked, and two main xylem categories can be distinguished: *diffuse-porous wood* and *ring-porous wood*. In diffuse-porous wood, the vessels are more or less uniform in size, whereas in ring-porous wood, the vessels produced at the beginning of the growth season are significantly wider than those produced at the end of the season (Evert 2006). Earlywood vessels in ring-porous trees can be huge (width of up to 500 μm and length of the tree itself) and therefore are very efficient in water conductance, even though they usually function for only one season (Tyree and Zimmermann 2002; Evert 2006).

The challenge to understand the mechanisms that have shaped these vessel patterns during the evolution of temperate deciduous hardwood trees requires elucidation of the roles of tissue sensitivity to auxin (Trewavas 1983; Bradford and Trewavas 1994; Barbez et al. 2012) and the specific hormonal signaling in these trees (Aloni 1991, 2001). It has been suggested that ring-porous trees have originated from diffuse-porous species (Aloni 1991; Wheeler and Baas 1991). The development of ring porosity has probably arisen independently multiple times during the diversification of angiosperms, and different lineages might therefore have modified mechanisms in different families. To explain how ring-porous wood has developed during evolution of hardwood trees, Aloni (1991) proposed the *limited-growth hypothesis*, suggesting that during the evolution of temperate deciduous hardwood trees, the ring-porous species have developed from diffuse-porous species under selective pressures in limiting environments which resulted in limited vegetative growth. It was further postulated that the natural selection for ring-porous wood has led to a decrease in the intensity of vegetative growth, accompanied by reduced auxin levels. The latter was followed by an increase in the sensitivity of the cambium to relatively low auxin stimulation;

these changes created the conditions which enable wide-earlywood-vessel differentiation (Aloni 1991), as will be clarified below.

Evidence that supports the hypothesis comes from observations that a diffuse-porous tree (*Populus euphratica*) and a ring-porous tree (*Quercus ithaburensis*) can change their porosity under opposite environmental conditions (Liphschitz 1995). Thus, under stress conditions when extension growth is suppressed, both tree species produced narrow annual rings characterized by ring-porous wood (as predicted by the hypothesis), whereas under favorable conditions when extensive growth is intensive, both species produce wide annual rings with diffuse-porous wood (Liphschitz 1995). Cytokinins from the root caps (Aloni et al. 2005, 2006a) increase the sensitivity of the cambium to the free-auxin signal originating in young leaves (Aloni 1995; Aloni et al. 2003, 2006b). Cytokinin prevents the usually rapid occurring IAA conjugation (Coenen and Lomax 1997); therefore, elevated CK concentration enables the transport of very low IAA concentrations via the cambium, which may explain the increased sensitivity of the cambium to free auxin. Experimental evidence from transformed plants (Zhang et al. 1995; Eklöf et al. 1997) supports the idea that reduced auxin concentrations can elevate cytokinin concentration, which would enhance tissue sensitivity to free auxin (Trewavas 1983; Aloni 1991; Bradford and Trewavas 1994; Barbez et al. 2012). The experiments demonstrate that auxin or cytokinin modifies the content of the other hormone by affecting its rate of synthesis. Reduced free-auxin concentration increases free cytokinin level (Palni et al. 1988; Zhang et al. 1995; Eklöf et al. 1997). This, in turn, enhances cambium sensitivity to extremely low-level IAA streams originating in swelling buds and creates the special physiological conditions that enable the differentiation of very wide earlywood vessels during a limited period of time in early spring (Aloni 1991, 2001).

The increased cambium sensitivity to IAA in ring-porous trees enables early cambium reactivation at the beginning of the growth season before bud break. This adaptation of ring-porous trees created the special internal conditions that enable them to respond to initial flows of extremely low IAA concentrations originating in dormant looking (before swelling) buds a few weeks before bud break (Aloni 1991; Aloni and Peterson 1997; Aloni et al. 1997), stimulating slow vessel differentiation which permits more cell expansion before secondary wall deposition, resulting in the formation of very wide earlywood vessels. Therefore, their first wide earlywood vessels are initiated six to two weeks before the onset of leaf expansion (Suzuki et al. 1996). Conversely, in diffuse-porous species, the first earlywood vessels are initiated two to seven weeks after the onset of leaf expansion (Suzuki et al. 1996), and because of its low sensitivity, their cambium requires high auxin levels (from fast growing young leaves) for reactivation. These results explain the old report of Priestley and Scott (1936) who found that in a deciduous ring-porous tree, the cambium undergoes extremely fast reactivation before bud break, which occurs almost simultaneously in the branches and along the trunk. This is why the bark of deciduous ring-porous trees may be peeled a few days before any bud swelling can be observed in spring. (The bark can be removed along the newly formed cell layer of reactivated cambium because it possesses new thin radial cell walls following early cambial cell divisions.) Conversely, a deciduous diffuse-porous species requires several weeks for a "wave" of cambial reactivation to extend from the twigs of a large tree downward to the base of its trunk (Priestley and Scott 1936).

Earlywood vessel width may be influenced by climatic signals. In the ring-porous chestnut trees grown on the Swiss Alps, the earlywood vessels were positively affected by early spring (April) temperatures, at the time of resumption of shoot growth, likely affecting cambial sensitivity to auxin (Fonti et al. 2007). Rising temperature before bud break increased the expression of genes involved in polar auxin transport (Schrader et al. 2003), providing evidence that sensitivity to auxin can be modulated by temperature. When triggered later in the season, this higher cell sensitivity to the IAA signal would result in smaller vessels as a consequence of an earlier and faster process of differentiation (Aloni and Zimmermann 1983).

Diffuse-porous species start the growth season a few weeks earlier than ring-porous trees and have a longer growth season which is characterized by continuous production of young leaves during a few months. Conversely, ring-porous trees which are late leafing trees (Lechowicz 1984) produce young leaves for only a short period of a few weeks, and later, they have mainly mature leaves (Aloni et al. 1997). Because young diffuse-porous trees possess greater growth intensity, they might produce more xylem per year than young ring-porous trees (Aloni et al. 1997). The continuous production of the IAA-producing-young leaves on diffuse-porous trees stimulates continuous production of vessels along the entire growth season with relatively thin-wall fibers. Whereas the mature leaves, which produce the gibberellin signal (Dayan et al. 2012), on the ring-porous trees induce the development of numerous well-developed hard lignified fibers during most of the growth season. These diverse earlywood and latewood properties in ring-porous wood, namely, the soft wide earlywood vessels versus the numerous hard latewood fibers, affect lumber stability and can have major effects on wood and fiber utilization.

## 8 Concluding Remarks

Understanding the hormonal mechanisms that control secondary xylem differentiation enables to improve wood production and quality by modifying tree growth, wood development, and response to stress. The hormonal regulation of trees can be improved with molecular genetic (Nieminen et al. 2012) and physiology methods, and the selected trees should produce superior wood for industry and agriculture. To improve forest trees, their endogenous hormonal concentrations and the sensitivity to hormones can be modified. For instance, increasing the endogenous bioactive gibberellin concentrations in transgenic forest trees and industrial crop plants improves stem elongation, induces numerous longer fiber and long tracheid production, and modifies lignin biosynthesis. Similarly, decreasing the sensitivity to ethylene in conifer trees is expected to promote tree growth and prevent the development of large traumatic resin cavities which damage the wood in response to wounding and stress. Likewise, fruit trees can be screened for reduced sensitivity to ethylene. In the selected trees the damage from tumorous crown galls (Aloni et al. 1998) will be prevented, resulting in healthy tree growth and high crop yield.

## References

Agusti J, Herold S, Schwarz M, Sanchez P, Ljung K, Dun EA, Brewer PB, Beveridge CA, Sieberer T, Sehr EM, Greb T (2011) Strigolactone signaling is required for auxin-dependent stimulation of secondary growth in plants. Proc Natl Acad Sci USA 108:20242–20247

Akiyama K, Matsuzaki K, Hayashi H (2005) Plant sesquiterpenes induce hyphal branching in arbuscular mycorrhizal fungi. Nature 435:824–827

Aloni R (1979) Role of auxin and gibberellin in differentiation of primary phloem fibers. Plant Physiol 63:609–614

Aloni R (1980) Role of auxin and sucrose in the differentiation of sieve and tracheary elements in plant tissue cultures. Planta 150:255–263

Aloni R (1982) Role of cytokinin in differentiation of secondary xylem fibers. Plant Physiol 70:1631–1633

Aloni R (1985) Plant growth method and composition. US Patent No 4507144

Aloni R (1987) Differentiation of vascular tissues. Annu Rev Plant Physiol 38:179–204

Aloni R (1991) Wood formation in deciduous hardwood trees. In: Raghavendra AS (ed) Physiology of trees. Wiley, New York, pp 175–197

Aloni R (1995) The induction of vascular tissues by auxin and cytokinin. In: Davies PJ (ed) Plant hormones: physiology, biochemistry and molecular biology. Kluwer, Dordrecht, pp 531–546

Aloni R (2001) Foliar and axial aspects of vascular differentiation – hypotheses and evidence. J Plant Growth Regul 20:22–34

Aloni R (2010) The induction of vascular tissues by auxin. In: Davies PJ (ed) Plant hormones: biosynthesis, signal transduction, action! Kluwer, Dordrecht, pp 485–506

Aloni R, Barnett JR (1996) The development of phloem anastomoses between vascular bundles and their role in xylem regeneration after wounding in *Cucurbita* and *Dahlia*. Planta 198:595–603

Aloni R, Peterson CA (1990) The functional significance of phloem anastomoses in stems of *Dahlia pinnata* Cav. Planta 182:583–590

Aloni R, Peterson CA (1997) Auxin promotes dormancy callose removal from the phloem of *Magnolia kobus* and callose accumulation and earlywood vessel differentiation in *Quercus robur*. J Plant Res 110:37–44

Aloni R, Sachs T (1973) The three-dimensional structure of primary phloem systems. Planta 113:343–353

Aloni R, Ullrich CI (2007) Biology of crown gall tumors. In: Tzfira T, Citovsky V (eds) Agrobacterium. Springer, Berlin, pp 525–549

Aloni R, Zimmermann MH (1983) The control of vessel size and density along the plant axis – a new hypothesis. Differentiation 24:203–208

Aloni R, Baum SF, Peterson CA (1990a) The role of cytokinin in sieve tube regeneration and callose production in wounded *Coleus* internodes. Plant Physiol 93:982–989

Aloni R, Tollier T, Monties B (1990b) The role of auxin and gibberellin in controlling lignin formation in primary phloem fibers and in xylem of *Coleus blumei* stems. Plant Physiol 94:1743–1747

Aloni R, Raviv A, Peterson CA (1991) The role of auxin in the removal of dormancy callose and resumption of phloem activity in *Vitis vinifera*. Can J Bot 69:1825–1832

Aloni R, Pradel KS, Ullrich CI (1995) The three-dimensional structure of vascular tissues in *Agrobacterium tumefaciens*-induced crown galls and in the host stem of *Ricinus communis* L. Planta 196:597–605

Aloni R, Alexander JD, Tyree MT (1997) Natural and experimentally altered hydraulic architecture of branch junctions in *Acer saccharum* Marsh. and *Quercus velutina* Lam. trees. Trees 11:255–264

Aloni R, Wolf A, Feigenbaum P, Avni A, Klee HJ (1998) The *Never ripe* mutant provides evidence that tumor-induced ethylene controls the morphogenesis of *Agrobacterium tumefaciens*-induced crown galls on tomato stems. Plant Physiol 117:841–847

Aloni R, Feigenbaum P, Kalev N, Rozovsky S (2000) Hormonal control of vascular differentiation in plants: the physiological basis of cambium ontogeny and xylem evolution. In: Savidge RA, Barnett JR, Napier R (eds) Cell and molecular biology of wood formation. BIOS Scientific Publishers, Oxford, pp 223–236

Aloni R, Schwalm K, Langhans M, Ullrich CI (2003) Gradual shifts in sites of free-auxin production during leaf-primordium development and their role in vascular differentiation and leaf morphogenesis in *Arabidopsis*. Planta 216:841–853

Aloni R, Langhans M, Aloni E, Ullrich CI (2004) Role of cytokinin in the regulation of root gravitropism. Planta 220:177–182

Aloni R, Langhans M, Aloni E, Dreieicher E, Ullrich CI (2005) Root-synthesized cytokinin in *Arabidopsis* is distributed in the shoot by the transpiration stream. J Exp Bot 56:1535–1544

Aloni R, Aloni E, Langhans M, Ullrich CI (2006a) Role of cytokinin and auxin in shaping root architecture: regulating vascular differentiation, lateral root initiation, root apical dominance and root gravitropism. Ann Bot 97:883–893

Aloni R, Aloni E, Langhans M, Ullrich CI (2006b) Role of auxin in regulating *Arabidopsis* flower development. Planta 223:315–328

Anfodillo T, Deslauriers A, Menardi R, Tedoldi L, Petit G, Rossi S (2012) Widening of xylem conduits in a conifer tree depends on the longer time of cell expansion downwards along the stem. J Exp Bot 63:837–845

Bailey IW (1944) The development of vessels in angiosperms in morphological research. Am J Bot 31:421–428

Bannan MW (1958) An occurrence of perforated tracheids in *Thuja occidentalis* L. New Phytol 57:132–134

Barbez E, Kubeš M, Rolčík J, Béziat C, Pěnčík A, Wang B, Rosquete MR, Zhu J, Dobrev PI, Lee Y, Zažímalovà E, Petrášek J, Geisler M, Friml J, Kleine-Vehn J (2012) A novel putative auxin carrier family regulates intracellular auxin homeostasis in plants. Nature 485:119–122

Barker-Bridgers M, Ribnicky DM, Cohen JD, Jones AM (1998) Red-light regulated growth. II. Changes in the abundance of indoleacetic acid in the maize mesocotyl. Planta 204:207–211

Baucher M, El Jaziri M, Vandeputte O (2007) From primary to secondary growth: origin and development of the vascular system. J Exp Bot 58:3485–3501

Baum SF, Aloni R, Peterson CA (1991) The role of cytokinin in vessel regeneration in wounded *Coleus* internodes. Ann Bot 67:543–548

Benková E, Michniewicz M, Sauer M, Teichmann T, Seifertova D, Jürgens G, Friml J (2003) Local, efflux-dependent auxin gradients as a common module for plant organ formation. Cell 115:591–602

Berleth T, Mattsson J, Hardtke CS (2000) Vascular continuity and auxin signals. Trends Plant Sci 5:387–393

Biemelt S, Tschiersch H, Sonnewald U (2004) Impact of altered gibberellin metabolism on biomass accumulation, lignin biosynthesis, and photosynthesis in transgenic tobacco plants. Plant Physiol 135:254–265

Bishopp A, Help H, El-Showk S, Weijers D, Scheres B, Friml J, Benková E, Mähönen AP, Helariutta Y (2011a) A mutually inhibitory interaction between auxin and cytokinin specifies vascular pattern in roots. Curr Biol 21:917–926

Bishopp A, Lehesranta S, Vatén A, Help H, El-Showk S, Scheres B, Helariutta K, Mähönen AP, Sakakibara H, Helariutta Y (2011b) Phloem-transported cytokinin regulates polar auxin transport and maintains vascular pattern in the root meristem. Curr Biol 21:927–932
Bollhöner B, Prestele J, Tuominen H (2012) Xylem cell death: emerging understanding of regulation and function. J Exp Bot 63:1081–1094
Booker J, Chatfield S, Leyser O (2003) Auxin acts in xylem-associated or medullary cells to mediate apical dominance. Plant Cell 15:495–507
Bradford KJ, Trewavas AJ (1994) Sensitivity thresholds and variable time scales in plant hormone action. Plant Physiol 105:1029–1036
Caño-Delgado A, Lee JY, Demura T (2010) Regulatory mechanisms for specification and patterning of plant vascular tissues. Annu Rev Cell Dev Biol 26:605–637
Carlsbecker A, Helariutta Y (2005) Phloem and xylem specifications: pieces of the puzzle emerge. Curr Opin Plant Biol 8:512–517
Chaffey N, Cholewa E, Regan S, Sundberg B (2002) Secondary xylem development in Arabidopsis: a model for wood formation. Physiol Plant 114:594–600
Clouse SD (2011) Brassinosteroids. In: The Arabidopsis book, vol 9. American Society of Plant Biologists, Rockville, MD. doi:10.1199/tab.0073, http://www.aspb.org/publications/arabidopsis/
Coenen C, Lomax TL (1997) Auxin-cytokinin interactions in higher plants: old problems and new tools. Trends Plant Sci 2:351–356
Cook CE, Whichard LP, Wall ME (1972) Germination stimulants. II. The structure of strigol – a potent seed germination stimulant for witchweed (*Striga lutea* Lour.). J Am Chem Soc 94:6198–6199
Dayan J, Schwarzkopf M, Avni A, Aloni R (2010) Enhancing plant growth and fiber production by silencing GA 2-oxidase. Plant Biotechnol J 8:425–435
Dayan J, Voronin N, Gong F, Sun TP, Hedden P, Fromm H, Aloni R (2012) Leaf-induced gibberellin signaling is essential for internode elongation, cambial activity, and fiber differentiation in tobacco stems. Plant Cell 24:66–79
Demura T, Fukuda H (2007) Transcriptional regulation in wood formation. Trends Plant Sci 12:64–70
Deslauriers A, Giovannelli A, Rossi S, Castro G, Fragnelli G, Traversi L (2009) Intra-annual cambial activity and carbon availability in stem of poplar. Tree Physiol 29:1223–1235
Dettmer J, Elo A, Helariutta Y (2009) Hormone interactions during vascular development. Plant Mol Biol 69:347–360
Dodd IC (2003) Hormonal interactions and stomatal responses. J Plant Growth Regul 22:32–46
Domagalska MA, Leyser O (2011) Signal integration in the control of shoot branching. Nat Rev Mol Cell Biol 12:211–221
Eklöf S, Åstot C, Blackwell J, Moritz T, Olsson O, Sandberg G (1997) Auxin-cytokinin interactions in wild-type and transgenic tobacco. Plant Cell Physiol 38:225–235
Erbilgin N, Gillette NE, Mori SR, Stein JD, Owen DR, Wood DL (2007) Acetophenone as an antiattractant for the western pine beetle, Dendroctonus brevicomis LeConte (Coleoptera: Scolytidae). J Chem Ecol 33:817–823
Eriksson ME, Moritz T (2002) Daylength and special expression of gibberellin 20-oxidase isolated from hybrid aspen (*Populus termulata* × *P. tremuloides* Michx.). Planta 214:920–930
Eriksson ME, Israelsson M, Olsson O, Moritz T (2000) Increased gibberellin biosynthesis in transgenic trees promotes growth, biomass production and xylem fiber length. Nat Biotechnol 18:784–788
Evert RF (2006) Esau's plant anatomy, meristems, cells, and tissues of the plant body – their structure, function, and development. Wiley, Hoboken, NJ
Fahn A (1964) Some anatomical adaptations of desert plants. Phytomorphology 14:93–102
Fahn A (1990) Plant anatomy, 4th edn. Pergamon Press, Oxford
Fahn A, Leshem B (1963) Wood fibres with living protoplasts. New Phytol 62:91–98

Fahn A, Zamski E (1970) The influence of pressure, wind, wounding and growth substances on the rate of resin duct formation in *Pinus halepensis* wood. Isr J Bot 19:429–446

Fonti P, Solomonoff N, García-González I (2007) Earlywood vessels of *Castanea sativa* record temperature before their formation. New Phytol 173:562–570

Friml J (2010) Subcellular trafficking of PIN auxin efflux carriers in auxin transport. Eur J Cell Biol 89:231–235

Friml J, Palme K (2002) Polar auxin transport – old questions and new concepts? Plant Mol Biol 49:273–284

Frydman VM, Wareing PF (1973) Phase change in *Hedera helix* L. II. The possible role of roots as a source of shoot gibberellin-like substances. J Exp Bot 24:1139–1148

Frydman VM, Wareing PF (1974) Phase change in *Hedera helix* L. III. The effects of gibberellins, abscisic acid and growth retardants on juvenile and adult ivy. J Exp Bot 25:420–429

Fu PL, Jiang YJ, Wang AY, Brodribb TJ, Zhang JL, Zhu SD, Cao KF (2012) Stem hydraulic traits and leaf water-stress tolerance are co-ordinated with the leaf phenology of angiosperm trees in an Asian tropical dry karst forest. Ann Bot 110(1):189–199

Fukuda H (1997) Tracheary element differentiation. Plant Cell 9:1147–1156

Fukuda H (2004) Signals that control vascular cell differentiation. Nat Rev Mol Cell Biol 5:379–391

Galiba G, Vagujfalvi A, Li CX, Soltesz A, Dubcovsky J (2009) Regulatory genes involved in the determination of frost tolerance in temperate cereals. Plant Sci 176:12–19

Gälweiler L, Guan C, Müller A, Wisman E, Mendgen K, Yephremov A, Palme K (1998) Regulation of polar auxin transport by AtPIN1 in *Arabidopsis* vascular tissue. Science 282:2226–2230

Geldner N, Anders N, Wolters H, Keicher J, Kornberger W, Muller P, Delbarre A, Ueda T, Nakano A, Jürgens G (2003) The *Arabidopsis* GNOM ARF-GEF mediates endosomal recycling, auxin transport, and auxin-dependent plant growth. Cell 112:219–230

Gerrienne P, Gensel PG, Strullu-Derrien C, Lardeux H, Steemans P, Prestianni C (2011) A simple type of wood in two early Devonian plants. Science 333:837

Gessler A, Kopriva S, Rennenberg H (2004) Regulation of nitrate uptake at the whole-tree level: interaction between nitrogen compounds, cytokinins and carbon metabolism. Tree Physiol 24:1313–1321

Goldschmidt EE, Samach A (2004) Aspects of flowering in fruit trees. In: Kang SM et al (eds) Proceedings of ths 9th IS on plant bioregulators. Acta Horticulture 653, pp 23–27

Goldsmith MHM, Catealdo DA, Karn J, Brenneman T, Trip P (1974) The nonpolar transport of auxin in the phloem of intact *Coleus* plants. Planta 116:301–317

Gomez-Roldan V, Fermas S, Brewer PB, Puech-Pagès V, Dun EA, Pillot JP, Letisse F, Matusova R, Danoun S, Portais JC, Bouwmeester H, Bécard G, Beveridge CA, Rameau C, Rochange SF (2008) Strigolactone inhibition of shoot branching. Nature 455:189–194

Gordon D, Karbat I, Ilan N, Cohen L, Kahn R, Gilles N, Dong K, Stuhmer W, Tytgat J, Gurevitz M (2007) The differential preference of scorpion alpha-toxins for insect or mammalian sodium channels: implications for improved insect control. Toxicon 49:452–472

Grace LJ, Charity JA, Gresham B, Kay N, Walter C (2005) Insect-resistant transgenic *Pinus radiata*. Plant Cell Rep 24:103–111

Groover AT (2005) What genes make a tree? Trends Plant Sci 10:210–214

Groover A, Robischon M (2006) Developmental mechanisms regulating secondary growth in woody plants. Curr Opin Plant Biol 9:55–58

Gurevitz M, Karbat I, Cohen L, Ilan N, Kahn R, Turkov M, Stankiewicz M, Stuhmer W, Dong K, Gordon D (2007) The insecticidal potential of scorpion beta-toxins. Toxicon 49:473–489

Hacke UG, Sperry JS, Wheeler JK, Castro L (2006) Scaling of angiosperm xylem structure with safety and efficiency. Tree Physiol 26:689–701

Hagen G, Martin G, Li Y, Guilfoyle TJ (1991) Auxin-induced expression of the soybean GH3 promoter in transgenic tobacco plants. Plant Mol Biol 17:567–579

Hargreaves CL, Grace LJ, van der Maas SA, Menzies MI, Kumar S, Holden DG, Foggo MN, Low CB, Dibley MJ (2005) Comparative *in vitro* and early nursery performance of adventitious and axillary shoots from epicotyls of same zygotic embryo of control-pollinated *Pinus radiata*. Can J For Res 35:2629–2641

Hellgren JM, Olofsson K, Sundberg B (2004) Patterns of auxin distribution during gravitational induction of reaction wood in poplar and pine. Plant Physiol 135:212–220

Hess T, Sachs T (1972) The influence of a mature leaf on xylem differentiation. New Phytol 71:903–914

Hirota A, Kato T, Fukaki H, Aida M, Tasaka M (2007) The auxin-regulated AP2/EREBP gene PUCHI is required for morphogenesis in the early lateral root primordium of Arabidopsis. Plant Cell 19:2156–2168

Horner HT, Lersten NR, Wirth CL (1994) Quantitative survey of sieve tube distribution in foliar terminal veins of ten dicot species. Am J Bot 81:1267–1274

Hou H-W, Zhou Y-T, Mwange K-N, Li W-F, He X-Q, Cui K-M (2006) ABP1 expression regulated by IAA and ABA is associated with the cambium periodicity in *Eucommia ulmoides* Oliv. J Exp Bot 57:3857–3867

Howe GA (2004) Jasmonates. In: Davies PJ (ed) Plant hormones: biosynthesis, signal transduction, action! Kluwer, Dordrecht, pp 610–634

Huber DP, Philippe RN, Madilao LL, Sturrock RN, Bohlmann J (2005) Changes in anatomy and terpene chemistry in roots of Douglas-fir seedlings following treatment with methyl jasmonate. Tree Physiol 25:1075–1083

Hudgins JW, Franceschi VR (2004) Methyl jasmonate-induced ethylene production is responsible for conifer phloem defense responses and reprogramming of stem cambial zone for traumatic resin duct formation. Plant Physiol 135:2134–2149

Hudgins JW, Christiansen E, Franceschi VR (2003) Methyl jasmonate induces changes mimicking anatomical defenses in diverse members of the Pinaceae. Tree Physiol 23:361–371

Hudgins JW, Ralph SG, Franceschi VR, Bohlmann J (2006) Ethylene in induced conifer defense: cDNA cloning, protein expression, and cellular and subcellular localization of 1-aminocyclopropane-1-carboxylate oxidase in resin duct and phenolic parenchyma cells. Planta 224:865–877

Israelsson M, Sundberg B, Moritz T (2005) Tissue-specific localization of gibberellins and expression of gibberellin-biosynthetic and signaling genes in wood-forming tissues in aspen. Plant J 44:494–504

Iwasaki T, Shibaoka H (1991) Brassinosteroids act as regulators of tracheary-element differentiation in isolated *Zinnia* mesophyll cells. Plant Cell Physiol 32:1007–1014

Jacobs WP (1952) The role of auxin in differentiation of xylem around a wound. Am J Bot 39:301–309

Jiang S, Xu K, Wang YZ, Ren YP, Gu S (2008) Role of GA3, GA4 and uniconazole-P in controlling gravitropism and tension wood formation in *Fraxinus mandshurica* Rupr. var. *japonica* Maxim. seedlings. J Integr Plant Biol 50:19–28

Jung JH, Park CM (2007) Vascular development in plants: specification of xylem and phloem tissues. J Plant Biol 50:301–305

Kalev N, Aloni R (1998) Role of auxin and gibberellin in regenerative differentiation of tracheids in *Pinus pinea* L. seedlings. New Phytol 138:461–468

Keeling CI, Bohlmann J (2006) Genes, enzymes and chemicals of terpenoid diversity in the constitutive and induced defence of conifers against insects and pathogens. New Phytol 170:657–675

Klee H (2008) Evidence points to the existence of a hitherto uncharacterized type of hormone that controls different aspects of plant growth and interaction. The hunt for that hormone is heating up. Nature 455:176–177

Klee HJ, Horsch RB, Hinchee MA, Hein MB, Hoffmann MB (1987) The effects of overproduction of two *Agrobacterium tumefaciens* T-DNA auxin biosynthetic gene products in transgenic petunia plants. Genes Dev 1:86–96

Kleine-Vehn J, Wabnik K, Martinière A, Łangowski Ł, Willig K, Naramoto S, Leitner J, Tanaka H, Jakobs S, Robert S, Luschnig C, Govaerts W, Hell SW, Runions J, Friml J (2011) Recycling, clustering, and endocytosis jointly maintain PIN auxin carrier polarity at the plasma membrane. Mol Syst Biol 7:540

Kohlen W, Charnikhova T, Liu Q, Bours R, Domagalska MA, Beguerie S, Verstappen F, Leyser O, Bouwmeester H, Ruyter-Spira C (2011) Strigolactones are transported through the xylem and play a key role in shoot architectural response to phosphate deficiency in nonarbuscular mycorrhizal host *Arabidopsis*. Plant Physiol 155:974–987

Koiwai H, Nakaminami K, Seo M, Mitsuhashi W, Toyomasu T, Koshiba T (2004) Tissue-specific localization of an abscisic acid biosynthetic enzyme, AAO3, in *Arabidopsis*. Plant Physiol 134:1697–1707

Kramer EM (2006) Wood grain pattern formation: a brief review. J Plant Growth Regul 25:290–301

Kramer EM, Borkowski MH (2004) Wood grain patterns at branch junctions: modeling and implications. Trees 18:493–500

Kudo T, Kiba T, Sakakibara H (2010) Metabolism and long-distance translocation of cytokinins. J Integr Plant Biol 52:53–60

Kuroha T, Satoh S (2007) Involvement of cytokinins in adventitious and lateral root formation. Plant Root 1:27–33

Larkin PJ, Gibson JM, Mathesius U, Weinmann JJ, Gärtner E, Hall E, Tanner GJ, Rolfe BG, Djordjevic MA (1996) Transgenic white clover. Studies with the auxin-responsive promoter, GH3, in root gravitropism and lateral root development. Transgenic Res 5:325–335

Larson PR (1969) Wood formation and the concept of wood quality. Yale University School of Forestry Bulletin 74, New Haven

Larson PR (1994) The vascular cambium: development and structure. Springer, Berlin

Lechowicz MJ (1984) Why do temperate deciduous trees leaf out at different times? Adaptation and ecology of forest communities. Am Nat 124:821–842

Leitch MA (2001) Vessel-element dimensions and frequency within the most current growth increment along the length of *Eucalyptus globules* stems. Trees 15:353–357

Lersten NR (1990) Sieve tubes in foliar vein endings: review and quantitative survey of *Rudbeckia laciniata* (Asteraceae). Am J Bot 77:1132–1141

Lersten NR, Curtis JD (1993) Paraveinal mesophyll in *Calliandra tweedii* and *C. emarginata* (Leguminosa; Mimosoideae). Am J Bot 80:561–568

Lev-Yadun S, Aloni R (1990) Vascular differentiation in branch junctions of trees: circular patterns and functional significance. Trees 4:49–54

Lev-Yadun S, Aloni R (1995) Differentiation of the ray system in woody plants. Bot Rev 61:45–84

Li S, Pezeshki SR, Shields FD (2006) Partial flooding enhances aeration in adventitious roots of black willow (*Salix nigra*) cuttings. J Plant Physiol 163:619–628

Liphschitz N (1995) Ecological wood anatomy: changes in xylem structure in Israeli trees. In: Shuming W (ed) Wood anatomy research 1995. Proceedings of the international symposium on tree anatomy and wood formation, Tianjin, China. International Academic Publishers, Beijing, pp 12–15

Ljung K, Hull AK, Kowalczyk M, Marchant A, Celenza J, Cohen JD, Sandberg G (2002) Biosynthesis, conjugation, catabolism and homeostasis of indol-3-acetic acid in *Arabidopsis*. Plant Mol Biol 50:309–332

López-Ráez JA, Charnikhova T, Gómez-Roldán V, Matusova R, Kohlen W, De Vos R, Verstappen F, Puech-Pages V, Bécard G, Mulder P, Bouwmeester H (2008) Tomato strigolactones are derived from carotenoids and their biosynthesis is promoted by phosphate starvation. New Phytol 178:863–874

Love J, Bjorklund S, Vahala J, Hertzberg M, Kangasjarvi J, Sundberg B (2009) Ethylene is an endogenous stimulator of cell division in the cambial meristem of *Populus*. Proc Natl Acad Sci USA 106:5984–5989

Ludwig-Müller J (2011) Auxin conjugates: their role for plant development and in the evolution of land plants. J Exp Bot 62:1757–1773
Matsumoto-Kitano M, Kusumoto T, Tarkowski P, Kinoshita-Tsujimura K, Václavíková K, Miyawaki K, Kakimoto T (2008) Cytokinins are central regulators of cambial activity. Proc Natl Acad Sci USA 105:20027–20031
Matte Risopatron JP, Sun Y, Jones BJ (2010) The vascular cambium: molecular control of cellular structure. Protoplasma 247:145–161
Mattsson J, Sung ZR, Berleth T (1999) Responses of plant vascular systems to auxin transport inhibition. Development 126:2979–2991
Mertens R, Eberle J, Arnscheidt A, Ledebur A, Weiler EW (1985) Monoclonal antibodies to plant growth regulators. II. Indole-3-acetic acid. Planta 166:389–393
Miyawaki K, Matsumoto-Kitano M, Kakimoto T (2004) Expression of cytokinin biosynthetic isopentenyltransferase genes in *Arabidopsis*: tissue specificity and regulation by auxin, cytokinin, and nitrate. Plant J 37:128–138
Morris DA, Kadir GO, Barry AJ (1973) Auxin transport in intact pea seedlings (*Pisum sativum* L.): the inhibition of transport by 2,3,5-triiodobenzoic acid. Planta 110:173–182
Moubayidin L, Di Mambro R, Sabatini S (2009) Cytokinin-auxin crosstalk. Trends Plant Sci 14:557–562
Muday GK, Rahman A, Binder BM (2012) Auxin and ethylene: collaborators or competitors? Trends Plant Sci 17:181–195
Müller A, Düchting P, Weiler EW (2002) A multiplex GC-MS/MS technique for the sensitivity and qualitative single-run analysis of acidic phytohormones and related compounds, and its application to *Arabidopsis thaliana*. Planta 216:44–56
Nieminen K, Immanen J, Laxell M, Kauppinen L, Tarkowski P, Dolezal K, Tähtiharju S, Elo A, Decourteix M, Ljung K, Bhalerao R, Keinonen K, Albert VA, Helariutta Y (2008) Cytokinin signaling regulates cambial development in poplar. Proc Natl Acad Sci USA 105:20032–20037
Nieminen K, Robischon M, Immanen J, Helariutta Y (2012) Towards optimizing wood development in bioenergy trees. New Phytol 194:46–53
Nugroho WD, Yamagishi Y, Nakaba S, Fukuhara S, Begum S, Sri Nugroho Marsoem SN, Ko J-H, Jin H-O, Funada R (2012) Gibberellin is required for the formation of tension wood and stem gravitropism in *Acacia mangium* seedlings. Ann Bot 110(4):887–895
Oda Y, Fukuda H (2012) Secondary cell wall patterning during xylem differentiation. Curr Opin Plant Biol 15:38–44
Olson ME, Rosell JA (2013) Vessel diameter-stem diameter scaling across woody angiosperms and the ecological causes of xylem vessel diameter variation. New Phytol 197:1204–1213
Palme P, Gälweiler L (1999) PIN-pointing the molecular basis of auxin transport. Curr Opin Plant Biol 2:375–381
Palni LMS, Burch L, Horgan R (1988) The effect of auxin concentration on cytokinin stability and metabolism. Planta 174:231–234
Pesquet E, Tuominen H (2011) Ethylene stimulates tracheary element differentiation in *Zinnia elegans* cell cultures. New Phytol 190:138–149
Priestley JH, Scott LI (1936) A note upon summer wood production in the tree. Proc Leeds Phil Soc 3:235–248
Ragni L, Nieminen K, Pacheco-Villalobos D, Sibout R, Schwechheimer C, Hardtke CS (2011) Mobile gibberellin directly stimulates *Arabidopsis* hypocotyl xylem expansion. Plant Cell 23:1322–1336
Rahayu YS, Walch-Liu P, Neumann G, Römheld V, von Wirén N, Bangerth F (2005) Root-derived cytokinins as long-distance signals for $NO_3$- -induced stimulation of leaf growth. J Exp Bot 56:1143–1152
Ralph SG, Hudgins JW, Jancsik S, Franceschi VR, Bohlmann J (2007) Aminocyclopropane carboxylic acid synthase is a regulated step in ethylene-dependent induced conifer defense. Full-length cDNA cloning of a multigene family, differential constitutive, and wound- and

insect-induced expression, and cellular and subcellular localization in spruce and Douglas fir. Plant Physiol 143:410–424
Rathgeber CB, Rossi S, Bontemps JD (2011) Cambial activity related to tree size in a mature silver-fir plantation. Ann Bot 108:429–438
Raven PH, Evert RF, Eichhorn SE (2005) Biology of plants, 7th edn. Freeman, New York
Reinhardt D, Pesce E-R, Stieger P, Mandel T, Baltensperger K, Bennett M, Traas J, Friml J, Kuhlemeier C (2003) Regulation of phyllotaxis by polar auxin transport. Nature 426:255–260
Roberts LW, Gahan BP, Aloni R (1988) Vascular differentiation and plant growth regulators. Springer, Berlin
Romano CP, Hein MB, Klee HJ (1991) Inactivation of auxin in tobacco transformed with the indoleacetic acid-lysine synthetase gene of *Pseudomonas savastanoi*. Genes Dev 5:438–446
Ruffel S, Krouk G, Ristova D, Shasha D, Birnbaum KD, Coruzzi GM (2011) Nitrogen economics of root foraging: transitive closure of the nitrate–cytokinin relay and distinct systemic signaling for N supply vs. demand. Proc Natl Acad Sci USA 108:18524–18529
Runions J, Friml J (2011) Recycling, clustering, and endocytosis jointly maintain PIN auxin carrier polarity at the plasma membrane. Mol Syst Biol 7:540
Sachs T (1981) The control of patterned differentiation of vascular tissues. Adv Bot Res 9:151–262
Sachs T (1991) Callus and tumor development. In: Sachs T (ed) Pattern formation in plant tissues. Cambridge University Press, Cambridge, pp 38–55
Sachs T (2000) Integrating cellular and organismal aspects of vascular differentiation. Plant Cell Physiol 41:649–656
Sachs T, Cohen D (1982) Circular vessels and the control of vascular differentiation in plants. Differentiation 21:22–26
Sakakibara H, Takei K, Hirose N (2006) Interactions between nitrogen and cytokinin in the regulation of metabolism and development. Trends Plant Sci 11:440–448
Saks Y, Feigenbaum P, Aloni R (1984) Regulatory effect of cytokinin on secondary xylem fiber formation in an *in vivo* system. Plant Physiol 76:638–642
Salleo S, LoGullo MA, Siracusano L (1984) Distribution of vessel ends in stems of some diffuse and ring-porous trees: the nodal region as ‘safety zones’ of the water conducting system. Ann Bot 54:543–552
Sauer M, Balla J, Luschnig C, Wisniewska J, Reinöhl V, Friml J, Benková E (2006) Canalization of auxin flow by Aux/IAA-ARF-dependent feedback regulation of PIN polarity. Genes Dev 20:2902–2911
Savaldi-Goldstein S, Peto C, Chory J (2007) The epidermis both drives and restricts plant shoot growth. Nature 446:199–202
Savidge RA (1996) Xylogenesis, genetic and environmental regulation. IAWA J 17:269–310
Scarpella E, Helariutta Y (2010) Vascular pattern formation in plants. Curr Top Dev Biol 91:221–265
Scarpella E, Marcos D, Friml J, Berleth T (2006) Control of leaf vascular patterning by polar auxin transport. Genes Dev 20:1015–1027
Scheres B, Xu J (2006) Polar auxin transport and patterning: grow with the flow. Genes Dev 20:922–926
Schmidt A, Nagel R, Krekling T, Christiansen E, Gershenzon J, Krokene P (2011) Induction of isoprenyl diphosphate synthases, plant hormones and defense signalling genes correlates with traumatic resin duct formation in Norway spruce (*Picea abies*). Plant Mol Biol 77:577–590
Schrader J, Baba K, May ST, Palme K, Bennet M, Bhalerao RP, Sandberg G (2003) Polar auxin transport in the wood-forming tissue of hybrid aspen is under simultaneous control of developmental and environmental signals. Proc Natl Acad Sci USA 100:10096–10101
Schwalm K, Aloni R, Langhans M, Heller W, Stich S, Ullrich CI (2003) Flavonoid-related regulation of auxin accumulation in *Agrobacterium tumefaciens*-induced plant tumors. Planta 218:163–178

Sehr EM, Agusti J, Lehner R, Farmer EE, Schwarz M, Greb T (2010) Analysis of secondary growth in the Arabidopsis shoot reveals a positive role of jasmonate signalling in cambium formation. Plant J 63:811–822

Sieburth LE (1999) Auxin is required for leaf vein pattern in *Arabidopsis*. Plant Physiol 121:1179–1190

Sieburth LE, Deyholos MK (2006) Vascular development: the long and winding road. Curr Opin Plant Biol 9:48–54

Spicer R, Groover A (2010) Evolution of development of vascular cambia and secondary growth. New Phytol 186:577–592

Sundberg B, Uggla C, Tuominen H (2000) Cambial growth and auxin gradients. In: Savidge RA, Barnett JR, Napier R (eds) Cell and molecular biology of wood formation. BIOS Scientific Publishers, Oxford, pp 169–188

Suzuki M, Yoda K, Suzuki H (1996) Phenological comparison of the onset of vessel formation between ring-porous and diffuse-porous deciduous trees in a Japanese temperate forest. IAWA J 17:431–444

Swarup R, Friml J, Marchant A, Ljung K, Sandberg G, Palme K, Bennett M (2001) Localization of the auxin permease AUX1 suggests two functionally distinct hormone transport pathways operate in the *Arabidopsis* root apex. Genes Dev 15:2648–2653

Taiz L, Zeiger E (2006) Plant physiology, 4th edn. Sinauer, Sunderland, MA

Takei K, Sakakibara H, Taniguchi M, Sugiyama T (2001) Nitrogen-dependent accumulation of cytokinins in root and the translocation to leaf: implication of cytokinin species that induces gene expression of maize response regulator. Plant Cell Physiol 42:85–93

Takei K, Ueda N, Aoki K, Kuromori T, Hirayama T, Shinozaki K, Yamaya T, Sakakibara H (2004) *AtIPT3* is a key determinant of nitrate-dependent cytokinin biosynthesis in *Arabidopsis*. Plant Cell Physiol 45:1053–1062

Tanaka M, Takei K, Kojima M, Sakakibara H, Mori H (2006) Auxin controls local cytokinin biosynthesis in the nodal stem in apical dominance. Plant J 45:1028–1036

Teale WD, Paponov IA, Palme K (2006) Auxin in action: signalling, transport and the control of plant growth and development. Nat Rev Mol Cell Biol 7:847–859

Thorpe MR, Ferrieri AP, Herth MM, Ferrieri RA (2007) 11C-imaging: methyl jasmonate moves in both phloem and xylem, promotes transport of jasmonate, and of photoassimilate even after proton transport is decoupled. Planta 226:541–551

Timell TE (1986) Compression wood in gymnosperms, vol 2. Springer, Berlin

Tokunaga N, Uchimura N, Sato Y (2006) Involvement of gibberellin in tracheary element differentiation and lignification in *Zinnia elegans* xylogenic culture. Protoplasma 228:179–187

Trewavas AJ (1983) Is plant development regulated by changes in concentration of growth substances or by changes in the sensitivity to growth substances? TIBS 8:354–357

Turner S, Sieburth LE (2003) Vascular patterning. In: Somerville CR, Meyerowitz EM (eds) The Arabidopsis book, vol 2. American Society of Plant Biologists, Rockville, MD. doi:10.1199/tab.0073, http://www.aspb.org/publications/arabidopsis/

Turner S, Gallois P, Brown D (2007) Tracheary element differentiation. Annu Rev Plant Biol 58:407–433

Tyree MT, Zimmermann MH (2002) Xylem structure and the ascent of Sap, 2nd edn. Springer, Berlin

Uggla C, Moritz T, Sandberg G, Sundberg B (1996) Auxin as a positional signal in pattern formation in plants. Proc Natl Acad Sci USA 93:9282–9286

Uggla C, Mellerowicz EJ, Sundberg B (1998) Indole-3-acetic acid controls cambial growth in Scots pine by positional signaling. Plant Physiol 117:113–121

Ullrich CI, Aloni R (2000) Vascularization is a general requirement for growth of plant and animal tumours. J Exp Bot 51:1951–1960

Ulmasov T, Murfett J, Hagen G, Guilfoyle TJ (1997) Aux/IAA proteins repress expression of reporter genes containing natural and highly active synthetic auxin response elements. Plant Cell 9:1963–1971

Umehara M, Hanada A, Yoshida S, Akiyama K, Arite T, Takeda-Kamiya N, Magome H, Kamiya Y, Shirasu K, Yoneyama K, Kyozuka J, Yamaguchi S (2008) Inhibition of shoot branching by new terpenoid plant hormones. Nature 455:195–200

Ursache R, Nieminen K, Helariutta Y (2012) Genetic and hormonal regulation of cambial development. Physiol Plant 147:36–45

Wabnik K, Kleine-Vehn J, Balla J, Sauer M, Naramoto S, Reinöhl V, Merks RM, Govaerts W, Friml J (2010) Emergence of tissue polarization from synergy of intracellular and extracellular auxin signaling. Mol Syst Biol 6:447

Wächter R, Fischer K, Gäbler R, Kühnemann F, Urban W, Bögemann GM, Voesenek LACJ, Blom CWPM, Ullrich CI (1999) Ethylene production and ACC-accumulation in *Agrobacterium tumefaciens*-induced plant tumours and their impact on tumour and host stem structure and function. Plant Cell Environ 22:1263–1273

Wheeler EA, Baas P (1991) A survey of the fossil record for dicotyledonous wood and its significance for evolutionary and ecological wood anatomy. IAWA Bull ns 12:275–332

Xue-Xuan X, Hong-Bo S, Yuan-Yuan M, Gang X, Jun-Na S, Dong-Gang G, Cheng-Jiang R (2010) Biotechnological implications from abscisic acid (ABA) roles in cold stress and leaf senescence as an important signal for improving plant sustainable survival under abiotic-stressed conditions. Crit Rev Biotechnol 30:222–230

Yamaguchi M, Goué N, Igarashi H, Ohtani M, Nakano Y, Mortimer JC, Nishikubo N, Kubo M, Katayama Y, Kakegawa K, Dupree P, Demura T (2010a) VASCULAR-RELATED NAC-DOMAIN6 and VASCULAR-RELATED NAC-DOMAIN7 effectively induce transdifferentiation into xylem vessel elements under control of an induction system. Plant Physiol 153:906–914

Yamaguchi M, Ohtani M, Mitsuda N, Kubo M, Ohme-Takagi M, Fukuda H, Demura T (2010b) VND-INTERACTING2, a NAC domain transcription factor, negatively regulates xylem vessel formation in *Arabidopsis*. Plant Cell 22:1249–1263

Yamaguchi M, Mitsuda N, Ohtani M, Ohme-Takagi M, Kato K, Demura T (2011) VASCULAR-RELATED NAC-DOMAIN7 directly regulates the expression of a broad range of genes for xylem vessel formation. Plant J 66:579–590

Yamamoto F, Angeles G, Kozlowski TT (1987) Effect of ethrel on stem anatomy of *Ulmus americana* seedlings. IAWA Bull NS 8:3–9

Yamamoto F, Kozlowski TT (1987) Effect of ethrel on growth and stem anatomy of *Pinus halepensis* seedlings. IAWA Bull ns 8:11–19

Yamamoto R, Demura T, Fukuda H (1997) Brassinosteroids induce entry into the final stage of tracheary element differentiation in cultured *Zinnia* cells. Plant Cell Physiol 38:980–983

Yamamoto R, Fujioka S, Demura T, Takatsuto S, Yoshida S, Fukuda H (2001) Brassinosteroid levels increase drastically prior to morphogenesis of tracheary elements. Plant Physiol 125:556–563

Zhang R, Zhang X, Wang J, Letham DS, McKinney SA, Higgins TJV (1995) The effect of auxin on cytokinin levels and metabolism in transgenic tobacco tissue expressing an *ipt* gene. Planta 196:84–94

Zhang S, Wei Y, Lu Y, Wang X (2009) Mechanisms of brassinosteroids interacting with multiple hormones. Plant Signal Behav 12:1117–1120

Zhang J, Elo A, Helariutta Y (2011) Arabidopsis as a model for wood formation. Curr Opin Biotechnol 22:293–299

Zimmermann MH (1983) Xylem structure and the ascent of sap. Springer, Berlin

Zimmermann MH, Jeje AA (1981) Vessel-length distribution in stems of some American woody plants. Can J Bot 59:1882–1892

Zobel BJ, van Buijtenen JP (1989) Wood variation, its causes and control. Springer, Berlin

# Transcriptional Regulation of Wood Formation in Tree Species

**Ruiqin Zhong and Zheng-Hua Ye**

**Abstract** Wood formation is mediated by a complex developmental program consisting of a number of sequentially occurred processes, including vascular cambial cell division, differentiation of xylem mother cells, cell elongation, secondary wall deposition, programmed cell death, and heartwood formation, all of which are proposed to be coordinated by transcriptional networks. Recent molecular and genetic studies have uncovered a transcriptional network regulating secondary wall biosynthesis during wood formation in tree species. This network encompasses a multileveled feed-forward loop regulatory structure in which the top master transcriptional switches, wood-associated NAC transcription factors (WNDs), together with WND-regulated transcription factors, regulate an array of downstream genes thereby activating the wood biosynthetic program. Genome-wide transcriptome analysis has revealed a number of wood-associated transcription factors, some of which have been proposed to be involved in vascular cambial cell division and secondary xylem differentiation. With the availability of the genome sequence data and the wood-associated gene expression data from tree species, it is expected that we will soon be able to reveal the transcriptional networks regulating various processes of wood formation, the knowledge of which could be applied to genetically engineer wood biomass composition tailored for diverse end uses such as biofuel production.

## 1 Introduction

Wood is the most abundant biomass produced by land plants. It is estimated that land plants fix up to 56 billion tons of $CO_2$ every year, bulk of which is stored in the wood (Field et al. 1998). Therefore, wood is important for the regulation of

R. Zhong • Z.-H. Ye (✉)
Department of Plant Biology, University of Georgia, Athens, GA 30602, USA
e-mail: zhye@plantbio.uga.edu

J. Fromm (ed.), *Cellular Aspects of Wood Formation*, Plant Cell Monographs 20,
DOI 10.1007/978-3-642-36491-4_5, 

greenhouse gas level by serving as the most abundant reservoir of fixed carbon in plants. Furthermore, wood is an important raw material for a myriad of applications in our daily life. It is widely used for energy by direct burning, building construction, pulping and papermaking, furniture, charcoal, musical instrument, and many others. Wood is also considered to be an abundant source of lignocellulosic biomass for the production of second generation of biofuels (Carroll and Somerville 2009). Because of its vast economic values, tremendous efforts have been invested to understand the process of wood formation at the anatomical, chemical, physiological, biochemical, cellular, molecular, genetic, and genomic levels. In this review, we will focus on the molecular and genetic understanding of how the process of wood formation is transcriptionally regulated.

## 2 Wood Formation Is a Complex Developmental Process Leading to the Massive Deposition of Secondary Walls

Wood (secondary xylem) is formed through a complex developmental program involving cambial cell division, differentiation of the vascular cambial into secondary xylem mother cells, cell elongation, massive secondary wall deposition, programmed cell death, and finally heartwood formation (Plomion et al. 2001). At maturity, wood is largely the remains of secondary walls composed of three major types of polymers, including cellulose, hemicelluloses, and lignin. The chemistry and biosynthesis of these wood components have been intensively studied in the past decades (Mellerowicz and Sundberg 2008). Cellulose is composed of long chains of β-1,4-linked glucosyl residues that are assembled into semicrystalline microfibrils. Its biosynthesis in secondary walls requires three major groups of cellulose synthase genes in Arabidopsis (Taylor et al. 2004). Their close orthologs are present in tree species and were found to be coordinately expressed during wood formation, indicating that as in Arabidopsis, cellulose biosynthesis during wood formation in tree species also involves the cooperative actions of three cellulose synthase subunits (Joshi and Mansfield 2007).

The major hemicelluloses in wood are xylan and glucomannan depending on tree species. Xylan is made of a linear chain of β-1,4-linked xylosyl residues that are often substituted with α-1,2-linked methylglucuronic acid and may also be highly acetylated at C-2 and/or C-3 (Timell 1967). The reducing end of xylan from both dicot wood and gymnosperm wood has been shown to include a unique tetrasaccharide sequence, β-D-Xyl*p*-(1→3)-α-L-Rha*p*-(1→2)-α-D-Gal*p*A-(1→4)-D-Xyl*p* (Shimizu et al. 1976; Johansson and Samuelson 1977; Andersson et al. 1983; Lee et al. 2009). A number of wood-associated glycosyltransferase genes (Aspeborg et al. 2005) have been demonstrated to play essential roles in xylan biosynthesis; these include family GT43 genes required for xylan backbone elongation (Zhou et al. 2007; Lee et al. 2011, 2012) and families GT8 and GT47 genes involved in the biosynthesis of xylan reducing end sequence (Zhou et al. 2006; Kong et al. 2009; Lee et al. 2009, 2011; Li et al. 2011).

Lignin is a complex phenylpropanoid polymer produced by dehydrogenative polymerization of three monolignols, *p*-coumaryl alcohol, coniferyl alcohol, and sinapyl alcohol (Boerjan et al. 2003). Lignin composition in wood varies depending on tree species. Gymnosperm wood mainly contains lignin composed of guaiacyl unit polymerized from coniferyl alcohol, whereas angiosperm wood includes lignin made of both guaiacyl and syringyl units polymerized from coniferyl alcohol and sinapyl alcohol, respectively. The biosynthetic pathway of monolignols starts with the general phenylpropanoid pathway leading to the production of hydroxycinnamoyl-CoA esters, which are further converted into three monolignols by two successive reductive steps. Monolignols are transported through plasma membrane-localized transporters into cell walls, where they are oxidized by peroxidases and laccases for polymerization into lignin (Vanholme et al. 2008). Many of lignin biosynthetic genes in tree species have been functionally characterized (Vanholme et al. 2008) as well as applied to downregulate the deposition of lignin in wood (Pilate et al. 2002).

It is conceivable that to make wood, all the genes involved in the biosynthesis of wood components need to be coordinately turned on. Furthermore, developmental signals and signal transductions pathways must be sequentially triggered at different stages of wood formation to regulate xylem mother cell differentiation, cell elongation, secondary wall deposition, programmed cell death, and heartwood formation. Indeed, genomic analysis of wood formation has revealed that thousands of genes implicated in signaling, transcriptional regulation, cytoskeletal organization, programmed cell death, and cell wall biosynthesis are preferentially activated (Prassinos et al. 2005; Andersson-Gunneras et al. 2006; Wang et al. 2009; Wilkins et al. 2009; Dharmawardhana et al. 2010). It is envisioned that transcriptional networks must play essential roles in thc coordinated regulation of genes controlling various stages of wood development. To better understand the molecular mechanisms underlying wood formation, it is critical to uncover the transcriptional factors and associated networks regulating wood formation. Recent molecular, genetic and genomic studies of wood formation have provided a glimpse of the complexity of how wood formation is regulated in tree species.

## 3 Promoter Analysis of Wood Biosynthetic Genes Reveals *cis*-Elements Critical for Wood-Specific Expression

Early work has employed gene promoter deletion analysis to find out which promoter fragment is required for wood-specific expression. By defining the wood-specific *cis*-elements, one may use them to identify corresponding transcription factors that are involved in the regulation of wood formation. Promoters of several lignin biosynthetic genes from poplar, Eucalyptus, and conifers have been shown to be able to drive the GUS reporter gene expression in developing woody tissues (Feuillet et al. 1995; Gray-Mitsumune et al. 1999; Lacombe et al. 2000; Lauvergeat et al. 2002; Rahantamalala et al. 2010), indicating that the promoter sequences used contain regulatory elements for wood-specific expression. Promoters of two Eucalyptus lignin genes, cinnamoyl-CoA reductase (EgCCR) and cinnamyl alcohol dehydrogenase (EgCAD2), were subjected to detailed molecular dissection in order to define minimal regulatory elements responsible for wood-specific expression. Promoter deletion analysis identified a 50-bp region located between −119 and −77 from the transcriptional start site of *EgCCR* as necessary and sufficient for expression in vascular tissues (Lacombe et al. 2000). Likewise, an 80-bp region located between −203 and −124 of the *EgCAD2* promoter was found to be sufficient for driving the GUS reporter gene expression in vascular tissues (Lauvergeat et al. 2002; Rahantamalala et al. 2010).

Rahantamalala et al. (2010) employed in vivo footprinting coupled with transcriptional activation assay to further define the vascular-specific *cis*-elements within the 80-bp *EgCAD2* promoter. It was found that four putative *cis*-elements might be involved in interaction with transcription factors. Two of these *cis*-elements have identical sequences, ACCTACC, corresponding to the well-known AC element that was first identified in the lignin biosynthetic gene, phenylalanine ammonia-lyase (PAL), from bean (Hatton et al. 1995). The 7-bp AC element [ACC(A/T)A(A/C)C] is present in most of the lignin biosynthetic genes and considered to be a common *cis*-element responsible for the coordinated expression of lignin biosynthetic genes in vascular tissues (Raes et al. 2003; Zhong and Ye 2009). The AC element was previously shown to be the binding site for some MYB (v-myb myeloblastosis viral oncogene homolog) transcription factors involved in the regulation of lignin biosynthetic genes, including Eucalyptus MYB2 (Goicoechea et al. 2005). Both electrophoretic mobility shift assay and transactivation activity study demonstrated that the two *cis*-elements harboring the AC element in the 80-bp *EgCAD2* promoter fragment are indeed required for the binding of EgMYB2 as well as the transcriptional activation by EgMYB2 (Rahantamalala et al. 2010). Likewise, the one AC element present in the 50-bp *EgCCR* promoter fragment is also responsible for the EgMYB2 binding and EgMYB2-induced transcriptional activation. In addition to the AC element, two additional putative *cis*-elements harboring the identical sequence CTGGTT are present in the 80-bp *EgCAD2* promoter fragment although the transcription factor that binds to this new element has not been identified. These studies provide molecular evidence supporting the hypothesis that the coordinated regulation of one of the wood biosynthetic pathways, lignification, is controlled through the AC element present in the promoters of lignin biosynthetic genes.

Although functional study of *cis*-elements regulating wood-specific expression was mainly done using lignin biosynthetic gene promoters, recent bioinformatic analysis of orthologous cellulose synthase promoters of poplar, Eucalyptus, and Arabidopsis has revealed a number of putative *cis*-regulatory elements that are evolutionarily conserved in angiosperms (Creux et al. 2008). The functional roles of these putative *cis*-elements in regulating the expression of cellulose synthase gene expression during wood formation remain to be investigated.

## 4 MYB Transcription Factors as Key Regulators of Wood Formation

As described above, the study of lignin biosynthetic gene promoters led to the discovery of the AC element as a common *cis*-element regulating lignin biosynthetic pathway (Hatton et al. 1995; Raes et al. 2003). The AC element sequences are similar to the DNA binding sequence [CC(T/A)ACC] identified by binding sequence selection for the maize MYB protein P (Grotewold et al. 1994), suggesting that MYB transcription factors may also be involved in the regulation of the lignin biosynthetic pathway via binding to the AC element. Study of wood-associated MYB genes from Eucalyptus (EgMYB2) and pine (PtMYB4) provides direct molecular evidence demonstrating the binding of the AC element by MYBs and roles of MYBs in regulating lignin biosynthesis. As discussed above, EgMYB2 was found to be able to bind to the AC element in the promoters of EgCCR and EgCAD and activate the EgCCR or EgCAD promoter-driven GUS reporter gene. When overexpressed in tobacco, EgMYB2 was capable of inducing the expression of a number of lignin biosynthetic genes and a concomitant alteration in lignin profiles (Goicoechea et al. 2005). Similarly, PtMYB4 was shown to be capable of binding to the AC element and activating the AC element-driven GUS reporter gene. Overexpression of PtMYB4 in tobacco was proven to be sufficient to induce lignin biosynthetic genes and an ectopic deposition of lignin in pith parenchyma cells (Patzlaff ct al. 2003a). These findings provide the first line of evidence demonstrating the involvement of MYBs in regulating lignification during wood formation in tree species.

In addition to their roles in regulating lignin biosynthesis, recent functional analysis has revealed that EgMYB2 and PtMYB4 are also involved in regulating the biosynthesis of cellulose and xylan (Zhong and Ye 2009; Zhong et al. 2010a). EgMYB2 and PtMYB4 are functional orthologs of Arabidopsis MYB46 and MYB83, which function redundantly as key transcriptional activators of entire secondary wall biosynthetic program (Zhong et al. 2007a; Ko et al. 2009; McCarthy et al. 2009). Similar to MYB46 and MYB83, EgMYB2 is able to activate the expression of biosynthetic genes of cellulose, xylan, and lignin and concomitantly induce the ectopic deposition of all three major secondary wall components, including cellulose, xylan, and lignin, when overexpressed in Arabidopsis (Zhong et al. 2010a). Expression of EgMYB2 or PtMYB4 driven by the Arabidopsis MYB46 promoter was capable of complementing the secondary wall thickening defect in the *myb46 myb83* double mutant (Zhong et al. 2010a). Similarly, poplar wood-associated PtrMYB3 and PtrMYB20, which are close homologs of MYB46 and MYB83, were able to activate the entire secondary wall biosynthetic pathways and complement the secondary wall thickening defect in the myb46 myb83 double mutant (McCarthy et al. 2010). Since EgMYB2, PtMYB4, PtrMYB3, and PtrMYB20 are able to complement the *myb46 myb83* mutant phenotype, this indicates that they are functional orthologs sharing the same DNA binding sequences and thereby activating the same downstream targets. Electrophoretic mobility shift assay coupled with transactivation analysis revealed that MYB46 and MYB83 bind to a 7-bp secondary wall MYB-responsive element (SMRE), ACC(A/T)A(A/C)(T/C) (Zhong and Ye 2012). This SMRE consensus sequence encompasses the AC element, ACC(A/T)A(A/C)C. We have recently found that PtrMYB3, PtrMYB20, EgMYB4, and PtMYB2 also bind to the SMRE sequences (Zhong et al. unpublished data), indicating that similar to MYB46 and MYB83, they regulate the expression of secondary wall biosynthetic genes via binding to the SMRE site. These findings indicate that PtrMYB3, PtrMYB20, EgMYB4, PtMYB2, and their other tree orthologs are key transcriptional regulators activating the entire secondary wall biosynthetic program during wood formation in tree species.

Two additional pine MYB genes, PtMYB1 and PtMYB8, have been shown to be important players in regulating wood formation (Bomal et al. 2008). PtMYB8 is a close homolog of PtMYB4, EgMYB2, and Arabidopsis MYB46/MYB83 (Bomal et al. 2008). When overexpressed in transgenic spruce, PtMYB8 is able to activate a number of biosynthetic genes for lignin as well as cellulose and xylan and cause an ectopic deposition of secondary walls, indicating that similar to PtMYB4, EgMYB2, and MYB46/MYB83, PtMYB8 is likely a transcriptional activator of the entire secondary wall biosynthetic program.

While PtMYB8 together with its close homologs, PtMYB4, EgMYB2, and MYB46/MYB83, are key regulators of the entire secondary wall biosynthetic program, PtMYB1 appears to be specific to lignin biosynthesis (Bomal et al. 2008). Overexpression of PtMYB1 in transgenic spruce leads to an elevated expression of lignin biosynthetic genes, ectopic lignin deposition, and an increased lignin content, suggesting that PtMYB1 is involved in regulating lignin biosynthesis. PtMYB1 is a close homolog of Arabidopsis MYB85, which was found to be a transcriptional regulator of lignin biosynthesis (Zhong et al. 2008). Overexpression of MYB85 in Arabidopsis results in an induction of expression of biosynthetic genes for lignin but not cellulose and xylan and an ectopic deposition of lignin. These findings indicate that pine PtMYB1 and Arabidopsis MYB85 are specific transcriptional activators controlling lignin biosynthesis.

Bedon et al. (2007) isolated and characterized 13 and 5 MYB cDNAs from white spruce (*Picea glauca*) and loblolly pine (*Pinus taeda*), respectively. Three of them, *PgMYB2*, *PgMYB4*, and *PgMYB8*, are preferentially expressed in secondary xylem and suggested to play a role in lignification of woody tissues.

The study of wood-associated Eucalyptus *EgMYB1* gene suggests that transcriptional regulation of wood formation may be controlled by not only activators but also repressors. EgMYB1 is a close homolog of Arabidopsis MYB4, *Antirrhinum* MYB308/MYB330, and maize MYB31/MYB42, which are transcriptional repressors of phenylpropanoid biosynthetic pathway (Tamagnone et al. 1998; Jin et al. 2000; Sonbol et al. 2009; Fornale et al. 2010). These MYBs contain an ethylene-responsive element binding factor-associated amphiphilic repression (EAR)-like repression motif at the carboxyl terminus (Kazan 2006). EgMYB1 binds specifically the AC element located in the promoters of lignin biosynthetic genes (Legay et al. 2007). When overexpressed in Arabidopsis and poplar, EgMYB1 causes a reduction in expression of biosynthetic genes for not only lignin but also cellulose and xylan and concomitantly a reduced secondary wall thickening (Legay et al. 2010). It was proposed that EgMYB1 may act as a repressor of secondary wall biosynthesis during wood formation although its proposed repressor function awaits further knockdown study in transgenic tree species. Another MYB gene, *PttMYB21a*, from poplar was suggested as a repressor of lignin biosynthesis as antisense knockdown of *PttMYB21a* results in an increased expression of a lignin biosynthetic gene, caffeoyl-CoA 3-*O*-methyltransferase (Karpinska et al. 2004).

# 5 NAC Transcription Factors as Master Switches of Wood Formation

In Arabidopsis, a group of secondary wall NAC domain transcription factors (collectively called SWNs), including VASCULAR-RELATED NAC-DOMAIN6 (VND6), VND7; NAC SECONDARY WALL THICKENING PROMOTING FACTOR1 (NST1), NST2; and SECONDARY WALL-ASSOCIATED NAC DOMAIN PROTEIN1 (SND1), were found to be master transcriptional switches of secondary wall biosynthesis (Kubo et al. 2005; Mitsuda et al. 2005, 2007; Zhong et al. 2006, 2007b, 2008; Yamaguchi et al. 2008; Zhong and Ye 2007). SWNs bind to a common *cis*-acting element, namely, secondary wall NAC binding element (SNBE), which is an imperfect palindromic 19-bp consensus sequence, (T/A)NN (C/T)(T/C/G)TNNNNNNNA(A/C)GN(A/C/T)(A/T). SWNs directly activate not only a number of transcription factors but also a suite of genes involved in secondary wall biosynthesis, cell wall modification, and programmed cell death via binding to the SNBE sites in the promoters of target genes (Zhong et al. 2010c). Among these direct target genes are MYB46 and MYB83 that function as second-level master switches regulating secondary wall biosynthesis.

Because EgMYB2 and PtMYB4 are functional orthologs of MYB46/MYB83, it is conceivable that SWN orthologs from tree species might also be upstream regulators of these wood-associated MYBs and involved in transcriptional regulation of wood formation. There are 6 pairs of SWN homologs in poplar (Zhong et al. 2010b; Zhong and Ye 2010; Ohtani et al. 2011); the members of each pair are likely originated from genome duplication (Tuskan et al. 2006). These poplar SWN homologs are expressed in developing wood and hence are named as wood-associated NAC domain transcription factors (PtrWNDs) (Zhong et al. 2010b). Overexpression of PtrWNDs was found to be sufficient to cause the induction of biosynthetic genes for all three major secondary wall components, including

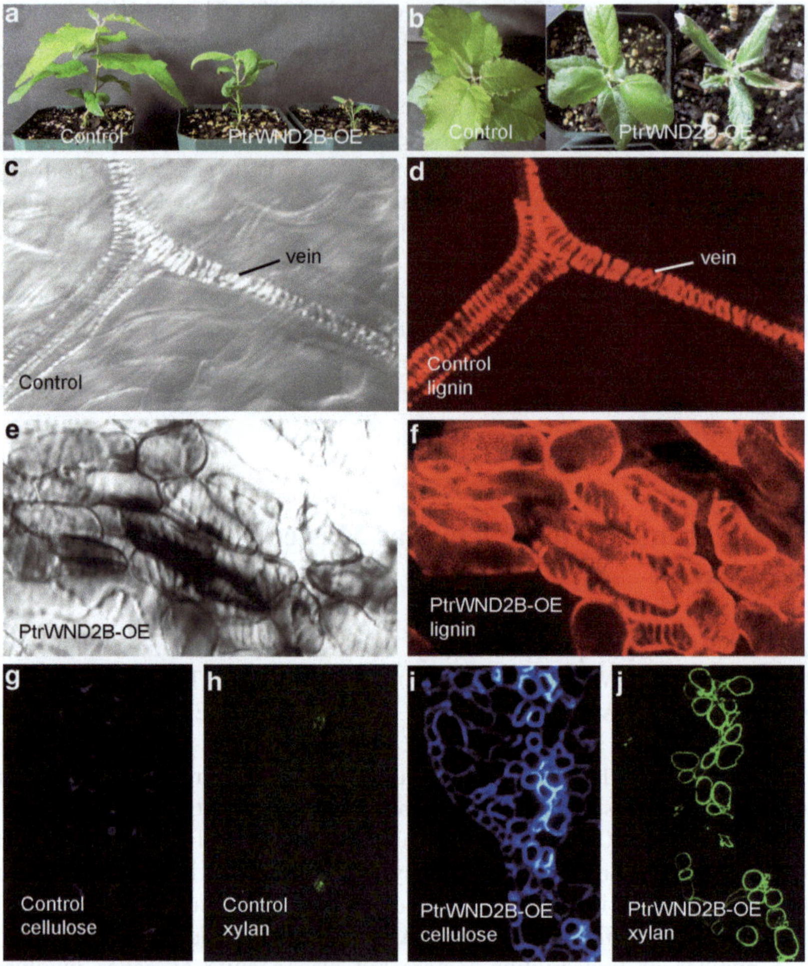


**Fig. 1** Overexpression of PtrWND2B (OE) leads to curly leaves (**a**, **b**) and ectopic production of woody cell walls, which are composed of cellulose (**i**), xylan (**j**), and lignin (**f**), in the leaf mesophyll cells of transgenic poplar plants. Note the ectopic helical secondary wall thickening in the mesophyll cells of PtrWND2B-OE (**e**, **f**). The control expressing an empty vector shows secondary wall deposition only in leaf veins (**c**, **d**, **g**, **h**)

cellulose, xylan, and lignin, and concomitantly results in ectopic deposition of secondary walls in both Arabidopsis and poplar (Fig. 1; Zhong et al. 2010b; Ohtani et al. 2011). Dominant repression of PtrWND function leads to a significant reduction in secondary wall thickness in both vessels and fibers in transgenic poplar wood (Zhong et al. 2011). PtrWNDs are able to bind to the SNBE sites and directly activate the expression of PtrMYB3 and PtrMYB20, which are functional orthologs of Arabidopsis MYB46/MYB83 (McCarthy et al. 2010; Zhong et al. 2011). In addition, wood-associated EgWND1 is also an Arabidopsis SWN ortholog capable of activating the entire secondary wall biosynthetic program (Zhong et al. 2010a). These findings provide strong evidence demonstrating that WNDs are master transcriptional switches regulating secondary wall biosynthesis during wood formation in tree species.

# 6 A Transcriptional Network Regulating Wood Formation

The finding that PtrWNDs are transcriptional master switches of wood formation provides an unprecedented tool to uncover all PtrWND-regulated transcription factors involved in wood formation. It has been shown that PtrWNDs activate the expression of over 40 wood-associated transcription factors (Zhong et al. 2010b; Ohtani et al. 2011). Among them, 29 are close homologs of Arabidopsis SWNs that are known to be part of the SWN-mediated transcriptional network regulating secondary wall biosynthesis. These include the *MYB46* homologs (*PtrMYB2/3/20/21*), *SND2* homologs (*PtrNAC154/156*; Grant et al. 2010), *SND3* homologs (*PtrNAC105/157*), *KNAT7* homolog (*PtrKNAT7*), *MYB20/43* homologs (*PtrMYB18*), *MYB42/85* homologs (*PtrMYB75/92/125/199*), *MYB52/54* homologs (*PtrMYB90/161/167/175*), *MYB58/63* (*PtrMYB28/192*), *MYB69* homologs (*PtrMYB26/31/158/189*), *MYB103* homologs (*PtrMYB10/128*), *LBD15* homolog (*PtrLBD15*), and *XND1* homolog (*PtrNAC118*; Grant et al. 2010). Both PtrWNDs and EgWND1 are able to bind and activate the SNBE sites in the promoters of PtrMYB3, PtrMYB21, PtrNAC157, PtrMYB128, PtrKNAT7, PtrLBD15, and PtrNAC118, indicating that WNDs directly regulate the expression of these downstream targets (Zhong et al. 2011). Furthermore, PtrWNDs activate the expression of 13 other transcription factors whose Arabidopsis close homologs are not known to be involved in the regulation of secondary wall biosynthesis. These include two NACs (*PtrNAC150/151*), two MYBs (*PtrMYB74/121*), two WRKYs (*PtrWRKY12/13*), one IAA (*PtrIAA11*), two WUSCHEL-related homeobox gene (*PtrWUS1* and *PtrWOX13*), one BEL1-like homeobox gene (*PtrBLH3*), one ULTRAPETALA-like gene (*PtrULT1*), and two zinc-finger transcription factors (*PtrZF1* and *PtrGATA8*). Identification of these PtrWND-regulated downstream transcription factors opens a new avenue to further dissect the transcriptional program regulating wood formation.

The discovery that PtrWNDs are the master switches activating a suite of wood-associated downstream transcription factors suggests a complexity of transcriptional regulation of wood formation. Currently, we are still far from a complete understanding of the interrelationships of these transcription factors and how they function together in the activation of wood biosynthetic genes. Analysis of PtrWND direct targets has shown that similar to Arabidopsis SWNs (Ohashi-Ito et al. 2010; Zhong et al. 2010c; Yamaguchi et al. 2011), PtrWNDs are able to not only directly induce the expression of a number of transcription factors but also directly bind and activate the SNBE sites in the promoters of a number of genes involved in secondary wall biosynthesis, cell wall modification, and programmed cell death

(Zhong et al. 2011). Among the PtrWND-regulated transcription factors, PtrMYB3 and PtrMYB20, which are orthologs of Arabidopsis MYB46/MYB83, function as the second-level master switches capable of activating the entire secondary cell wall biosynthetic program (McCarthy et al. 2010). Furthermore, a set of additional transcription factors, such as PtrNAC150, PtrNAC156, PtrNAC157, PtrMYB18, PtrMYB74, PtrMYB75, PtrMYB121, PtrMYB128, PtrZF1, and PtrGATA1, are also able to activate the promoters of genes for all three secondary wall biosynthetic pathways. Since only SWNs and MYB46/MYB83 are capable of activating the entire secondary wall biosynthetic program in Arabidopsis, it was proposed that tree species might have evolved a much more complex transcriptional network consisting additional multiple levels of master controls to ensure secondary wall biosynthesis during wood formation, which requires the deposition of massive amount of secondary wall components.

In addition to those master switches capable of activating all three major wood biosynthetic pathway genes, several PtrWND-regulated transcription factors are involved in activating genes of specific wood biosynthetic pathways. For example, PtrMYB26, PtrMYB90, and PtrMYB28 were found to specifically activate the promoters of lignin biosynthetic genes (Zhong and Ye 2009; Zhong et al. 2011), whereas PtrNAC156, PtrNAC157, and PtrLBD15 were shown to only activate the promoters of one or more specific wood biosynthetic pathway genes. These findings indicate that some transcription factors may only regulate the expression of one or a few secondary wall biosynthetic genes instead of all genes in a specific biosynthetic pathway.

The functional characterization of wood-associated transcription factors has led to the proposal that a transcriptional network encompassing the top-level master switches WNDs together with the WNDs-regulated downstream transcription factors is involved in regulating wood formation. Since many of these transcription factors are able to activate the expression of all three major wood biosynthetic pathway genes, it is unlikely that the transcriptional network regulating wood formation involves a linear cascade of transcriptional regulation in which the top-level WNDs activate downstream transcription factors, which in turn activate wood biosynthetic genes. Instead, the available evidence suggests that similar to the Arabidopsis SWN-mediated transcriptional network (Zhong and Ye 2012), the transcriptional network regulating wood formation may also utilize a multiple-leveled feed-forward loop regulatory structure in which WNDs regulate other transcription factors, and they together activate downstream targets (Fig. 2).

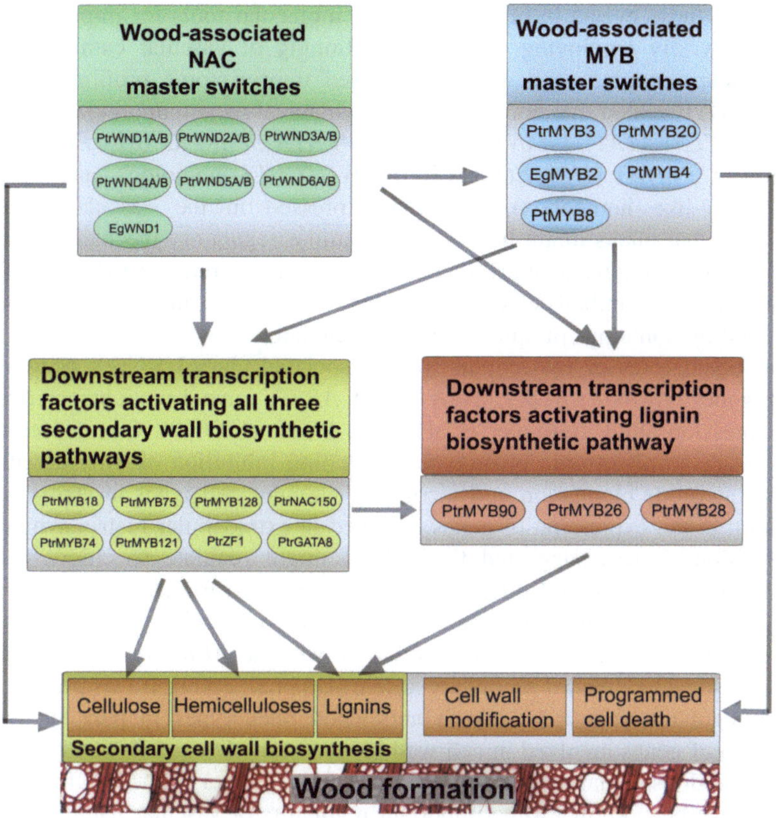


**Fig. 2** Diagram of the transcriptional regulatory network controlling wood formation. The wood-associated NAC transcription factors, WNDs, function as the first-level master switches that directly activate a number of downstream transcription factors as well as many genes involved in secondary wall biosynthesis, cell wall modification, and programmed cell death. A number of the PtrWND-regulated transcription factors act as second-level master switches capable of activating the entire secondary wall program. It is proposed that the transcriptional network regulating wood formation may utilize a multiple-leveled feed-forward loop regulatory structure in which WNDs regulate other transcription factors, and they together activate downstream targets and subsequent wood biosynthetic program

# 7 Transcriptional Regulation of Cambial Activity and Secondary Xylem Differentiation

Compared to the breakthrough discovery of many key transcriptional regulators controlling secondary wall biosynthesis during wood formation, we have just begun to identify a few potential regulators of cambial activity and secondary xylem differentiation (Du and Groover 2010). Cambial cell division and differentiation into secondary xylem mother cells are important events determining the rate of wood formation. Therefore, it is critical to uncover the transcriptional networks controlling these events in order to provide molecular tools for improving wood productivity. By employing tangential sectioning of vascular cambial cells of poplar stems coupled with microarray-based transcriptome profiling, Schrader et al. (2004) revealed that homologs of known apical meristem regulators, such as *PttCLV1* (*CLAVATA1*), *PttANT* (*AINTEGUMENTA*), *PttRLK3* (*RECEPTOR-LIKE KINASE3*), *PttHB3* (*HOMEOBOX3*; the *WUSCHEL* homolog), and *PttKNOX* (*KNOTTED*-like homeobox), are also expressed in the vascular cambium, suggesting that plants might have evolved similar regulatory mechanisms in both the vascular cambium and apical meristems. However, it remains to be investigated whether vascular cambial cell division and maintenance during wood formation are controlled by homologs of apical meristem regulators.

Functional roles of two poplar wood-associated class I KNOX transcription factor genes, *ARBORKNOX1* (*ARK1*) and *ARK2*, have been characterized by overexpression study (Groover et al. 2006; Du et al. 2009). Overexpression of *ARK1* leads to pleiotropic phenotypes, including inhibition of differentiation of leaves, internode elongation and secondary vascular cell types in stems, and an alteration in the expression of genes involved in extracellular matrix synthesis or modification. It was suggested that ARK1 might regulate specific aspects of cambial functions and cell differentiation during wood formation (Groover et al. 2006). *ARK2* overexpression was found to result in a wider cambial zone and delayed cambial daughter cell differentiation, whereas *ARK2* knockdown leads to premature development of secondary xylem (Du et al. 2009). It was proposed that ARK2 regulates secondary growth and wood phenotypes in part by negative regulation of genes involved in cell wall biosynthesis (Du and Groover 2010).

In Arabidopsis, class III homeodomain-leucine zipper (HD-ZIP) transcription factors have been known to be important regulators of vascular patterning, organ polarity, and fiber differentiation (Zhong and Ye 1999 and 2004; Prigge et al. 2005). Poplar homologs of class III HD-ZIP genes, including *popREVOLUTA*, *popCORONA*, and *PtaHB1*, are expressed in developing wood (Ko et al. 2006; Du et al. 2011; Robischon et al. 2011). Overexpression of a microRNA-resistant form of *popREVOLUTA* affects secondary growth and results in abnormal formation of cambia within cortical parenchyma. It was suggested that popREVOLUTA is involved in regulating the initiation of the vascular cambium as well as the patterning of secondary vascular tissues (Robischon et al. 2011). Functional characterization of popCORONA revealed that overexpression of a miRNA-resistant *popCORONA* causes a delay in wood lignification, whereas synthetic miRNA knockdown of *popCORONA* leads to abnormal lignification in pith cells (Du et al. 2011). Further investigation of their downstream targets will help understand how these vascular cambium-associated KNOX and HD-ZIP III genes regulate the division and differentiation of the vascular cambium.

# 8 Genomic Analysis of Transcriptional Regulation of Wood Formation

Although the studies of transcriptional regulation of secondary wall biosynthesis and cambial activity have uncovered a number of key regulators controlling these processes, it is conceivable that additional transcriptional regulators and networks are involved in different aspects of wood formation considering that wood development involves tightly controlled sequential processes. With the availability of the sequenced poplar genome and vast cDNA sequence and expressed sequence tag data from other tree species (Tuskan et al. 2006; Demura and Fukuda 2007; Pavy et al. 2008; Wang et al. 2009; Rigault et al. 2011), it is now possible to identify and functionally characterize transcriptional networks involved in every step of wood development, including cambial cell division and differentiation into secondary xylem mother cells, cell elongation, secondary wall thickening, programmed cell death, and heartwood formation. Using the Affymetrix Poplar Genome Arrays, Wilkins et al. (2009) performed whole-genome transcriptome profiling in seedlings grown under different light regimes, young leaves, mature leaves, roots, xylem, female catkins, and male catkins. The compendium of data derived from this transcriptome profiling is referred to as the *Populus* Gene Expression (PopGenExpress) data set. The whole PopGenExpress transcript abundance data is presented as a simple, graphical format with a Web-based tool, namely, the *Populus* Electronic Fluorescent Pictograph (eFP) browser (http://www.bar.utoronto.ca/efpop/cgi-bin/efpWeb.cgi). For the *Populus* eFP browser, diagrams of poplar tissues, organs, or growth conditions are shaded, with colors corresponding to the quantity of transcript for a given gene under that condition. Thus, one can easily find out which transcription factors are preferentially expressed in the xylem (or wood) and then investigate the roles of these wood-associated transcription factors in the regulation of wood formation. Wilkins et al. (2009) demonstrated the effectiveness of using the PopGenExpress data to identify MYB genes expressed in developing wood. Among 180 R2R3-MYB genes, 23 of them, including PtrMYB2/3/20/21 that are orthologs of EgMYB2 and PtMYB4, showed the highest level of transcript abundance in differentiating xylem. PtrWNDs and their downstream transcription factors were also found to be highly expressed in developing wood based on the *Populus* eFP browser profiles (Zhong et al. 2011). It is expected that the *Populus* eFP browser will be a powerful tool to uncover all the transcriptional regulators involved in the regulation of wood formation.

Additional genome-wide transcriptome analysis has been applied to study genetic regulation of secondary growth and revealed a number of transcription factors that are potentially involved in the regulation of wood formation (Prassinos et al. 2005; Andersson-Gunneras et al. 2006; Wang et al. 2009; Dharmawardhana et al. 2010). Functional characterization of all the wood-associated transcription factors will likely lead to uncovering their roles in specific processes of wood formation.

# 9 Concluding Remarks

With the rapidly increasing genomic sequence and transcriptome profiling data available in tree species, it is now an exciting time to tackle the long-standing question regarding how plants make wood, the most abundantly stored biomass by plants. We have now begun to unveil the transcriptional network regulating secondary wall biosynthesis during wood formation (Fig. 2). We have also uncovered a few transcriptional regulators potentially involved in vascular cambial cell division and secondary xylem differentiation. The next challenging step is to use the available genomic data to identify and functionally characterize the transcriptional networks governing each process of wood formation, including vascular cambial cell division and differentiation into secondary xylem, cell elongation, secondary wall thickening, programmed cell death, and heartwood formation. The knowledge gained from transcriptional regulation of wood formation will likely help develop strategies to custom design wood biomass composition better suited for diverse end uses, such as biofuel production.

**Acknowledgments** Work in our laboratory was supported by grants from the National Science Foundation (ISO-1051900) and the US Department of Agriculture National Institute of Food and Agriculture [AFRI Plant Biology program (#2010-65116-20468)].

## References

Andersson S-I, Samuelson O, Ishihara M, Shimizu K (1983) Structure of the reducing end-groups in spruce xylan. Carbohydr Res 111:283–288

Andersson-Gunneras S, Mellerowicz EJ, Love J, Segerman B, Ohmiya Y, Coutinho PM, Nilsson P, Henrissat B, Moritz T, Sundberg B (2006) Biosynthesis of cellulose-enriched tension wood in *Populus*: global analysis of transcripts and metabolites identifies biochemical and developmental regulators in secondary wall biosynthesis. Plant J 45:144–165

Aspeborg H, Schrader J, Coutinho PM, Stam M, Kallas A, Djerbi S, Nilsson P, Denman S, Amini B, Sterky F, Master E, Sandberg G, Mellerowicz E, Sundberg B, Henrissat B, Teeri TT (2005) Carbohydrate-active enzymes involved in the secondary cell wall biogenesis in hybrid aspen. Plant Physiol 137:983–997

Bedon F, Grima-Pettenati J, Mackay J (2007) Conifer R2R3-MYB transcription factors: sequence analyses and gene expression in wood-forming tissues of white spruce (*Picea glauca*). BMC Plant Biol 7:17

Boerjan W, Ralph J, Baucher M (2003) Lignin biosynthesis. Annu Rev Plant Biol 54:519–546

Bomal C, Bedon F, Caron S, Mansfield SD, Levasseur C, Cooke JE, Blais S, Tremblay L, Morency MJ, Pavy N, Grima-Pettenati J, Séguin A, Mackay J (2008) Involvement of *Pinus taeda* MYB1 and MYB8 in phenylpropanoid metabolism and secondary cell wall biogenesis: a comparative in planta analysis. J Exp Bot 59:3925–3939

Carroll A, Somerville C (2009) Cellulosic biofuels. Annu Rev Plant Biol 60:165–182

Creux NM, Ranik M, Berger DK, Myburg AA (2008) Comparative analysis of orthologous cellulose synthase promoters from *Arabidopsis*, *Populus* and *Eucalyptus*: evidence of conserved regulatory elements in angiosperms. New Phytol 179:722–737

Demura T, Fukuda H (2007) Transcriptional regulation in wood formation. Trends Plant Sci 12:64–70

Dharmawardhana P, Brunner AM, Strauss SH (2010) Genome-wide transcriptome analysis of the transition from primary to secondary stem development in *Populus trichocarpa*. BMC Genomics 11:150

Du J, Groover A (2010) Transcriptional regulation of secondary growth and wood formation. J Integr Plant Biol 52:17–27

Du J, Mansfield SD, Groover AT (2009) The *Populus* homeobox gene *ARBORKNOX2* regulates cell differentiation during secondary growth. Plant J 60:1000–1014

Du J, Miura E, Robischon M, Martinez C, Groover A (2011) The *Populus* Class III HD ZIP transcription factor *POPCORONA* affects cell differentiation during secondary growth of woody stems. PLoS One 6:e17458

Feuillet C, Lauvergeat V, Deswarte C, Pilate G, Boudet A, Grima-Pettenati J (1995) Tissue- and cell-specific expression of a cinnamyl alcohol dehydrogenase promoter in transgenic poplar plants. Plant Mol Biol 27:651–667

Field CB, Behrenfeld MJ, Randerson JT, Falkowski P (1998) Primary production of the biosphere: integrating terrestrial and oceanic components. Science 281:237–240

Fornalé S, Shi X, Chai C, Encina A, Irar S, Capellades M, Fuguet E, Torres JL, Rovira P, Puigdomènech P, Rigau J, Grotewold E, Gray J, Caparrós-Ruiz D (2010) ZmMYB31 directly represses maize lignin genes and redirects the phenylpropanoid metabolic flux. Plant J 64:633–644

Goicoechea M, Lacombe E, Legay S, Mihaljevic S, Rech P, Jauneau A, Lapierre C, Pollet B, Verhaegen D, Chaubet-Gigot N, Grima-Pettenati J (2005) EgMYB2, a new transcriptional activator from *Eucalyptus xylem*, regulates secondary cell wall formation and lignin biosynthesis. Plant J 43:553–567

Grant EH, Fujino T, Beers EP, Brunner AM (2010) Characterization of NAC domain transcription factors implicated in control of vascular cell differentiation in Arabidopsis and *Populus*. Planta 232:337–352

Gray-Mitsumune M, Molitor EK, Cukovic D, Carlson JE, Douglas CJ (1999) Developmentally regulated patterns of expression directed by poplar *PAL* promoters in transgenic tobacco and poplar. Plant Mol Biol 39:657–669

Groover AT, Mansfield SD, DiFazio SP, Dupper G, Fontana JR, Millar R, Wang Y (2006) The *Populus* homeobox gene *ARBORKNOX1* reveals overlapping mechanisms regulating the shoot apical meristem and the vascular cambium. Plant Mol Biol 61:917–932

Grotewold E, Drummond BJ, Bowen B, Peterson T (1994) The myb-homologous P gene controls phlobaphene pigmentation in maize floral organs by directly activating a flavonoid biosynthetic gene subset. Cell 76:543–553

Hatton D, Sablowski R, Yung M-H, Smith C, Schuch W, Bevan M (1995) Two classes of *cis* sequences contribute to tissue-specific expression of a *PAL2* promoter in transgenic tobacco. Plant J 7:859–876

Jin H, Cominelli E, Bailey P, Parr A, Mehrtens F, Jones J, Tonelli C, Weisshaar B, Martin C (2000) Transcriptional repression by AtMYB4 controls production of UV-protecting sunscreens in Arabidopsis. EMBO J 19:6150–6161

Johansson MH, Samuelson O (1977) Reducing end groups in birch xylan and their alkaline degradation. Wood Sci Technol 11:251–263

Joshi CP, Mansfield SD (2007) The cellulose paradox – simple molecule, complex biosynthesis. Curr Opin Plant Biol 10:220–226

Karpinska B, Karlsson M, Srivastava M, Stenberg A, Schrader J, Sterky F, Bhalerao R, Wingsle G (2004) MYB transcription factors are differentially expressed and regulated during secondary vascular tissue development in hybrid aspen. Plant Mol Biol 56:255–270

Kazan K (2006) Negative regulation of defense and stress genes by EAR-motif-containing repressors. Trends Plant Sci 11:109–112

Ko JH, Prassinos C, Han KH (2006) Developmental and seasonal expression of *PtaHB1*, a *Populus* gene encoding a class III HD-Zip protein, is closely associated with secondary growth and inversely correlated with the level of microRNA (miR166). New Phytol 169:469–478

Ko JH, Kim WC, Han KH (2009) Ectopic expression of MYB46 identifies transcriptional regulatory genes involved in secondary wall biosynthesis in Arabidopsis. Plant J 60:649–665

Kong Y, Zhou G, Avci U, Gu X, Jones C, Yin Y, Xu Y, Hahn MG (2009) Two poplar glycosyltransferase genes, *PdGATL1.1* and *PdGATL1.2*, are functional orthologs to *PARVUS/AtGATL1* in Arabidopsis. Mol Plant 2:1040–1050

Kubo M, Udagawa M, Nishikubo N, Horiguchi G, Yamaguchi M, Ito J, Mimura T, Fukuda H, Demura T (2005) Transcription switches for protoxylem and metaxylem vessel formation. Genes Dev 19:1855–1860

Lacombe E, Doorsselaere JV, Boerjan W, Boudet AM, Grima-Pettenati J (2000) Characterization of *cis*-elements required for vascular expression of the *Cinnamoyl CoA Reductase* gene and for protein-DNA complex formation. Plant J 23:663–676

Lauvergeat V, Rech P, Jauneau A, Guez C, Coutos-Thevenot P, Grima-Pettenati J (2002) The vascular expression pattern directed by the *Eucalyptus gunnii* cinnamyl alcohol dehydrogenase *EgCAD2* promoter is conserved among woody and herbaceous plant species. Plant Mol Biol 50:497–509

Lee C, Teng Q, Huang W, Zhong R, Ye Z-H (2009) Down-regulation of PoGT47C expression in poplar results in a reduced glucuronoxylan content and an increased wood digestibility by cellulase. Plant Cell Physiol 50:1075–1089

Lee C, Teng Q, Zhong R, Ye Z-H (2011) Molecular dissection of xylan biosynthesis during wood formation in poplar. Mol Plant 4:730–747

Lee C, Zhong R, Ye Z-H (2012) Arabidopsis family GT43 members are xylan xylosyltransferases required for the elongation of the xylan backbone. Plant Cell Physiol 53:135–143

Legay S, Lacombe E, Goicoechea M, Briere C, Seguin A, Mackay J, Grima-Pettenati J (2007) Molecular characterization of *EgMYB1*, a putative transcriptional repressor of the lignin biosynthetic pathway. Plant Sci 173:542–549

Legay S, Sivadon P, Blervacq AS, Pavy N, Baghdady A, Tremblay L, Levasseur C, Ladouce N, Lapierre C, Séguin A, Hawkins S, Mackay J, Grima-Pettenati J (2010) EgMYB1, an R2R3 MYB transcription factor from eucalyptus negatively regulates secondary cell wall formation in Arabidopsis and poplar. New Phytol 188:774–786

Li Q, Min D, Wang JP, Peszlen I, Horvath L, Horvath B, Nishimura Y, Jameel H, Chang HM, Chiang VL (2011) Down-regulation of glycosyltransferase 8D genes in *Populus trichocarpa* caused reduced mechanical strength and xylan content in wood. Tree Physiol 31:226–236

McCarthy RL, Zhong R, Ye Z-H (2009) MYB83 is a direct target of SND1 and acts redundantly with MYB46 in the regulation of secondary cell wall biosynthesis in Arabidopsis. Plant Cell Physiol 50:1950–1964

McCarthy RL, Zhong R, Fowler S, Lyskowski D, Piyasena H, Carleton K, Spicer C, Ye ZH (2010) The poplar MYB transcription factors, PtrMYB3 and PtrMYB20, are involved in the regulation of secondary wall biosynthesis. Plant Cell Physiol 51:1084–1090

Mellerowicz EJ, Sundberg B (2008) Wood cell walls: biosynthesis, developmental dynamics and their implications for wood properties. Curr Opin Plant Biol 11:293–300

Mitsuda N, Seki M, Shinozaki K, Ohme-Takagi M (2005) The NAC transcription factors NST1 and NST2 of *Arabidopsis* regulates secondary wall thickening and are required for anther dehiscence. Plant Cell 17:2993–3006

Mitsuda N, Iwase A, Yamamoto H, Yoshida M, Seki M, Shinozaki K, Ohme-Takagi M (2007) NAC transcription factors, NST1 and NST3, are key regulators of the formation of secondary walls in woody tissues of *Arabidopsis*. Plant Cell 19:270–280

Ohashi-Ito K, Oda Y, Fukuda H (2010) Arabidopsis VASCULAR-RELATED NAC-DOMAIN6 directly regulates the genes that govern programmed cell death and secondary wall formation during xylem differentiation. Plant Cell 22:3461–3473

Ohtani M, Nishikubo N, Xu B, Yamaguchi M, Mitsuda N, Goué N, Shi F, Ohme-Takagi M, Demura T (2011) A NAC domain protein family contributing to the regulation of wood formation in poplar. Plant J 67:499–512

Patzlaff A, McInnis S, Courtenay A, Surman C, Newman LJ, Smith C, Bevan MW, Mansfield S, Whetten RW, Sederoff RR, Campbell MM (2003a) Characterization of a pine MYB that regulates lignification. Plant J 36:743–754

Patzlaff A, Newman LJ, Dubos C, Whetten RW, Smith C, McInnis S, Bevan MW, Sederoff RR, Campbell MM (2003b) Characterization of PtMYB1, an R2R3-MYB from pine xylem. Plant Mol Biol 53:597–608

Pavy N, Boyle B, Nelson C, Paule C, Giguère I, Caron S, Parsons LS, Dallaire N, Bedon F, Bérubé H, Cooke J, Mackay J (2008) Identification of conserved core xylem gene sets: conifer cDNA microarray development, transcript profiling and computational analyses. New Phytol 180:766–786

Pilate G, Guiney E, Holt K, Petit-Conil M, Lapierre C, Leplé JC, Pollet B, Mila I, Webster EA, Marstorp HG, Hopkins DW, Jouanin L, Boerjan W, Schuch W, Cornu D, Halpin C (2002) Field and pulping performances of transgenic trees with altered lignification. Nat Biotechnol 20:607–612

Plomion C, Leprovost G, Stokes A (2001) Wood formation in trees. Plant Physiol 127:1513–1523

Prassinos C, Ko JH, Yang J, Han KH (2005) Transcriptome profiling of vertical stem segments provides insights into the genetic regulation of secondary growth in hybrid aspen trees. Plant Cell Physiol 46:1213–1225

Prigge MJ, Otsuga D, Alonso JM, Ecker JR, Drews GN, Clark SE (2005) Class III homeodomain-leucine zipper gene family members have overlapping, antagonistic, and distinct roles in Arabidopsis development. Plant Cell 17:61–76

Raes J, Rohde A, Christensen JH, Peer YV, Boerjan W (2003) Genome-wide characterization of the lignification toolbox in Arabidopsis. Plant Physiol 133:1051–1071

Rahantamalala A, Rech P, Martinez Y, Chaubet-Gigot N, Grima-Pettenati J, Pacquit V (2010) Coordinated transcriptional regulation of two key genes in the lignin branch pathway–CAD and CCR–is mediated through MYB- binding sites. BMC Plant Biol 10:130

Rigault P, Boyle B, Lepage P, Cooke JE, Bousquet J, MacKay JJ (2011) A white spruce gene catalog for conifer genome analyses. Plant Physiol 157:14–28

Robischon M, Du J, Miura E, Groover A (2011) The Populus class III HD ZIP, popREVOLUTA, influences cambium initiation and patterning of woody stems. Plant Physiol 155:1214–1225

Schrader J, Nilsson J, Mellerowicz E, Berglund A, Nilsson P, Hertzberg M, Sandberg G (2004) A high-resolution transcript profile across the wood-forming meristem of poplar identifies potential regulators of cambial stem cell identity. Plant Cell 16:2278–2292

Shimizu K, Ishihara M, Ishihara T (1976) Hemicellulases of brown rotting fungus, *Tyromyces palustris*. II. The oligosaccharides from the hydrolysate of a hardwood xylan by the intracellular xylanase. Mokuzai Gaikkashi 22:618–625

Sonbol FM, Fornalé S, Capellades M, Encina A, Touriño S, Torres JL, Rovira P, Ruel K, Puigdomènech P, Rigau J, Caparrós-Ruiz D (2009) The maize ZmMYB42 represses the phenylpropanoid pathway and affects the cell wall structure, composition and degradability in *Arabidopsis thaliana*. Plant Mol Biol 70:283–296

Tamagnone L, Merida A, Parr A, Mackay S, Culianez-Macia FA, Roberts K, Martin C (1998) The AmMYB308 and AmMYB330 transcription factors from *Antirrhinum* regulate phenylpropanoid and lignin biosynthesis in transgenic tobacco. Plant Cell 10:135–154

Taylor NG, Gardiner JC, Whiteman R, Turner SR (2004) Cellulose synthesis in the Arabidopsis secondary cell wall. Cellulose 11:329–338

Timell TE (1967) Recent progress in the chemistry of wood hemicelluloses. Wood Sci Technol 1:45–70

Tuskan GA, Difazio S, Jansson S et al (2006) The genome of black cottonwood, *Populus trichocarpa* (Torr. & Gray). Science 313:1596–1604

Vanholme R, Morreel K, Ralph J, Boerjan W (2008) Lignin engineering. Curr Opin Plant Biol 11:278–285

Wang M, Qi X, Zhao S, Zhang S, Lu MZ (2009) Dynamic changes in transcripts during regeneration of the secondary vascular system in *Populus tomentosa* Carr. revealed by cDNA microarrays. BMC Genomics 10:215

Wilkins O, Nahal H, Foong J, Provart NJ, Campbell MM (2009) Expansion and diversification of the *Populus* R2R3-MYB family of transcription factors. Plant Physiol 149:981–993

Yamaguchi M, Kubo M, Fukuda H, Demura T (2008) Vascular-related NAC-DOMAIN7 is involved in the differentiation of all types of xylem vessels in *Arabidopsis* roots and shoots. Plant J 55:652–664

Yamaguchi M, Mitsuda N, Ohtani M, Ohme-Takagi M, Kato K, Demura T (2011) VASCULAR-RELATED NAC-DOMAIN7 directly regulates the expression of a broad range of genes for xylem vessel formation. Plant J 66:579–590

Zhong R, Ye Z-H (1999) *IFL1*, a gene regulating interfascicular fiber differentiation in Arabidopsis, encodes a homeodomain-leucine zipper protein. Plant Cell 11:2139–2152

Zhong R, Ye ZH (2004) *Amphivasal vascular bundle 1*, a gain-of-function mutation of the *IFL1/REV* gene, is associated with alterations in the polarity of leaves, stems and carpels. Plant Cell Physiol 45:369–385

Zhong R, Ye Z-H (2007) Regulation of cell wall biosynthesis. Curr Opin Plant Biol 10:564–572

Zhong R, Ye Z-H (2009) Transcriptional regulation of lignin biosynthesis. Plant Signal Behav 4:1028–1034

Zhong R, Ye Z-H (2010) The poplar PtrWNDs are transcriptional activators of secondary cell wall biosynthesis. Plant Signal Behav 5:469–472

Zhong R, Ye Z-H (2012) MYB46 and MYB83 bind to the SMRE sites and directly activate a suite of transcription factors and secondary wall biosynthetic genes. Plant Cell Physiol 53:368–380

Zhong R, Demura T, Ye Z-H (2006) SND1, a NAC domain transcription factor, is a key regulator of secondary wall synthesis in fibers of *Arabidopsis*. Plant Cell 18:3158–3170

Zhong R, Richardson EA, Ye Z-H (2007a) The MYB46 transcription factor is a direct target of SND1 and regulates secondary wall biosynthesis in Arabidopsis. Plant Cell 19:2776–2792

Zhong R, Richardson EA, Ye Z-H (2007b) Two NAC domain transcription factors, SND1 and NST1, function redundantly in regulation of secondary wall synthesis in fibers of Arabidopsis. Planta 225:1603–1611

Zhong R, Lee C, Zhou J, McCarthy RL, Ye Z-H (2008) A battery of transcription factors involved in the regulation of secondary cell wall biosynthesis in Arabidopsis. Plant Cell 20:2763–2782

Zhong R, Lee C, Ye Z-H (2010a) Evolutionary conservation of the transcriptional network regulating secondary cell wall biosynthesis. Trends Plant Sci 15:625–631

Zhong R, Lee C, Ye ZH (2010b) Functional characterization of poplar wood-associated NAC domain transcription factors. Plant Physiol 152:1044–1055

Zhong R, Lee C, Ye Z-H (2010c) Global analysis of direct targets of secondary wall NAC master switches in *Arabidopsis*. Mol Plant 3:1087–1103

Zhong R, McCarthy RL, Lee C, Ye Z-H (2011) Dissection of the transcriptional program regulating secondary wall biosynthesis during wood formation in poplar. Plant Physiol 157:1452–1468

Zhou GK, Zhong R, Richardson EA, Morrison WH 3rd, Nairn CJ, Wood-Jones A, Ye Z-H (2006) The poplar glycosyltransferase GT47C is functionally conserved with *Arabidopsis fragile fiber8*. Plant Cell Physiol 47:1229–1240

Zhou GK, Zhong R, Richardson EA, Himmelsbach DS, McPhail BT, Ye Z-H (2007) Molecular characterization of PoGT8D and PoGT43B, two secondary wall-associated glycosyltransferases in poplar. Plant Cell Physiol 48:689–699

# Climate Control of Wood Formation: Illustrated for Scots Pine at Its Northern Distribution Limit

**Jeong-Wook Seo, Dieter Eckstein, Andrea Olbrich, Risto Jalkanen, Hannu Salminen, Uwe Schmitt, and Jörg Fromm**

**Abstract** The growth of trees is a spectacular and exposed process based on a highly interlinked complex of hidden and cryptic metabolic and signaling pathways not yet fully understood. In this chapter, we focus on a sequence of studies on Scots pine as an example tree species during the past 10 years in the north of Finland. We particularly compare annual height growth and annual growth in girth in the long term. Moreover, we give attention to the chronological coherence between the growth in height and girth during a growing season. Finally, we go down on the cellular level and screen various variables of the water conducting cells for their suitability as climatic proxies.

Girth growth is promoted by a warm current summer and height growth by a warm preceding summer. Within a growing season, growth in height and girth culminates in the second half of June, clearly before the warmest period of the year in the second half of July. On the cellular level, it is concluded that diameter and wall thickness of earlywood tracheids are independent from one another and from tree-ring width and in consequence contain different climatic signals. These encouraging findings provide a strong rationale for further studies.

J.-W. Seo (✉)
University of Greifswald, Institute of Botany and Landscape Ecology, 17489 Greifswald, Germany
e-mail: jeongwook.seo@uni-greifswald.de

D. Eckstein • A. Olbrich • J. Fromm
University of Hamburg, Dept. of Wood Science, Division Wood Biology, 21031 Hamburg, Germany

R. Jalkanen • H. Salminen
Finnish Forest Research Institute, Northern Regional Unit, 96301 Rovaniemi, Finland

U. Schmitt
Thünen Institute of Wood Research, 21-31 Hamburg, Germany

J. Fromm (ed.), *Cellular Aspects of Wood Formation*, Plant Cell Monographs 20, DOI 10.1007/978-3-642-36491-4_6, © Springer-Verlag Berlin Heidelberg 2013

# 1 Introduction and Background

A tree—as a three-dimensional body—grows in length and girth, with some of its parts resting while others are growing, by accumulating new cells of which a small fraction (meristematic tissue) remains alive, whereas most of the other cells (xylem and phloem) have a life span of only days or weeks. Growth in length results from terminally located meristems distributed over the stem, branches, and roots. Growth in thickness results from the vascular cambium located between bark and wood and extending from the tips of the twigs continuously along the stem into the roots. It gives rise to new xylem cells to the inside and new phloem cells to the outside. Although the growth in length and girth in a tree is closely interlinked, both processes differ in their onset, duration, and end (Kozlowski 1971).

Whereas shoot growth is an externally visible process, the activity of the cambium remains hidden. Nevertheless, by the mid-nineteenth century, the existence and basic nature of the cambium (Larson 1994), and in consequence of the growth in girth, were correctly identified. Attempts to measure the increase in girth can be traced back to at least the end of the nineteenth century (e.g., Hartig 1885). At around the same time, coherence between growth in length and girth was implied (e.g., Jost 1893). Among the most interesting features of growth is the fact that trees divide their lifetime into annual growth layers provided that there is a regular alternation between a growing and a dormant season within 1 year. These layers "archive" information on "time and environment" entering a tree through its foliage and roots. Through metabolic processes, this information is transformed into new cells by the cambium (Fig. 1). The metabolic pathways involved can be traced through time and on various structural levels of a tree. But only on the anatomical, cellular, and chemical level are these effects of exogenous influences permanently "archived," provided that an influence had been limiting tree growth for a certain period of time. Environmental information can be found in the tree-ring width (e.g., Sirén 1961; Esper et al. 2012), earlywood and latewood widths (e.g., Eckstein and Schmidt 1974; Nola 1996; Tardif 1996), wood density (e.g., Parker and Hennoch 1971; Schweingruber et al. 1978; Grudd 2008), wood structure (e.g., Eckstein et al. 1977; Vaganov et al. 1985; Xu et al. 2012a), chemical composition of the cell walls (e.g., Schleser et al. 1999; McCarroll et al. 2011), and in shoot length (e.g., Jalkanen and Tuovinen 2001; Salminen and Jalkanen 2004). All these measurable quantities, at best, record a different detail of the tree's environment.

Until the mid-twentieth century, a great deal of information on tree growth has already been brought together (e.g., Zahner 1963; Zimmermann 1964; Kozlowski 1971; Carlquist 1975). But from that time on, a new quality of observations and interpretations of tree growth gradually emerged due to numerous technical progressions in microscopy, image analysis, data acquisition, and processing. The driving force behind this development was the ambition to push the frontiers of tree-ring research forward by finding new variables in the annually formed wood and by realizing a time resolution of higher than 1 year.

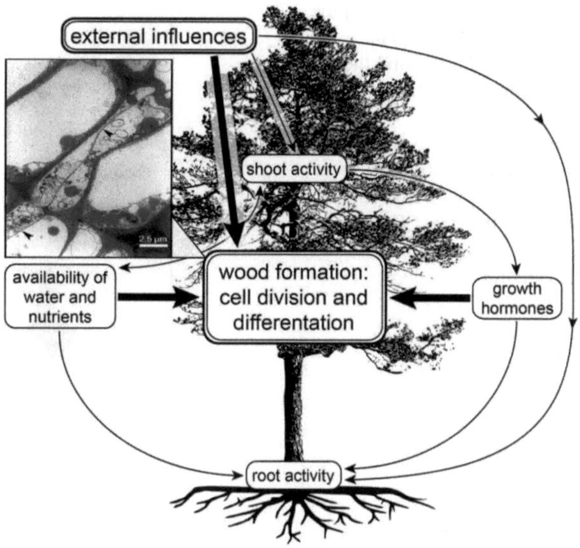


**Fig. 1** Interaction between external influences and shoot and root activity as well as wood formation (simplified after a model by Denne and Dodd 1981); from the internal factors, such as tree age and genetic makeup, only growth hormones are explicitly shown; *inset* in the left corner, dividing cambium cell; *arrow heads* point to new cell walls

In the meantime, but independently, the understanding of the metabolic pathways and of the subcellular processes by which xylem is produced and how it is controlled has been considerably improved (e.g., Barnett 1981; Bauch 1993; Savidge et al. 2000; Fromm 2010; Aloni 2013); but these aspects are not subject of our contribution.

In the following, we mainly reflect upon the outcome of a 10-year period of studies on Scots pine (*Pinus sylvestris* L.) in northern Finland (Fig. 2), based on the two aforementioned topics, viz., new variables and high time resolution. The forests in northern Finland are part of the circumpolar boreal forest belt covering altogether about one-third of the forested area of the world. At their northern limit, they are highly sensitive against environmental changes, caused by climate and/or human impacts, and have therefore increasingly attracted scientific attention (e.g., Hicks et al. 2000; McCarroll et al. 2003; Wilmking et al. 2004; Taulavuori et al. 2010). These forests are characterized by a short growing season and low air and soil temperatures (e.g., Venäläinen et al. 2001; Kirdyanov et al. 2003). Their growth processes and geographical distribution border largely depend on summer temperature (e.g., Jalkanen et al. 2007; Wilmking et al. 2012). That is why these trees have proven as useful proxies for summer temperature far back into the past (e.g., Briffa et al. 1990; Vaganov et al. 1996; Kirchhefer 2001; Watson and Luckman 2004; Pensa et al. 2005; Helama et al. 2009; Lindholm et al. 2011). The same awareness for timberline-associated ecosystems can be observed for the European Alps (e.g., Wieser et al. 2009), the Himalayas (e.g., Liu et al. 2013), and other high-altitude forests.

Despite these valuable achievements, a higher time resolution of tree growth and the search for new tree-ring variables have been aspired over the years, beginning with the easy separation between early- and latewood width of conifers as early as in the 1920s (Mork 1928). Later on, wood density profiles across the tree rings of

**Fig. 2** Location of Scots pine sites in northern Finland. Kevo: ~60 km beyond the Scots pine tree line. Laanila/Ivalo and Vanttauskoski/Rovaniemi: approximately 80 and 300 km, respectively, south from the tree line. The Arctic Circle goes through Rovaniemi

conifers using X-ray densitometry became feasible (Polge 1963). Around the same time, the focus started to be also put on the cellular ring structure (Vaganov and Terskov 1977). These early approaches have been further improved and are increasingly applied in practice (e.g., Loris 1981; Eckstein 1983; Park 1990; Bäucker et al. 1998; Yasue et al. 2000; Panyushkina et al. 2003; Deslauriers and Morin 2005; Eilmann et al. 2006; Gurskaya and Shiyatov 2006; Liang and Eckstein 2006; García-Gonzáles and Fonti 2008; Prislan et al. 2009; Gurskaya et al. 2012; Xu et al. 2012a).

## 2 Outline

After a very brief description and illustration of various techniques of sampling and data acquisition, we first of all give attention to the chronological coherence between the intra-annual dynamics of growth in height and girth during three growing seasons. Then we compare time series of annual height growth (shoot length) and of annual growth in girth (tree-ring width) in the long term. Subsequently, we focus on a 5-year period of intra-annual radial increment measurements in comparison with the entire increments of the same years (tree-ring width) and discuss the equality vs. inequality between both of them. Finally, we go down to the cellular level and screen various variables of earlywood tracheids (water conducting cells) for their potential as climatic proxies. The growth dynamics of roots has not been dealt with. Our contribution ends with some conclusions and perspectives.

# 3 Methods Applied

This subchapter may be omitted by those readers who are familiar with the various techniques of data acquisition for the objectives of this contribution.

## *3.1 Identification of Bud Break and Measurement of Height Growth on Saplings, Pole-Stage Trees, and Adult Trees*

On saplings, bud break after winter dormancy is achieved when a bud has elongated at least by 1 mm. The subsequent height growth can then be measured to the nearest millimeter once a week between the apex of the leader shoot and a pin, permanently inserted as a reference point in the previous annual shoot of the main stem (Fig. 3a). Onset and end of height growth are defined as the points in time when shoot growth accomplishes 5 and 95 %, respectively, of the total shoot length (Salminen and Jalkanen 2007).

On pole-stage trees, shoot growth during a growing season can be photographed from an experimental tower by a permanently attached camera (Fig. 3b).

On adult trees, bud break on the leader shoot has to be observed from the ground with a binocular (Fig. 4a). After felling, their annual height increments are measured to the nearest centimeter between the branch whorls which become visible after three sides of the trunk are worked off and the innermost tree rings revealed (Fig. 4b).

## *3.2 Monitoring Techniques for Intra-annual Wood Formation*

Mäkinen et al. (2008) compared various monitoring techniques for intra-annual wood formation, among them pinning and micro-coring. With the pinning technique, introduced by Wolter (1968) and later applied in a number of studies in various climatic environments, including the tropics, a thin needle (diameter of 1.2 mm) is inserted through the bark into the outer xylem and immediately withdrawn. By this action, a wound is set in the cambium and, as a consequence, wood formation next to the pinning canal stops. Further apart from it, the cambium produces modified cells as a response to wounding. The study tree is either felled and the stem section with the pinnings excised, or the pinning canal with some

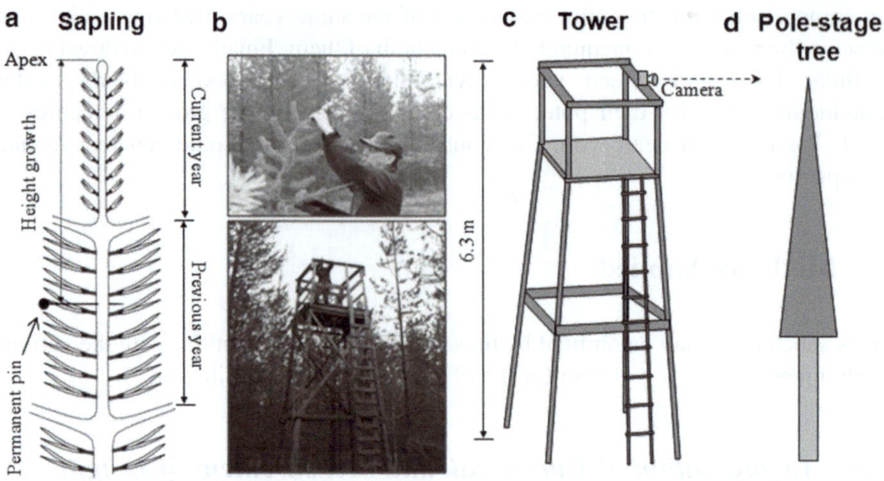


**Fig. 3** Monitoring of intra-annual height growth of saplings (**a**, top of **b**) and of pole-stage trees (bottom of **b** and **c**, **d**) (from Seo et al. 2012a)

**Fig. 4** The entire stem of an old, already round-topped Scots pine (**a**) and the samples prepared for measuring the lengths of the annual shoots, resulting in a 142-year long chronology (**b**); the most recent shoots are already rather short

tissue around is removed from the standing tree using a chisel (Seo et al. 2007). Finally, cross sections ≤18 μm in thickness are cut through the pinning canal on a sliding microtome. By microscopic observation, wood formation since the event of pinning can thus be traced and measured (Fig. 5, left). Pinning can be done easily and quickly in the forest, but the preparation of thin sections and their microscopic evaluation require special skills.

Taking wood samples of considerable sizes directly from living trees goes back to Mariaux (1967–1968) who in this way studied cambium activity in tropical tree species; in the meantime, a diameter of not more than 1.2 mm has been achieved by micro-coring (Rossi et al. 2006) (Fig. 5, right); micro-coring is less labor intensive

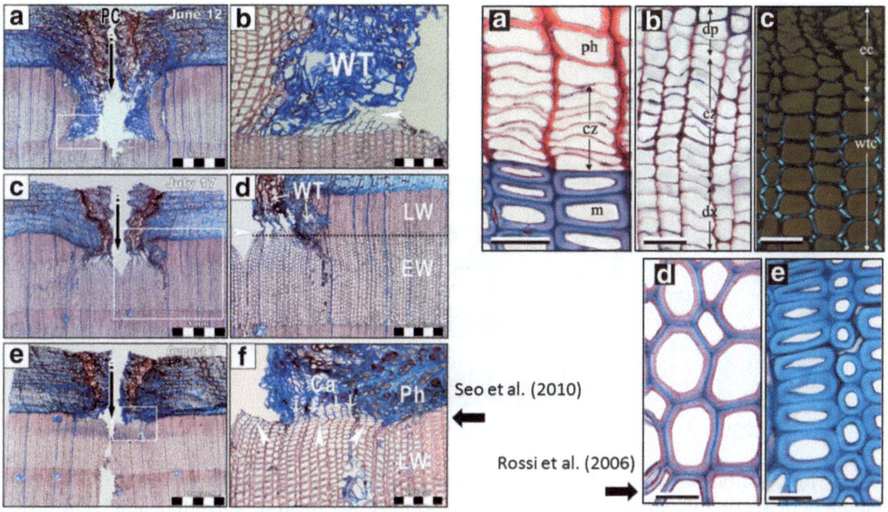


**Fig. 5** Monitoring of intra-annual wood formation using pinning (*left*) or micro-coring (right). *Left*, Scots pine; (**a**, **b**) pinning at June 12, first tracheids (*white arrow head* in **b**) after winter dormancy are visible; (**c**, **d**) pinning at July 17, tracheids (*white arrow head* in **d**) formed just after the transition from early- to latewood (*dotted horizontal line*); (**e**, **f**) pinning at August 7, tracheids formed at the end of wood formation (*white arrow heads* in **f**), their secondary wall is not yet completed. PC and *vertical black arrow*: pinning canal; *WT* wound tissue, *EW* earlywood, *LW* latewood, *Ca* cambium, *Ph* phloem. *Right*, (**A**) transverse thin section of Stone pine in winter; (**B**) cambial zone of Norway spruce (initial cells and derivatives) during the growing season; (**C**) wall thickening and enlarging cells of Norway spruce under polarized light; (**D**) earlywood of European larch with partially lignified cells; (**E**) mature latewood cells of European larch; *m* mature tracheids, *cz* cambial zone, *ph* phloem cells, *dx* developing xylem, *dp* developing phloem, *wtc* wall thickening cells, *ec* enlarging cells; scale bars = 20 μm (from Seo et al. 2010 and Rossi et al. 2006)

than pinning, and the trees have not to be felled; moreover, the reproducibility of its output is more reliable.

Sometimes, intact tissue sampling is still preferred to pinning or micro-coring (Gričar et al. 2007); the sample blocks of 30 × 10 × 10 $mm^3$ contain the inner part of living bark, the cambium, the newly forming xylem, and one fully formed tree ring. Intact tissue sampling is recommended if various development stages of cells in the xylem, cambium, and phloem are to be confidently distinguished.

### *3.3 Sampling of Increment Cores for Conventional Tree-Ring Research*

Standard coring for conventional tree-ring research has often been described, most recently by Speer (2010), and will therefore not be repeated here.

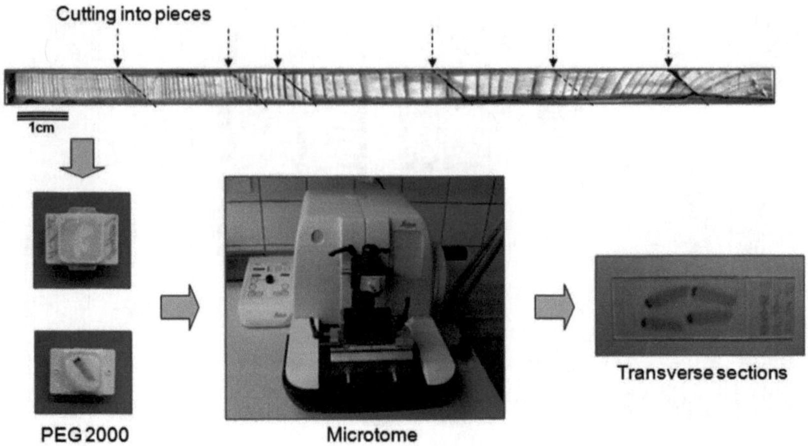


**Fig. 6** Preparation of thin sections from a standard increment core of a diameter of 4–6 mm: obliquely dissecting the core into shorter pieces, embedding them in polyethylene glycol (PEG) 2000, and cutting transverse sections of 7–9 μm in thickness by means of a rotary microtome (from Seo et al. 2012b)

## *3.4 Preparation of Samples for Measuring Variables of Cellular Anatomy*

The standard increment cores are divided into pieces of ≤2.5 cm in length in order to fit in the embedding forms filled with liquid polyethylene glycol (PEG) 2000 (Fig. 6). Then, 7–9 μm thick cross sections are cut with a rotary microtome and stained with aqueous safranin to enhance the contrast between cell walls and lumina. These separating cuts are made obliquely to enable a later, error-free, assembly of the image files, taken by a camera, into a continuous time series of all tree rings. From this series of digital images, various variables of the tracheids, such as cell diameter and cell-wall thickness (Fig. 7), are measured year by year, all in radial direction, using WinCELL™ (Regent Instruments Inc.). These measurements are taken along three (or more) radial transects through each tree ring at a magnification of 100×. The attribution of the tracheids to early- or latewood is made by WinCELL™ following Mork's formula (Mork 1928).

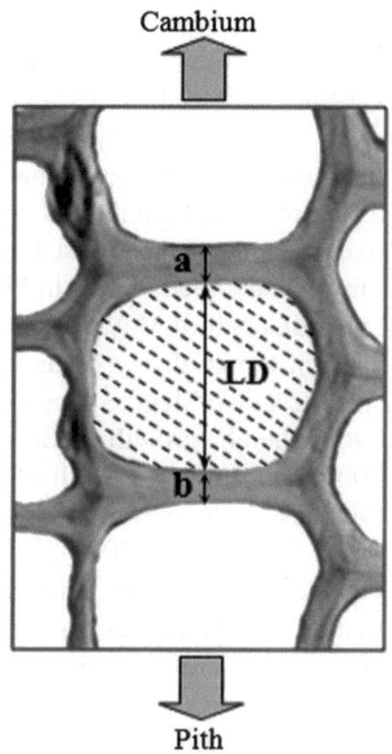


**Fig. 7** Definition of potential cell anatomical variables: lumen diameter (LD), cell-wall thickness (CW = $a/2 + b/2$), cell diameter (CD = $a/2$ + LD + $b/2$), and lumen area LA (hatched) (from Seo et al. 2012b)

# 4 Overall Context, Case Studies, and Implications

## *4.1 Chronological Coherence Between Intra-annual Growth in Height and Girth*

In the boreal zone, bud swelling is the first event of tree growth after winter dormancy, visible to the naked eye. Other phenological events, such as height growth and growth in girth, follow during the year up to the next winter dormancy. The timing of these phases within a given area, essential for the viability of a tree population (Kimmins 1987), is the result of a long-term acclimation of the tree species to the prevailing environmental conditions, particularly to the large diurnal and seasonal temperature fluctuations (e.g., Sarvas 1972; Repo et al. 2000). It is assumed that trees in cold environments generally synchronize their annual phenological development with the average annual temperature cycle (e.g., Heide 1985; Partanen et al. 1998) to protect their meristematic tissues from frost (Häkkinen et al. 1995). A compromise between maximizing the period of photosynthetic activity and minimizing the risk of damage is reached by regulating the onset and end of growth (Linkosalo et al. 2000). This is why favorable weather conditions at the beginning and end of the growing season are not fully used by the trees. In terms of carbon economy, it may be more advantageous to start growing as early as possible rather than to prolong the photosynthetic season further into the autumn (Karlsson 1989) as the trees need enough time, e.g., to complete cell-wall formation and lignification before winter dormancy (Rossi et al. 2006). However, the intricate linkages between the recurring phenological events during the annual cycle are not yet fully explored.

In a case study by Seo et al. (2010), the phenophases of Scots pine were monitored during three climatically different vegetation seasons at two climatically different sites, Laanila and Vanttauskoski, approximately 80 and 300 km, respectively, south of the forest border for Scots pine in northern Finland (see Fig. 2). From the detailed results in Fig. 8, it can be all in all concluded that the buds break in the first half of May. Height growth starts in the second half of May when the heat sum, in terms of degree days, has accumulated to ~5.3 % of the long-term, site-specific sum of degree days (Salminen and Jalkanen 2007). As compared to the cool 1960s, height growth of Scots pine in recent years has accelerated (Pensa et al. 2005) and volume growth of trees has increased (Tomppo et al. 2005). Growth in thickness follows around end of May/early June when the heat sum has reached ~12.5 % (Seo et al. 2008). Bud break and onset of growth in height and girth differ between years, significantly or nearly so, proving a flexible and immediate response to the annually changing temperature. By this capability, the trees take advantage of an above-average warm spring to improve their site dominance (Bailey and Harrington 2006) but to avoid the risk of late frost damage (Hannerz 1999). Growth in height and girth culminate in the second half of June, clearly before the warmest period of the year which is in the second half of July. Height growth finishes by the end of June/early July, when the heat sum has accumulated to ~41 % (Salminen and Jalkanen 2007). Soon after, the earlywood passes into latewood. Growth in thickness ceases by the end of July/mid-August with a heat sum of ~80 % of the long-term heat sum; the variability of this percentage value between years is considerably higher than for the onset of radial growth, thus supporting that for growth cessation temperature is not the only trigger.

In conclusion, tree growth from bud break to the end of radial growth takes 96 days at the southern site and 79 days at the northern site. However, the annual shoot length is independent both from the onset date and the duration of height growth; equally, the tree-ring width is independent from the onset date and the duration of cambium activity.

To preempt any criticism that the height growth was monitored with young trees and the growth in girth with adult trees, a subsequent study on bud break and intra-annual height growth with saplings and pole-stage Scots pines in northern Finland confirmed that both phenophases at the tops of the young and adult trees coincided with each other (Fig. 9) (Seo et al. 2012a). Hence, there is a possibility of transferring observations made on easily accessible saplings to hardly accessible adult trees.

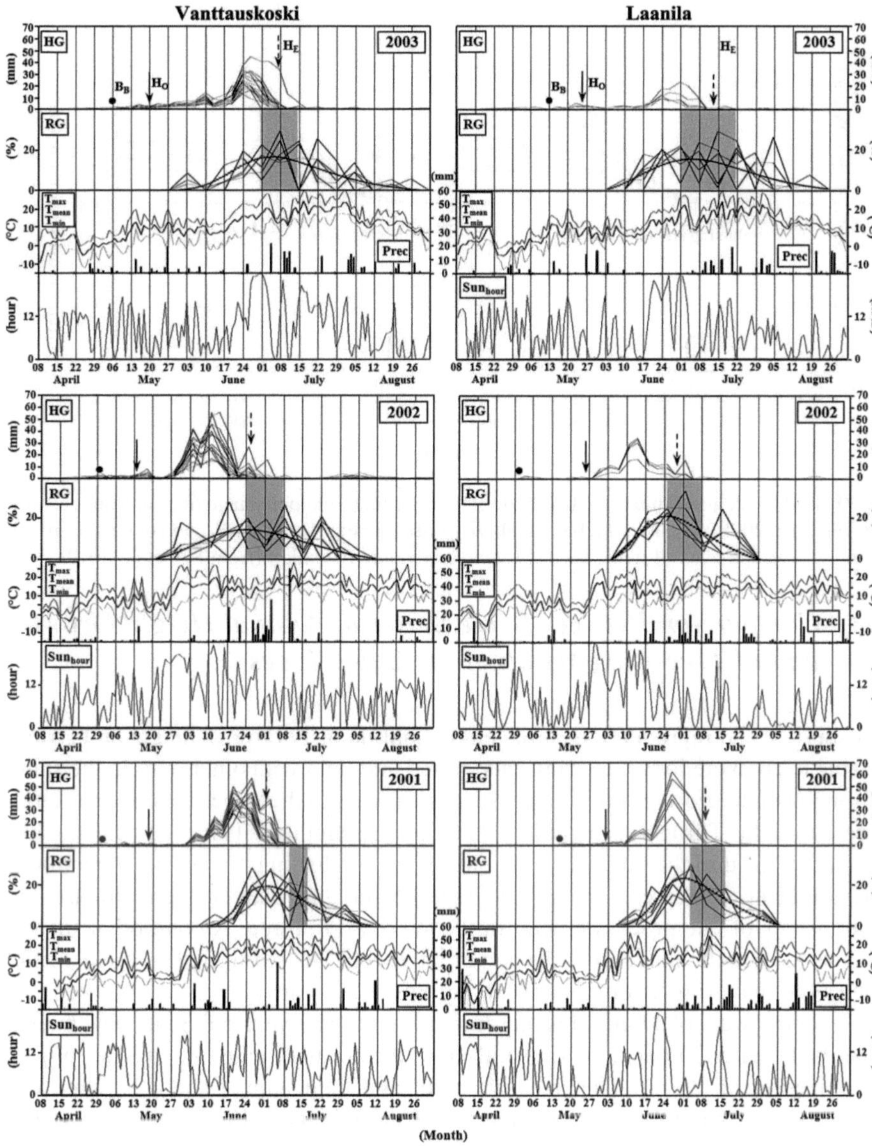


**Fig. 8** Succession of phenological events from bud break to the end of radial growth of Scots pine during three growing seasons at Vanttauskoski and Laanila (see Fig. 2); HG, intra-annual rate of height growth; RG, intra-annual rate of radial growth; daily maximum ($T_{max}$), mean ($T_{mean}$) and minimum ($T_{min}$) temperatures, daily total precipitation (Prec), and daily hours of sunshine ($Sun_{hour}$). $B_B$, median date of bud break; $H_O$, median date when 5 % of annual shoot length were accomplished; $H_E$, median date when 95 % of annual shoot length were accomplished; *shaded background*, transition from early- to latewood (from Seo et al. 2010)

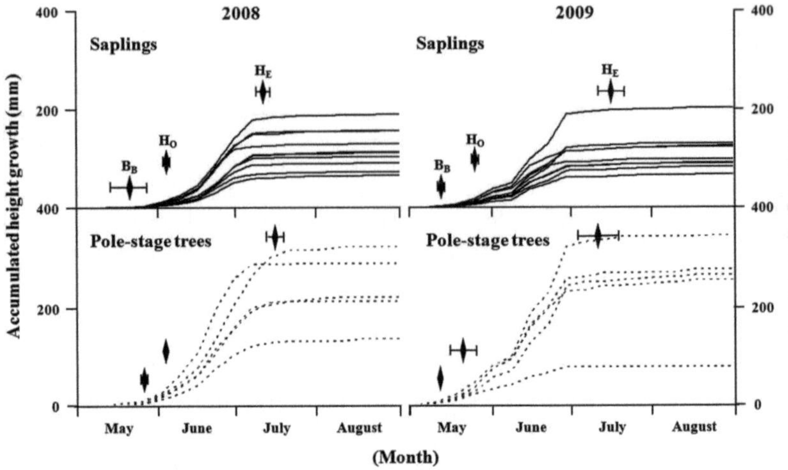


**Fig. 9** Bud break and cumulative height growth of saplings and pole-stage trees at Vanttauskoski (see Fig. 2) in 2008 and 2009. $B_B$, bud break; $H_O$, onset of height growth; $H_E$, end of height growth; diamond, average over all trees; *horizontal bars*, standard deviation (from Seo et al. 2012a)

## *4.2 Comparison Between Time Series of Annual Height Growth and of Tree-Ring Width of the Same Trees*

For survival and reproduction, a tree has to continuously grow in height and girth to be able to compete for space, nutrients, water, and light with other trees in the stand. Through height growth, a tree is moving its crown into the upper canopy for getting access to light. Through growth in girth, a tree strengthens its mechanical stability to stay in an upright position and renews its conducting tissues to maintain its physiological processes. So we see that height and diameter growth are two components of one process crucial to the survival of a tree. Among other factors, such as the genetic makeup or the age of a tree, climate strongly determines which one of the two growth components at any time has priority.

Recently, we looked into this question using ten 150-year-old Scots pine trees in Laanila, northern Finland, 80 km south from the northern pine tree line (see Fig. 2) (Salminen et al. 2009). The trees were cut and their annual shoot lengths were measured [for details, see Fig. 4b and Aalto and Jalkanen (1998)]; moreover, their tree-ring widths were measured from breast height disks. From both variables, chronologies were assembled applying the commonly used dendrochronological techniques and compared with the monthly mean temperatures since 1802, after the considerably high autocorrelation had been removed from the time series (Fig. 10).

Autocorrelation is clearly stronger and longer lasting in tree-ring width series than in height growth series; this means that one warm or one cool summer, respectively, furthers or impedes radial growth over more years than height growth.

It turned out that temperature of the current summer influences the girth growth and of the preceding summer the height growth. That is why height growth

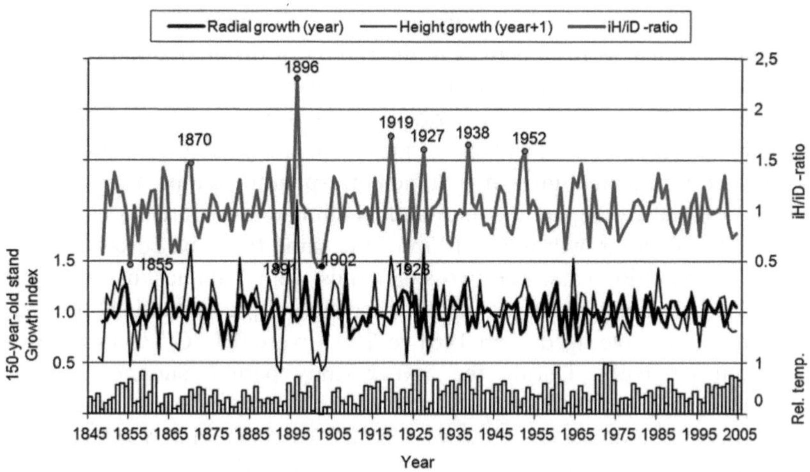


**Fig. 10** Chronologies of tree-ring width and shoot length (autocorrelation eliminated) as well as of the ratio between shoot length and tree-ring width of Scots pine trees (shoot-length chronologies offset by 1 year); summer temperature as bar chart (modified after Salminen et al. 2009)

significantly correlates with radial growth at lag 1 year ($r = +0.23$). There is a clear connection between warm growing seasons and intense growth and between cool growing seasons and slow growth. In years when the ratio between height growth of year $t + 1$ and radial growth of year $t$ is high, i.e., a long annual shoot and a medium-wide or narrow tree ring (example years are 1870, 1896, etc.), the levels of growth and of July temperature are higher than average. When this ratio is low, i.e., a short annual shoot and a medium-wide or narrow tree ring (example years are 1855, 1891, etc.), the levels of growth and of July temperature are lower than average. It can be concluded that height growth is more sensitive to summer temperature variations than radial growth. If this dependence is stable over time, a rising summer temperature would result in an increasing height and slenderness of Scots pine at tree line.

## 4.3 Intra-annual Wood Formation Dynamics

Despite the valuable climatic/environmental information "archived" in the tree-ring width, we should be aware that this information is integrated by a tree over an entire year and even from one or more previous years (Fritts 1976). Therefore, time was ready to aspire to a higher resolution of the annual growth by monitoring tree growth in weekly intervals during several growing seasons. An 8-year intra-annual monitoring of the growth of black spruce in north-eastern Canada by Rossi et al. (2012) is, for now, the culmination; according to the authors, the successive phenological phases of wood formation are statistically closely interconnected by complex cause-and-effect relationships, with the onset of cell differentiation being the main factor, directly or indirectly, triggering all successive phases of xylem maturation; in other words, changes in the timing of differentiation of earlywood can induce changes in the timing of maturation of latewood.

The aims of the case study in northern Finland by Seo et al. (2011) were to identify the onset, dynamics, and end of growth over 5 years from 2000 to 2004, to associate this intra-annual growth with the corresponding temperature and precipitation, and to finally compare the intra-annual climate/growth relationship with the inter-annual climate/growth relationship of the same trees, based on the tree-ring widths over about 40 years. The two study sites were Vanttauskoski and Laanila (see Fig. 2). The annual mean temperature and sum of precipitation at the southern and northern sites, averaged from 1961 to 2004, were 0.1 °C/538 mm and −0.7/448 mm, respectively. During the 5-year study period, summer temperature (May–September) was continuously above the long-term mean (southern site: 11.3 vs. 10.4 °C; northern site: 10.0 vs. 9.2 °C). Also the winters (October–April) were above-average warm, except in 2003. Precipitation in winter, also above average, had been continuously decreasing after a peak value in 2000.

Depending on these actual weather conditions during the 5-year study period, the trees at the southern site start radial growth between end of May and mid-June; earlywood passes into latewood during the first half of July and amounts to about 75 % of the total annual tree-ring width at both sites. Radial growth ends between end of July and mid-August. At the northern site, radial growth starts significantly later and ends slightly, but insignificantly earlier (Fig. 11). Thus, the cambium of Scots pine is active for around 9 weeks at the southern site and 7 weeks at the northern site.

The accumulated intra-annual growth dynamics follows an S-shaped function (Fig. 12). The growth rate at both sites is highest in the second half of June and first half of July; during these 4 weeks, two thirds of the total annual growth is formed; this applies even for the unusually cool and moist summer 1996 at the same two sites (Schmitt et al. 2004). According to Rossi et al. (2008), maximum growth appears to converge towards the summer solstice so that trees can safely complete cell-wall formation before an untimely frost may happen in the early autumn.

The intra-annual tree growth during the 5-year study period was, in view of the two sites, differently associated with summer temperature, mainly positively at the southern site but either negatively or indifferently at the northern site. This short-term episode of contrasting behavior is also reflected in the two tree-ring width site chronologies of the same trees which highly covaried with each other from 1961 to 1999 but then from 2000 to 2004, run off in different directions (Fig. 13). This is the first example that a highly time-resolved vs. a multi-year average climate/growth association can be inconsistent with one another, at least for a short period of a few years.

There is evidence from the early to the recent past that summer temperature is the most important growth factor for boreal forests at high latitudes (e.g., Mikola 1962;

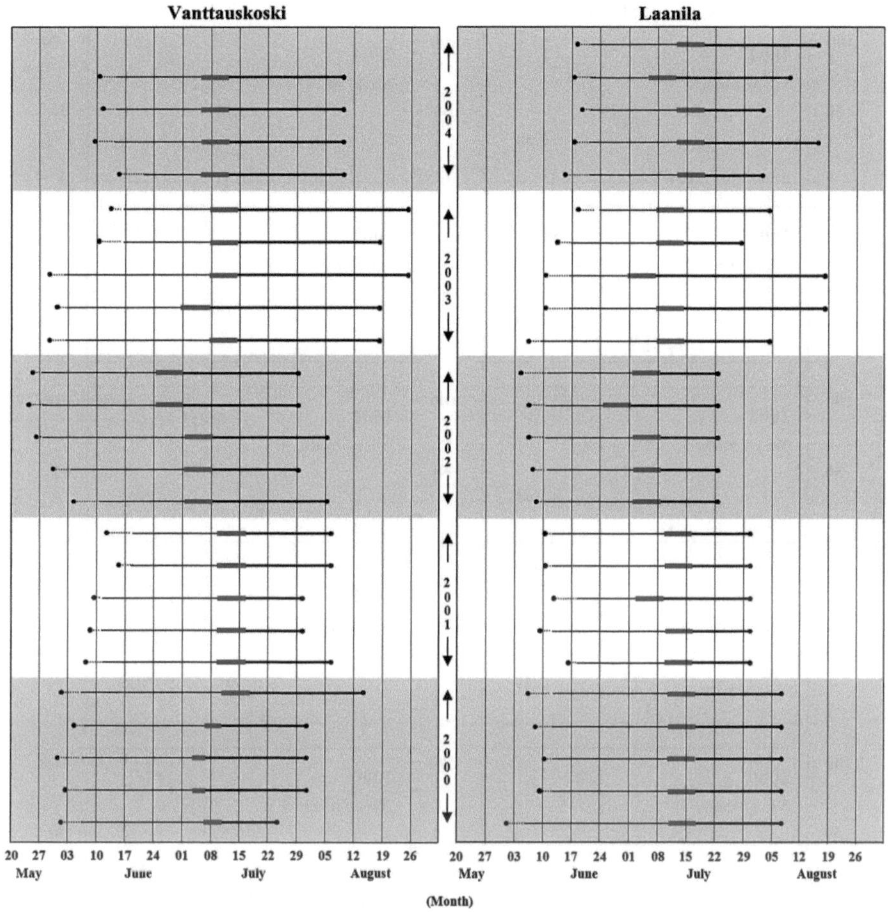


**Fig. 11** Duration of wood formation of Scots pine at Vanttauskoski and Laanila (see Fig. 2) during five vegetation periods, 2000–2004; *dotted line*, estimated range for the onset of wood formation; *thin line*, earlywood formation; *gray thick line*, transition from early- to latewood; *black line*, latewood formation (from Seo et al. 2011)

Kalela-Brundin 1999; Helama et al. 2009; Tuovinen et al. 2009). Against this background, an opposed growth response, as in our study, asks for an explanation. Most likely it was a temperature-induced drought stress resulting from a moisture deficit due to a lack of snow fall in May combined with an above-average warm period from May–July (Seo et al. 2011). A change in climatic control of tree growth due to locally and temporarily operative influences has repeatedly been observed during early forest decline studies (e.g., Eckstein and Krause 1989) and recently by Wilson and Elling (2004), Yonenobu and Eckstein (2006), and Kern et al. (2009).

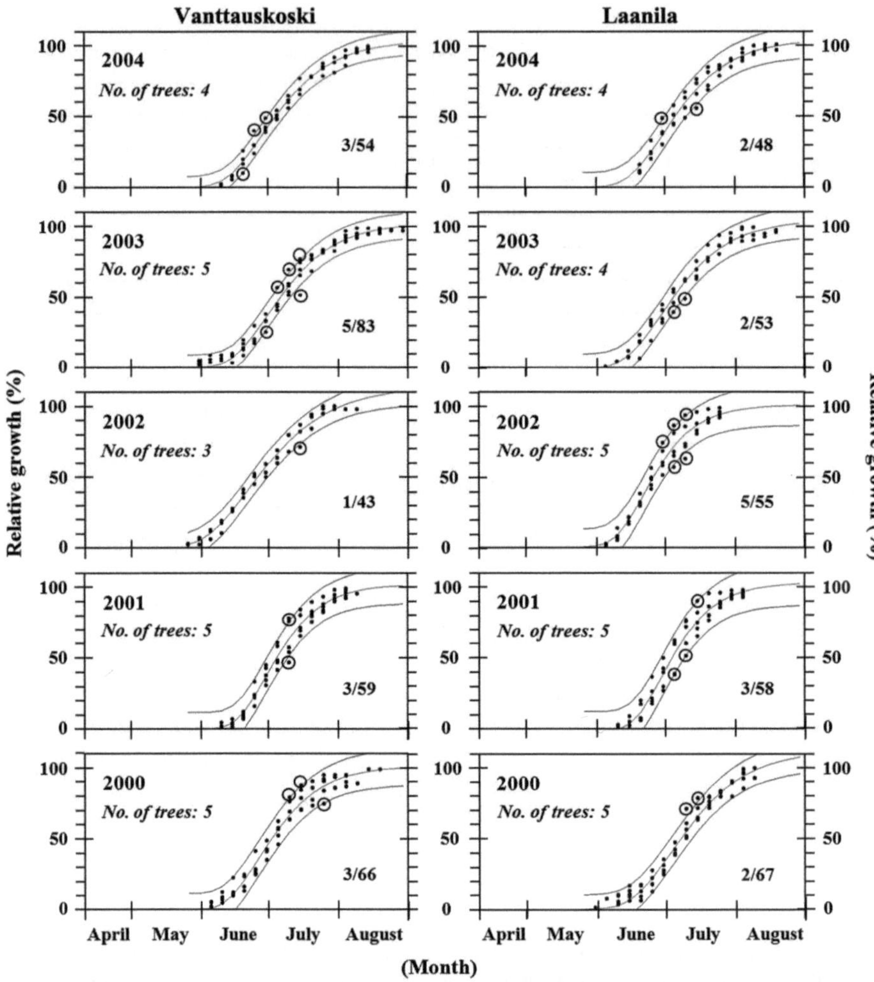


**Fig. 12** S-shaped clouds of data points of intra-annual growth of Scots pine at Vanttauskoski and Laanila during five consecutive growing seasons and superimposed Gompertz functions with their upper and lower 95 % confidence limits; data points outside these limits (*circled*) were defined as outliers, so that they were eliminated from further analysis; for example, 3/54 means three outliers out of 54 data points (from Seo et al. 2011)

A loss of thermal response of tree growth during the recent three decades is preferentially related to high-latitude forests (e.g., Driscoll et al. 2005; Wilmking and Juday 2005), but Carrer and Urbinati (2006) and Oberhuber et al. (2008) observed the same divergence also at Alpine sites; this phenomenon has become popular during the "global warming" discussion as "divergence problem" (e.g., Briffa et al. 1998; D'Arrigo et al. 2008).

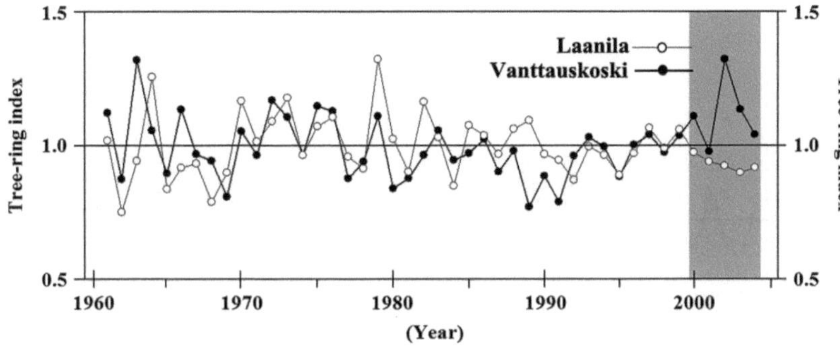


**Fig. 13** Tree-ring index chronologies of Scots pine at Vanttauskoski and Laanila (see Fig. 2) from 1961 to 2004; *shaded background*, monitoring period for intra-annual growth (from Seo et al. 2011)

## 4.4 Cell Anatomical Variables

On the cellular level, wood formation includes cell division, cell enlargement, cell-wall thickening, and lignification (Wodzicki 1971; Plomion et al. 2001; Fromm 2013). The search for wood anatomical variables as environmental proxies, at the beginning largely unsystematic and erratic, can be traced back, at least, by half a century (e.g., Knigge and Schulz 1961; Eckstein and Liese 1975; Vaganov and Terskov 1977). Already 30 years ago, Denne and Dodd (1981) had collected a vast quantity of literature concerned with environmental effects on cell dimensions, based on a confusing variety of experimental results and on a diversity of interpretations.

Due to the increasing capability and efficiency of automatic image analysis systems (e.g., Munro et al. 1996), wood anatomy-related tree-ring research became more and more visible in the 1990s (e.g., Woodcock 1989; von Wilpert 1991; Sass and Eckstein 1992; Antonova and Stasova 1997; García-Gonzáles and Eckstein 2003). Meanwhile, there are numerous publications and a diversity of viewpoints concerned with the effects of environment on wood anatomy (e.g., Wimmer 2002; Schweingruber 2007) so that Fonti et al. (2010) felt the necessity to distill some general trends and perspectives. At present, the water-conducting cells are in the foreground of interest whose arrangement, frequency, length, diameter, wall thickness, and pit size reflect a balance between efficiency and safety of water transport for an optimal tree growth (Fonti and Jansen 2012). The rationale behind is the assumption that trees have to adjust their xylem structure to the ecological setting and to the year-to-year climatic variability in their environment (DeSoto et al. 2011; Eilmann et al. 2011).

The longest chronology for tracheid dimensions, so far, has been assembled for *Larix cajanderi* Mayr. in northeastern Siberia by Panyushkina et al. (2003) who highlighted as an advantage that wood anatomical variables do not show any trends related to the age and size of the trees; the cell chronologies carried a strong signal

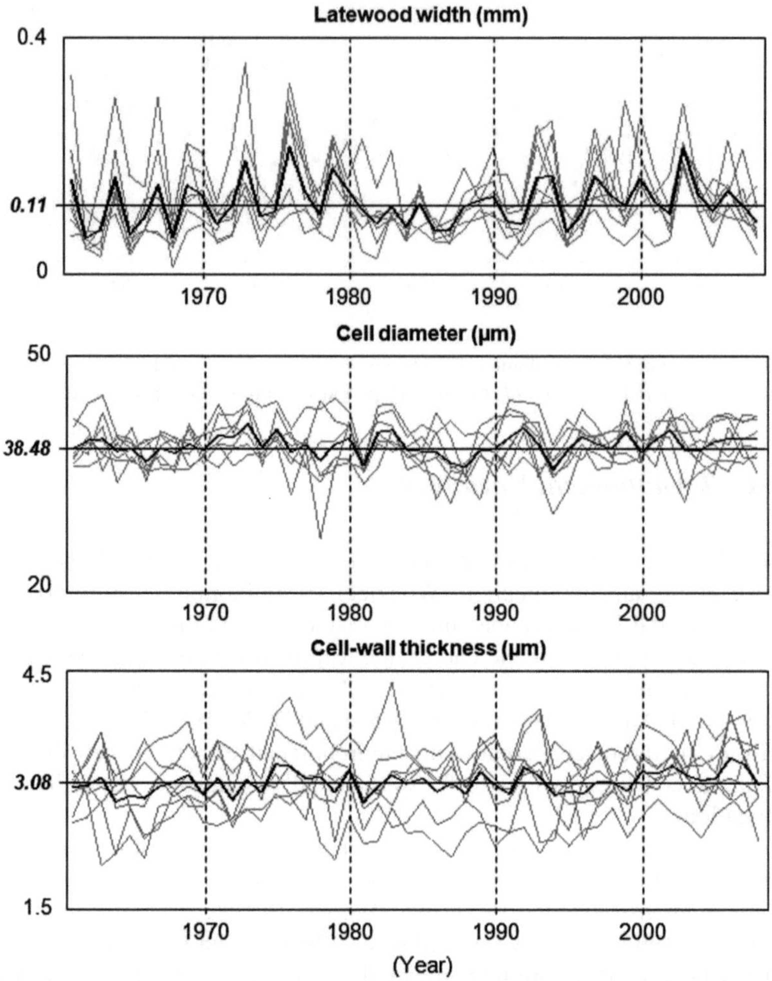


**Fig. 14** Raw measurements of latewood width as well as of cell diameter and cell-wall thickness in the earlywood (mean curves in *bold*) from 1961 to 2008; horizontal lines go through the mean values (from Seo et al. 2012b)

of the growing-season temperature used to reconstruct summer temperature in northeastern Siberia back to 1642.

In an exploratory study on Scots pine at Kevo in subarctic Finland (see Fig. 2) (Seo et al. 2012b), time series for various tree-ring and tracheidal variables were established back to 1961 and qualified for their potential as climate proxies (Fig. 14); it was concluded that the cell anatomical variables are statistically inferior to tree-ring width variables, as already described by Yasue et al. (2000) for *Pinus glehnii* (F. Schmidt) Voss. in northern Hokkaido/Japan and by García-Gonzáles and Fonti (2008) for broadleaved tree species. Nevertheless, it turned out

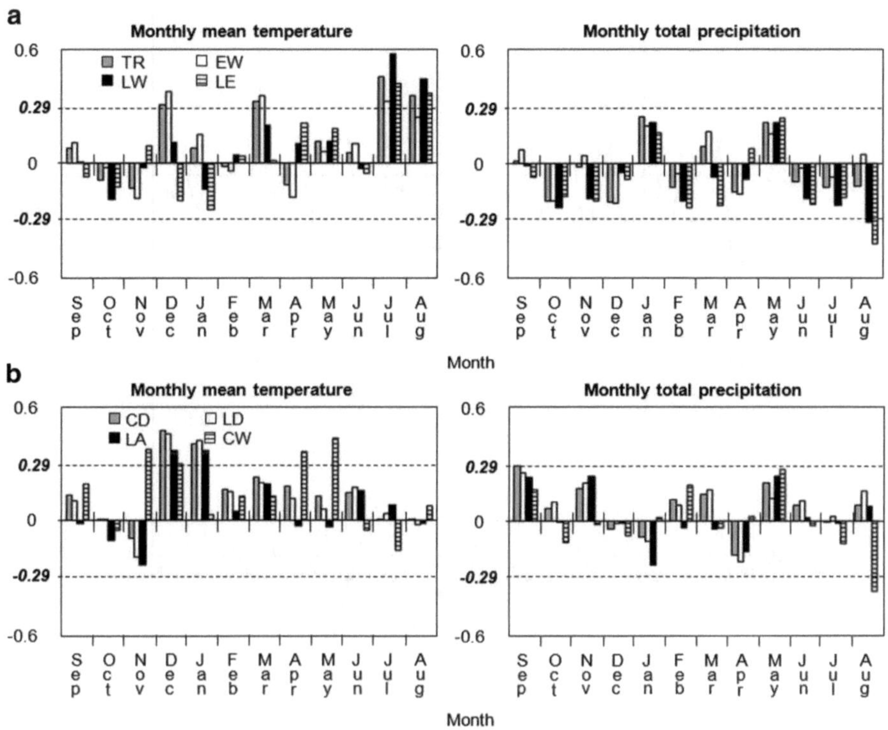


**Fig. 15** Correlation of latewood width (LW) as well as of cell diameter (CD) and cell-wall thickness (CW) in the earlywood with monthly mean temperature and precipitation from 1962–2008; correlation coefficients beyond the *dashed lines* are significant at the 95 % level (from Seo et al. 2012b) (EW earlywood index, LA lumen area, LD lumen diameter, LE late-/earlywood width, TR tree-ring index)

that diameter and wall thickness of the earlywood tracheids are independent from one another and contain different climatic signals (Fig. 15): Cell diameter is associated with the December/January temperature before the onset of cambium activity, and cell-wall thickness weakly correlates with April/May temperature (Fig. 16). This means that cell diameter and wall thickness respond to temperature already months before the earlywood tracheids come into existence. In fact, this is surprising although not unprecedented. Yasue et al. (2000), for example, have described for the cell-wall thickness of the three last-formed rows of tracheids of *P. glehnii* a positive association with temperature in February and March and a negative association with precipitation during the previous November; there are also associations with temperature and rainfall during the current growing period. Also Fonti et al. (2007) have observed that previous fall and spring temperature positively affect the earlywood vessel size of chestnut. The metabolic pathways for such delayed effects have not yet been experimentally fully explored. Temperature may affect wood formation with delay, for example, because of the time needed for synthesis and far-distance transport of hormones (e.g., Oribe et al. 2003;

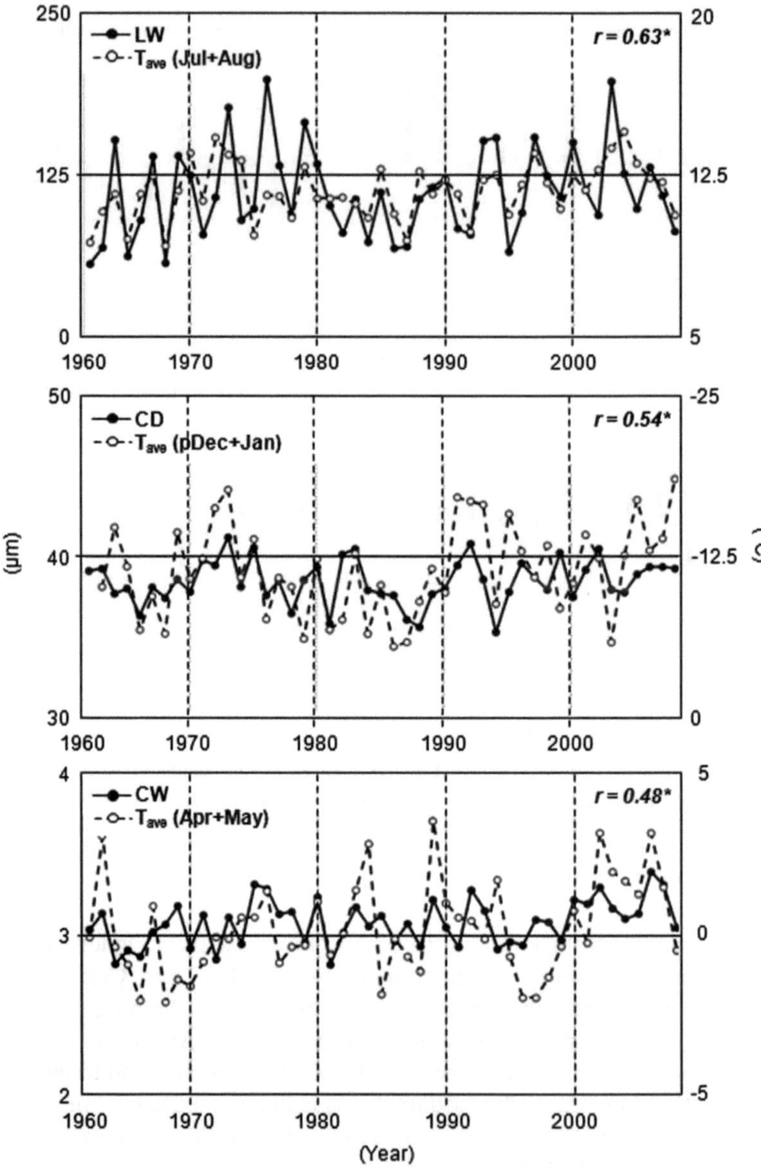


**Fig. 16** Chronologies of latewood width (LW) as well as of cell diameter (CD) and cell-wall thickness (CW) in the earlywood vs. significant temperature variables obtained from the climate/growth association in Fig. 15 (from Seo et al. 2012b)

Uggla et al. 2001). In contrast, water is effective through its influence on the turgidity during cell expansion (e.g., Gindl 2001; Abe et al. 2003) whereby, however, the sensitivity of cell formation to drought is much stronger in early than late summer (Arend and Fromm 2007).

## 4.5 Outlook

This 10-year retrospect describes the momentary state of the art of an ongoing process of research and development in a straightforward set of questions. In which direction could this process continue? There are still gaps of knowledge on the tree level as recently evidenced by Schulte (2012) by considering the length and diameter of tracheids as well as the number of pits per tracheid and the diameter of the pit membranes along the entire pathway from the roots to the needles. On the cellular level, we are even more on uncharted territory. Apart from the already ongoing research on the climatic signals in tracheids and vessels, the varying lignin content in the cell walls (e.g., Gindl 2001) and the variation of the angle of cellulose microfibrils (e.g., Xu et al. 2012b) deserve our attention. Just recently, Olano et al. (2013) came up with a surprisingly new anatomical xylem tissue, ray parenchyma, as climatic proxy. Gartner et al. (2002) have been calling dendrochronologists, wood anatomists, ecophysiologists, and others for "working in multi-disciplinary groups and becoming cross-trained in a wide range of scientific and technological areas." This call is still valid.

**Acknowledgements** The studies were funded by the EU-projects PINE "Predicting Impacts on Natural Ecotones (EVK2 CT-2002-00136) and Millenium (Contract no: 017008) as well as by projects of the German Science Foundation (DFG) (Project nos FR 955/16-1 and WI 2680/2-1) and the Academy of Finland (SA138937).

## References

Aalto T, Jalkanen R (1998) The needle trace method. Finnish Forest Research Institute, Research Papers 681, 36 p

Abe H, Nakai T, Utsuma Y, Kagawa A (2003) Temporal water stress and wood formation in *Cryptomeria japonica*. Tree Physiol 23:859–863

Aloni R (2013) The role of hormones in controlling vascular differentiation. In: Fromm J (ed) Cellular aspects of wood formation. Springer, Heidelberg

Antonova GF, Stasova VV (1997) Effects of environmental factors on wood formation in larch (*Larix sibirica* Ldb) stems. Trees 11:462–468

Arend M, Fromm J (2007) Seasonal change in the drought response of wood cell development in poplar. Tree Physiol 27:985–992

Bailey JD, Harrington CA (2006) Temperature regulation of bud-burst phenology within and among years in a young Douglas-fir (*Pseudotsuga menziesii*) plantation in western Washington, USA. Tree Physiol 26:421–430

Barnett JR (ed) (1981) Xylem cell development. Castle House Publications, Turnbridge Wells

Bauch J (1993) Mineralelementversorgung von Nadelbäumen und ihre Bedeutung für das Wachstum, vol 172. Mitt Bundesforschungsanst Forst-/Holzwirtschaft, Hamburg, pp 75–84

Bäucker E, Bues C-T, Vogel M (1998) Radial growth dynamics of spruce (*Picea abies*) measured by micro-cores. IAWA J 19:301–309

Briffa KR, Bartholin TS, Eckstein D, Jones PD, Karlen W, Schweingruber FH, Zetterberg P (1990) A 1400-year tree-ring record of summer temperature in Fennoscandia. Nature 346:434–439

Briffa KR, Schweingruber FH, Jones PD, Osborn TJ, Shivatov SG, Vaganov EA (1998) Reduced sensitivity of recent tree-growth to temperature at high northern latitudes. Nature 391:678–682

Carlquist S (1975) Ecological strategies of xylem evolution. University of California Press, Berkeley, CA
Carrer M, Urbinati C (2006) Long-term change in the sensitivity of tree-ring growth to climate forcing in *Larix decidua*. New Phytol 170:861–872
D'Arrigo R, Wilson R, Liepert B, Cherubini P (2008) On the 'divergence problem' in northern forests: a review of the tree-ring evidence and possible causes. Glob Planet Change 60:289–305
Denne MP, Dodd RS (1981) The environmental control of xylem differentiation. In: Barnett JR (ed) Xylem cell development. Castle House Publications, Turnbridge Wells, pp 236–255
Deslauriers A, Morin H (2005) Intra-annual tracheid production in balsam fir stems and the effect of meteorological variables. Trees 19:402–408
DeSoto L, De la Cruz M, Fonti P (2011) Intra-annual patterns of tracheid size in the Mediterranean tree *Juniperus thurifera* as an indicator of seasonal water stress. Can J For Res 41:1280–1294
Driscoll W, Wiles GC, D'Arrigo RD, Wilmking M (2005) Divergent tree growth response to recent climatic warming, Lake Clark National Park and Preserve, Alaska. Geophys Res Lett 32:L20703. doi:10.1029/2005GL024258
Eckstein D (1983) Biological basis of dendrochronology, vol 141. Mitt Bundesforschungsanst Forst-/Holzwirtschaft, Hamburg, pp 11–20
Eckstein D, Krause C (1989) Dendroecological studies on spruce trees to monitor environmental changes around Hamburg. IAWA Bull 10:175–182
Eckstein D, Liese W (1975) Veränderungen der Holzstruktur bei rauchgeschädigten Fichten. IX Intern IUFRO-meeting on air pollution and forestry, Mariánská Lázne, Czechoslovakia, 15–18 Oct 1974; Prague 1975, pp 205–214
Eckstein D, Schmidt B (1974) Dendroklimatologische Untersuchungen an Stieleichen aus dem maritimen Klimagebiet Schleswig-Holsteins. Angew Bot 48:371–383
Eckstein D, Frisse E, Quiehl F (1977) Holzanatomische Untersuchungen zum Nachweis anthropogener Einflüsse auf die Umweltbedingungen einer Rotbuche. Angew Bot 51:47–56
Eilmann B, Weber P, Rigling A, Eckstein D (2006) Growth reaction of *Pinus sylvestris* L. and *Quercus pubescens* Willd. to drought years on a dry site in Valais, Switzerland. Dendrochronologia 23:121–132
Eilmann B, Zweifel R, Buchmann N, Graf-Pannatier W, Rigling A (2011) Drought alters timing, quantity, and quality of wood formation in Scots pine. J Exp Bot 62:2763–2771
Esper J, Büntgen U, Timonen M, Frank DC (2012) Variability and extremes of northern Scandinavian summer temperatures over the past two millennia. Glob Planet Change 88–89:1–9
Fonti P, Jansen S (2012) Xylem plasticity in response to climate. New Phytol 195:734–736
Fonti P, Solomonoff N, García-Gonzáles I (2007) Earlywood vessels of *Castanea sativa* record temperature before their formation. New Phytol 173:562–570
Fonti P, von Arx G, García-Gonzáles I, Eilmann B, Sass-Klaassen U, Gärtner H, Eckstein D (2010) Studying global change through investigation of the plastic responses of xylem anatomy in tree rings. New Phytol 185:42–53
Fritts HC (1976) Tree rings and climate. Academic, London
Fromm J (2010) Wood formation of trees in relation to potassium and calcium nutrition. Tree Physiol 30:1140–1147
Fromm J (2013) Xylem development in trees: from cambial divisions to mature wood cells. In: Fromm J (ed) Cellular aspects of wood formation. Springer, Heidelberg
García-Gonzáles I, Eckstein D (2003) Climate signal of earlywood vessels of oak on a maritime site. Tree Physiol 23:497–504
García-Gonzáles I, Fonti P (2008) Ensuring a representative sample of earlywood vessels for dendroecological studies: an example from two ring-porous species. Trees 22:237–244
Gartner BL, Aloni R, Funada R, Lichtfuss-Gautier AN, Roig FA (2002) Clues for dendrochronology from studies of wood structure and function. Dendrochronologia 20:53–61
Gindl W (2001) Cell-wall lignin content related to tracheid dimensions in drought sensitive Austrian pine (*Pinus nigra* Arnold). IAWA J 22:113–120

Gričar J, Zupančič M, Čufar K, Oven P (2007) Wood formation in Norway spruce (*Picea abies*) studied by pinning and intact tissue sampling method. Wood Research 52:1–10

Grudd H (2008) Torneträsk tree-ring width and density AD 500–240: a test of climatic sensitivity and a new 1500-year reconstruction of north Fennoscandian summers. Clim Dyn 31:843–857

Gurskaya MA, Shiyatov SG (2006) Distribution of frost injuries in the wood of conifers. Russ J Ecol 37:7–12

Gurskaya MA, Hallinger M, Eckstein D, Wilmking M (2012) Extreme cold summers in western Siberia, concluded from light-rings in the wood of conifers. Phyton 52:101–119

Häkkinen R, Linkosalo T, Hari P (1995) Methods for combining phenological time series: application to bud burst in birch (*Betula pendula*) in Central Finland for the period 1896–1995. Tree Physiol 15:721–726

Hannerz M (1999) Evaluation of temperature models for predicting bud burst in Norway spruce. Can J For Res 29:9–19

Hartig R (1885) Das Holz der deutschen Nadelwaldbäume. Verlag von Julius Springer, Berlin, 147 p

Heide OM (1985) Physiological aspects of climatic adaptation in plants with special reference to high-latitude environments. In: Kaurin Å, Junttila O, Nilsen J (eds) Plant production in the North. Norwegian University Press, Tromsø, pp 1–12

Helama S, Timonen M, Holopainen J, Ogurtsov MG, Mielikäinen K, Eronen M, Lindholm M, Meriläinen J (2009) Summer temperature variations in Lapland during the medieval warm period and the Little Ice Age relative to natural instability of thermohaline circulation on multi-decadal and multi-centennial scales. J Quat Sci 24:450–456

Hicks S, Eckstein D, Schmitt U, Tuovinen M, Jalkanen R, McCarroll D, Pawellek F, Edouard J-L, Gagen M, Birks H, Serres R, Hyvärinen H, Nivala V (2000) Forest response to environmental stress at timberlines: sensitivity of Northern Alpine and Mediterranean forests to climate. In: European climate science conference, Vienna, Austria, 19–23 Oct 1998, 9 p

Jalkanen R, Tuovinen M (2001) Annual needle production and height growth: better climate predictors than radial growth at treeline? Dendrochronologia 19:39–44

Jalkanen R, Pensa M, Salminen H (2007) Development of Scots pine in the changing environment of the northern boreal zone in Finland. In: Taulavuori E, Taulavuori K (eds) Physiology of northern plants under changing environment. Research Signpost, Kerala, India, pp 271–289

Jost L (1893) Ueber Beziehungen zwischen der Blattentwicklung und der Gefässbildung in der Pflanze. Bot Ztg 51:89–138

Kalela-Brundin M (1999) Climatic information from tree-rings of *Pinus sylvestris* L and a reconstruction of summer temperatures back to AD 1500 in Femundsmarka, eastern Norway, using partial least squares regression (PLS) analysis. Holocene 9:59–77

Karlsson PS (1989) In situ photosynthetic performance of four coexisting dwarf shrubs in relation to light in a subarctic woodland. Funct Ecol 3:481–487

Kern Z, Popa I, Varga Z, Széles É (2009) Degraded temperature sensitivity of a stone pine chronology explained by dendrochemical evidences. Dendrochronologia 27:121–128

Kimmins JP (1987) Forest ecology. Macmillan, New York, pp 130–147

Kirchhefer AJ (2001) Reconstruction of summer temperatures from tree rings of Scots pine (*Pinus sylvestris* L.) in coastal northern Norway. Holocene 11:41–52

Kirdyanov A, Hughes M, Vaganov E, Schweingruber F, Silkin P (2003) The importance of early summer temperature and date of snow melt for tree growth in the Siberian Subarctic. Trees 17:61–69

Knigge W, Schulz H (1961) Einfluss der Jahreswitterung 1959 auf Zellartenverteilung, Faserlänge und Gefäßweite verschiedener Holzarten. Holz Roh-/Werkstoff 19:293–303

Kozlowski TT (1971) Growth and development of trees, vol I and II. Academic, New York

Larson PR (1994) The vascular cambium – development and structure. Springer, Berlin

Liang E, Eckstein D (2006) Light rings in Chinese pine (*Pinus tabulaeformis*) in semiarid areas of north China and their palaeo-climatological potential. New Phytol 171:783–791

Lindholm M, Aalto T, Jalkanen R, Salminen H, Ogurtsov M (2011) The height-increment record of summer temperature extended over the last millennium in Fennoscandia. Holocene 21:319–326

Linkosalo T, Carter TR, Häkkinen R, Hari P (2000) Predicting spring phenology and frost damage risk of *Betula* spp. under climatic warming: a comparison of two models. Tree Physiol 20:1175–1182

Liu B, Li Y, Eckstein D, Zhu L, Dawadi B, Liang E (2013) Has an extending growing season any effect on the growth of Smith fir at the timberline on the southeastern Tibetan Plateau. Trees 27:1432–2285. doi:10.1007/s00468-012-0819-z

Loris K (1981) Dickenwachstum von Zirbe, Fichte und Lärche an der alpinen Waldgrenze/ Patscherkofel. Mitt Forstl Bundesvers Wien 152:417–441

Mäkinen H, Seo J-W, Nöid P, Schmitt U, Jalkanen R (2008) Seasonal dynamics of wood formation: a comparison between pinning, microcoring and dendrometer measurements. Eur J For Res 127:235–245

Mariaux A (1967–1968) Les cernes dans les bois tropicaux africains, nature et périodicité. Bois et Forêts des Tropiques No 113:3–14; No 114:23–37

McCarroll D, Jalkanen R, Hicks S, Tuovinen M, Gagen M, Pawellek F, Eckstein D, Schmitt U, Autio J, Heikkinen O (2003) Multiproxy dendroclimatology: a pilot study in northern Finland. Holocene 13:831–841

McCarroll D, Tuovinen M, Campbell R, Gagen M, Grudd H, Jalkanen R, Loader NJ, Robertson I (2011) A critical evaluation of multi-proxy dendroclimatology in northern Finland. J Quat Sci 26:7–14

Mikola P (1962) Temperature and tree growth near the northern timberline. In: Kozlowski TT (ed) Tree growth. Ronald, New York, pp 265–287

Mork E (1928) Die Qualität des Fichtenholzes unter besonderer Rücksichtnahme auf Schleif- und Papierholz. Der Papier-Fabrikant 26:741–747

Munro MAR, Brown PM, Hughes MK, Garcia EMR (1996) Image-analysis of tracheid dimensions for dendrochronological use. In: Dean JS, Meko DM, Swetnam TW (eds) Tree rings, environment, and humanity. Proceedings of an international conference, Tucson, AZ, 17–21 May 1994. Radiocarbon, Department of Geosciences, University of Arizona, pp 843–851

Nola P (1996) Climatic signal in earlywood and latewood of deciduous oaks in northern Italy. In: Dean JS, Meko DM, Swetnam TW (eds) Tree rings, environment, and humanity. Proceedings of an international conference, Tucson, AZ, 17–21 May 1994. Radiocarbon, Department of Geosciences, University of Arizona, pp 249–258

Oberhuber W, Kofler W, Pfeifer K, Seeber A, Gruber A, Wieser G (2008) Long-term changes in tree-ring/climate relationships at Mt. Patscherkofel (Tyrol, Austria) since the mid-1980s. Trees 22:31–40

Olano JM, Arzac A, García-Cervigón AI, von Arx G, Rozas V (2013) New star on the stage: amount of ray parenchyma in tree rings shows a link to climate. New Phytol 188

Oribe Y, Funada R, Kubo T (2003) Relationships between cambial activity, cell differentiation and the localization of starch in storage tissues around the cambium in locally heated stems of *Abies sachalinensis* (Schmidt) Masters. Trees 17:185–192

Panyushkina P, Hughes M, Vaganov E, Munro M (2003) Summer temperature in northeastern Siberia since 1642 reconstructed from tracheid dimensions and cell numbers of *Larix cajanderi*. Can J For Res 33:1905–1914

Park WK (1990) Development of anatomical tree-ring chronologies from southern Arizona conifers using image analysis. PhD dissertation, University of Arizona, Tucson, AZ

Parker ML, Hennoch WES (1971) The use of Engelmann spruce latewood density for dendrochronological purposes. Can J For Res 1:90–98

Partanen J, Koski V, Hänninen H (1998) Effects of photoperiod and temperature on the timing of bud burst in Norway spruce (*Picea abies*). Tree Physiol 18:811–816

Pensa M, Salminen H, Jalkanen R (2005) A 250-year-long height-increment chronology for *Pinus sylvestris* at the northern coniferous timberline: a novel tool for reconstructing past summer temperature? Dendrochronologia 22:75–81

Plomion C, Leprovost G, Stokes A (2001) Wood formation in trees. Plant Physiol 127:1513–1523

Polge H (1963) L'analyse densitométrique de clichés radiographiques: Une nouvelle méthode de détermination de la texture du bois. Annales de l'Ecole Nationale des Eaux et Forêts de la Station de Recherches et Experiences 20:530–581

Prislan P, Koch G, Čufar K, Gričar J, Schmitt U (2009) Topochemical investigations of cell walls in developing xylem of beech (*Fagus sylvatica* L.). Holzforschung 63:482–490

Repo T, Zhang G, Ryyppö A, Rikala R, Vuorinen M (2000) The relation between growth cessation and frost hardening in Scots pines of different origins. Trees 14:456–464

Rossi S, Deslauriers A, Anfodillo T, Morin H, Saracino A, Motta R, Borghetti R, Borghetti M (2006) Conifers in cold environments synchronize maximum growth rate of tree-ring formation with day length. New Phytol 170:301–310

Rossi S, Deslauriers A, Gričar J, Seo J-W, Rathgeber CBK, Anfodillo T, Morin H, Levanič T, Oven P, Jalkanen R (2008) Critical temperatures for xylogenesis in conifers of cold climates. Glob Ecol Biogeogr 17:696–707

Rossi S, Morin H, Deslauriers A (2012) Causes and correlations in cambium phenology: towards an integrated framework of xylogenesis. Environ Exp Bot 63:2117–2163

Salminen H, Jalkanen R (2004) Does current summer temperature contribute to the final shoot length on *Pinus sylvestris*? A case study at the northern conifer timberline. Dendrochronologia 21:79–84

Salminen H, Jalkanen R (2007) Intra-annual height increment of *Pinus sylvestris* at high latitudes in Finland. Tree Physiol 27:1347–1353

Salminen H, Jalkanen R, Lindholm M (2009) Summer temperature affects the ratio of radial and height growth of Scots pine in northern Finland. Ann For Sci 66:810

Sarvas R (1972) Investigations on the annual cycle of development of forest trees. Active period. Comm Inst For Fenn 76:1–110

Sass U, Eckstein D (1992) The annual vessel area of beech as an ecological indicator. Lundqua Rep 34:281–285

Savidge RA, Barnett JR, Napier R (eds) (2000) Cell and molecular biology of wood formation. BIOS Scientific Publishers, Oxford

Schleser GH, Helle G, Lücke A, Vos H (1999) Isotope signals as climate proxies: the role of transfer functions in the study of terrestrial archives. Quat Sci Rev 18:927–943

Schmitt U, Jalkanen R, Eckstein D (2004) Cambium dynamics of *Pinus sylvestris* and *Betula* spp. in the northern boreal forest in Finland. Silva Fenn 38:167–178

Schulte PJ (2012) Vertical and radial profiles in tracheid characteristics along the trunk of Douglas-fir trees with implications for water transport. Trees 26:421–433

Schweingruber FH (2007) Wood structure and environment. Springer, Berlin

Schweingruber FH, Fritts HC, Bräker OU, Drew LG, Schär E (1978) The x-ray technique as applied to dendroclimatology. Tree Ring Bull 38:61–91

Seo J-W, Eckstein D, Schmitt U (2007) The pinning method: from pinning to data preparation. Dendrochronologia 25:79–86

Seo J-W, Eckstein D, Jalkanen R, Rickebusch S, Schmitt U (2008) Estimating the onset of cambial activity in Scots pine in northern Finland by means of the heat-sum approach. Tree Physiol 28:105–112

Seo J-W, Salminen H, Jalkanen R, Eckstein D (2010) Chronological coherence between intra-annual height and radial growth of Scots pine (*Pinus sylvestris* L.) in the northern boreal zone of Finland. Balt For 16:57–65

Seo J-W, Eckstein D, Jalkanen R, Schmitt U (2011) Climatic control of intra- and inter-annual wood-formation dynamics of Scots pine in northern Finland. Environ Exp Bot 72:422–431

Seo J-W, Aalto T, Jalkanen R, Eckstein D, Schmitt U, Fromm J (2012a) Bud break and intra-annual height growth dynamics of saplings and pole-stage trees of Scots pine: case study for a boreal forest in northern Finland. Balt For 18:144–149

Seo J-W, Eckstein D, Jalkanen R (2012b) Screening various variables of cellular anatomy of Scots pines in subarctic Finland for climatic signals. IAWA J 33:417–429

Sirén G (1961) Skogsgränstallen som indikator för klimafluktuationerna in norra Fennoskandien under historisk tid. Comm Inst For Fenn 53:1–66

Speer JH (2010) Fundamentals of tree-ring research. University of Arizona Press, Tucson, AZ

Tardif J (1996) Earlywood, latewood and total ring width of a ring-porous species (*Fraxinus nigra* Marsh) in relation to climatic and hydrologic factors. In: Dean JS, Meko DM, Swetnam TW (eds) Tree rings, environment, and humanity. Proceedings of an international conference, Tucson, AZ, 17–21 May 1994. Radiocarbon, Department of Geosciences, University of Arizona, pp 315–324

Taulavuori K, Sarala M, Taulavuori E (2010) Growth response of trees to Arctic light environment. Prog Bot 71:157–168

Tomppo E, Tuomainen T, Heikkinen J, Henttonen H, Ihalainen A, Korhonen KT, Mäkelä H, Tonteri T (2005) Lapin metsäkeskuksen alueen metsävarat 1970–2003 [Forest resources in Lapland, 1970–2003]. Metsätieteen aikakauskija 2B:199–287

Tuovinen M, McCarroll D, Grudd H, Jalkanen R, Los S (2009) Spatial and temporal stability of the climatic signal in northern Fennoscandian pine tree-ring width and maximum density. Boreas 38:1–12

Uggla C, Magel E, Moritz T, Sundberg B (2001) Function and dynamics of auxin and carbohydrates during earlywood/latewood transition in Scots pine. Plant Physiol 125:2029–2039

Vaganov EA, Terskov IA (1977) Tree growth analysis by tree-ring structure. Nauka Publishing House, Novosibirsk, USSR (in Russian)

Vaganov EA, Shashkin AV, Sviderskaya IV, Vysotskaya LG (1985) Histometrical analysis of woody plant growth. Nauka Publishing House, Novosibirsk, USSR (in Russian)

Vaganov EA, Naurazhaev MM, Schweingruber FH, Briffa KR, Moell M (1996) An 840-year tree-ring width chronology for Taimir as an indicator of summer temperature changes. Dendrochronologia 14:193–205

Venäläinen A, Tuomenvirta H, Heikinheimo M, Kellomäki S, Peltola H, Strandman H, Väisänen H (2001) Impact of climate change on soil frost under snow cover in a forested landscape. Clim Res 17:63–72

von Wilpert K (1991) Intra-annual variation of radial tracheid diameters as monitor of site specific water stress. Dendrochronologia 9:95–113

Watson E, Luckman BH (2004) Tree-ring-based mass-balance estimates for the past 300 years at Peyto Glacier, Alberta, Canada. Quat Res 62:9–18

Wieser G, Matyssek R, Luzian R, Zwerger P, Pindur P, Oberhuber W, Gruber A (2009) Effects of atmospheric and climate change at the timberline of the Central European Alps. Ann For Sci 66:402

Wilmking M, Juday GP (2005) Longitudinal variation of radial growth of Alaska's northern treeline – recent changes and possible scenarios for the 21st century. Glob Planet Change 47:282–300

Wilmking M, Juday GP, Barber VA, Zald HSJ (2004) Recent climate warming forces opposite growth responses of white spruce at treeline in Alaska through temperature threshold. Glob Change Biol 10:1724–1736

Wilmking M, Sanders TGM, Zhang Y, Kenter S, Holzkämper S, Grittenden PD (2012) Effects of climate, site conditions, and seed quality on recent treeline dynamics in NW Russia: permafrost and lack of reproductive success hamper treeline advance? Ecosystems 15:1053. doi:10.1007/s10021-012-9565-8

Wilson R, Elling W (2004) Temporal instability in tree-growth/climate response in the lower Bavarian Forest region: implications for dendroclimatic reconstruction. Trees 18:19–28

Wimmer R (2002) Wood anatomical features in tree rings as indicators of environmental change. Dendrochronologia 20:21–36

Wodzicki TJ (1971) Mechanism of xylem differentiation in *Pinus sylvestris* L. J Exp Bot 22:670–687

Wolter KE (1968) A new method for marking xylem growth. For Sci 14:102–104
Woodcock DW (1989) Climate sensitivity of wood-anatomical features of bur oak (*Quercus macrocarpa*). Can J For Res 19:639–644
Xu J, Lu J, Bao F, Evans R, Downes GM (2012a) Climate response of cell characteristics in tree rings of *Picea crassifolia*. Holzforschung. doi:10.1515/hf-2011-0144
Xu J, Lu J, Bao F, Evans R, Downes G, Huang R, Zhao Y (2012b) Cellulose microfibril angle variation in *Picea crassifolia* tree rings improves climate signals on the Tibetan Plateau. Trees 26:1007–1016
Yasue K, Funada R, Kobayashi O, Ohtani J (2000) The effects of tracheid dimensions on variations in maximum density of *Picea glehnii* and relationships to climatic factors. Trees 14:223–229
Yonenobu H, Eckstein D (2006) Reconstruction of early spring temperature for central Japan from the tree-ring widths of Hinoki cypress and its verification by other proxy records. Geophys Res Lett 33:L10701. doi:10.1029/2006GL026170
Zahner R (1963) Internal moisture stress and wood formation in conifers. For Prod J 13:240–247
Zimmermann MH (1964) The formation of wood in forest trees. Academic, New York

# Wood Formation Under Drought Stress and Salinity

**Silke Lautner**

**Abstract** As our environment changes, salinity and drought are becoming increasingly significant abiotic stress factors. Salinity reduces the ability to take up water, and this in turn causes alterations in wood formation; changes that are very often found to correspond closely to those caused by water deficiency. For example, both abiotic stress factors lead to a reduction in the extent of year ring increment, and they also affect xylem element architecture, leading to alterations in the hydraulic properties, as well as the chemical composition of the woody body. The intensity of the response is found to be dependent not only on the intensity of the stress but also on tree species, intraspecific variety, and even on provenances.

## 1 Drought and Salinity Influence Wood Formation as a Result of a Changing Environment

Since the last ice age, 10,000 years ago, changing climate and temperatures have greatly influenced the world's forests, while human activity has also had an increasing impact. There is evidence that recent anthropogenic acceleration of climate change has profoundly affected forest systems all over the world. In particular, desertification is one of the world's most alarming processes of environmental degradation. Today, it affects about two-thirds of the countries of the world and more than one-third of the earth's surface (more than four billion hectares; FAO 2007). Existing forest stands may survive these changing climate conditions for some time, but long-term responses might depend on the capability of species to adapt. This capability can be determined by the variation within and between species in their physiological responses to changes such as water deficiency and

S. Lautner (✉)
Department of Wood Science, Universität Hamburg, Leuschnerstrasse 91d, Hamburg 21031, Germany
e-mail: silke.lautner@uni-hamburg.de

J. Fromm (ed.), *Cellular Aspects of Wood Formation*, Plant Cell Monographs 20, DOI 10.1007/978-3-642-36491-4_7, © Springer-Verlag Berlin Heidelberg 2013 

salinity. Also, in many regions, agricultural irrigation results in aquifer depletion, leading to changes in the quality of groundwater, especially through increasing mineral content. Increased nutrient leaching and soil salinity affects not only agricultural land, however, but also forests. Soil salinity is characterised by a high concentration of soluble salts, and more than 800 million hectares of land worldwide are salt affected, accounting for more than 6 % of the world's total land area (Munns and Tester 2008).

Since salinity a priori reduces the ability of plants to take up water, early physiological responses to drought stress and salinity have much in common (Munns 2002). Independent of the volume changes that results from loss of water, plants try to osmotically adjust by net increase in solute concentration per cell, mainly by increases in concentration of various common solutes, including sugars, organic acids, amino acids, and inorganic ions. Osmotic adjustment promotes dehydration tolerance and occurs in roots and in leaves, but with salinity, also secondary events can be observed that inhibit plant function, arising from disruption of cell membrane integrity and cell metabolism. Of course, these effects are also applicable to trees, such as the lowering of the osmotic potential in roots in order to ensure water uptake, as well as the compartmentalisation of minerals in the vacuole or the apoplast region. Osmotic stresses derived from drought and salinity, therefore, reduce tree growth and lead to salt accumulation in different plant tissues (Chen and Polle 2010). Being perennial plants, however, trees also have to form wood in order to sustain their stability and facilitate water transport from root to shoot, despite any abiotic stress factors in place. The effects of drought stress and salinity in trees, therefore, extend to the physiological processes in the wood-forming meristem: the cambium. Hence, the xylem originating from a drought or salinity affected meristem also exhibits modifications in its anatomical and chemical aspects.

Compared to other crop plants, however, our knowledge of tree growth under abiotic stress factors is still somewhat limited, although a growing awareness of the role of trees and forests in processes such as carbon sequestration, environmental protection in agricultural shelter-belts or phytoremediation, to name but a few, and the role of wood as an energy resource and raw material for multiple utilisations, is increasingly turning the focus of research onto tree growth and wood formation. This chapter, therefore, presents the findings of recent research into the effects of drought stress and salinity on wood formation and the resulting xylem.

## 2 Impact on the Cambial Zone

Since xylem cell expansion is a turgor-driven process depending on cellular water uptake and on solute accumulation (Langer et al. 2002), drought stress can affect wood growth directly through its implications for the cambial meristem and the developing wood cells. Investigations on trees exposed to drought have confirmed loss of turgor pressure in the expanding cambial cell derivatives (Dünisch and

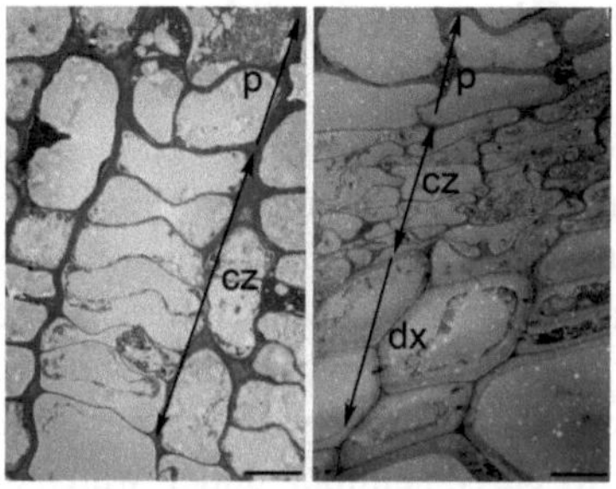


**Fig. 1** Transmission electron microscopy images of a stem cross section of young *Populus* × *canescens* tress grown under different exposure to salinity. *Left*: cambial zone of a control tree showing about eight cambial cells in radial direction which are highly vacuolated and form a consolidated meristematic tissue. *Right*: cambial zone is strongly reduced after 2 weeks of exposure to 75 mM NaCl, cambial tissue compound seems disintegrated in comparison to control tissue; secondary cell wall formation starts in close proximity to cambial zone. *p* phloem, *cz* cambial zone, *dx* developing xylem; bars represent 25 μm

Bauch 1994; Abe et al. 2003; Abe and Nakai 1999) as well as influences on cell wall metabolism, as shown on isolated stem tissue from pine (Whitmore and Zahner 1967). In poplar species, it has been shown that drought implies a significantly decreased concentration of osmotically active solutes in the cambial zone, which are closely related to predawn leaf water potentials (Arend and Fromm 2007). Microscopic analysis of the cambial zone has revealed a seasonal codependency of drought stress on the number of cambial cell derivatives, with only one or two enlarging cambial derivatives in early summer and none in late summer. No distinction could be detected, however, in the radial appearance of the undifferentiated cambial cells during drought stress compared to control trees during the growth season. Contrary to these findings in poplar, research in saplings of the Mediterranean pine species *Pinus halepensis* has revealed a significant and positive effect of irrigation on the number of cambial cells in spring (de Luis et al. 2011). This effect, however, levelled off during mid-season only to be proven effective again in late season. Closely related to the cambial dimension, the number of expanding tracheids decreased in water-deficient conditions during the spring months and remained at a very low level throughout the whole season if the water stress was maintained.

In saline conditions, the cambial zone also experiences significant changes. During the active growth period, trees exposed to salt stress form a reduced cambial zone, appearing also rather disorganised compared to the well-developed cambial zone in control trees (Fig. 1). Moreover, the cytoplasm of the cambial cells altered from being highly vacuolated towards exhibiting multiple smaller vacuoles, suggesting a shift in the osmotic balance within the wood-forming cells (Escalante-Perez et al. 2009).

## 3 Impact on Xylem Element Development

Anatomical analysis of xylem fibre and vessel elements of poplar has revealed a distinct alteration under drought stress conditions (Arend and Fromm 2007; Cocozza et al. 2011). Both the cross-sectional area of fibres, as well as their length, decreased in young poplar shoots which were exposed to drought in early summer, whereas, in late summer, no clear changes could be detected compared to control plants (Arend and Fromm 2007). Similarly, the vessel lumen area was only found to be significantly reduced in early summer under water-deficient conditions. This reduction was compensated for by an increased number of vessel elements, keeping the overall vessel-area: fibre-area ratio comparable with the values found in control trees. This compensatory increase in the number of vessel elements when trees were exposed to drought diminished towards the end of the growing season. The reduced vessel diameter under drought conditions in early summer can be explained in view of the fact that cell expansion is a turgor-driven process, and thus, the reduced volume of xylem element formation is linked to lower concentrations of osmotically active solutes in the xylem forming zone (Fereres et al. 1978; Meyer and Boyer 1972). The fact that this effect was overcome in late summer points to factors other than osmotic potential becoming more dominant, since well-watered trees also reveal a tendency towards reduction of vessel diameter during latewood formation (Arend and Fromm 2007).

In a similar way, the length of the vessel elements formed under drought stress was found to decrease slightly in early summer but to regain similar values to those of the control trees in late summer. Similar results concerning reduced vessel size and increased vessel density under water stress conditions have also been reported in different oak species (Villar-Salvador et al. 1997; Garcia-Gonzalez and Eckstein 2003; Corcuera et al. 2004; Eilmann et al. 2006; Gea-Izquierdo et al. 2012) and olive tree (Bacelar et al. 2007). Anatomical analyses of *Rhamnus* species have also revealed intraspecific variations in provision against drought conditions, with wood traits including elevated vessel frequency, smaller vessel diameter, as well as imperforate tracheary elements with spiral thickenings; the variety of these measures and their possible combination points to a potentially versatile adaptation under drought stress (Baas and Schweingruber 1987; Carlquist and Hoekman 1985). The effects of water deficiency on xylem element formation can also be detected in gymnosperm trees. Analysis of various pine species grown under different water regimes has revealed an increase in fibre lumen diameter under water deficiency compared to control trees (Eilmann et al. 2009, 2011; Esteban et al. 2012; Maherali and DeLucia 2000). These findings reveal a tendency towards the formation of an optimised water-conducting system in tandem with restricted year ring formation.

In contrast to this theory, however, other investigations have reported increasing lumen area along with increasing water supply (Nicholls and Waring 1977; Sheriff and Whitehead 1984; Sterck et al. 2008). Under water deficiency, reduction in the lumen area of water-conducting tracheids hence results in a reduction of stem water conductivity and, along with an observed decline in leaf area under drought conditions, favours the hypothesis that trees maintain a homeostatic water pressure gradient (Sterck et al. 2008).

Taking both theories into consideration, it seems that tracheid lumen area formation under water deficiency is a physiological process balancing mechanical support requirements and water stress resistance while at the same time also being strongly influenced by other environmental growth conditions, as well as by the region of provenance of the respective species (Esteban et al. 2012; Eilmann et al. 2011). Tracheid lumen diameter is also positively correlated with crossfield pit diameter and, hence, may have a direct impact on the cell physiology of living xylem elements, such as ray parenchyma cells and the epithelial cells of the resin ducts. Indeed, both ray frequency and radial resin duct frequency are elevated under drought conditions (Esteban et al. 2012; Rigling et al. 2003). While cumulative resin ducts are most often formed locally after mechanical damage or fungal or herbivore attack (when they are referred to as traumatic resin ducts), the function of enhanced resin duct formation under water deficiency is not yet clear. Presumably, such enhanced resin ducts constitute a preformed defence system serving to protect against attacks or diseases that are more likely to occur on trees weakened by previous stresses. Apart from the elevated occurrence of resin ducts, equipped with resin-producing epithelial cells, an elevated amount of resin pockets can also be observed under water-limited growth conditions (Frey-Wissling 1942; Seifert et al. 2010; Clifton 1969; Cown 1973). Interestingly, in pine trees grown under severe drought stress, increased formation of surface shelling fractures has been observed in the outer part of the stem (Donaldson 2002). These shelling fissures could be the anatomical precondition to the formation of resin pockets. Furthermore, tracheid cells along the fracture surfaces have been shown to be brighter than adjacent intact wood, pointing towards an altered lignin feature. Indeed, microscopic analysis of pine trees grown under drought stress has revealed significantly altered lignin deposition within the xylem cell walls (Donaldson 2002). In particular, the middle lamella is only poorly lignified causing deteriorated adhesion in the tracheids and resulting in intercellular checking, especially on radial cell walls. Within the cell wall structure itself, the $S_1$ region remains uniformly lignified but, in the $S_2$ layer, concentric layers of variable lignification intensities can be found, resulting in either increased or strongly decreased lignifications in a confined space. In fact, cell wall lignifications in the $S_2$ layer of tracheids can be reduced down to levels that cause their collapse. The terminal $S_3$ region, however, does not seem to be greatly affected by the drought-induced alterations in lignin deposition, but instead, occasionally, an increased level of development can be detected, possibly in order to compensate for the $S_2$ reductions by providing additional compression strength (Donaldson 2002).

Apart from drought, salinity also impairs tree water status and hence may increase tension in the water-conducting system, promoting cavitations and subsequent embolisms. Similar to angiosperm trees exposed to drought, vessel diameter decreases when trees are exposed to salinity (Fig. 2; Baas et al. 1983; Escalante-Perez et al. 2009; Junghans et al. 2006). At the same time, vessel

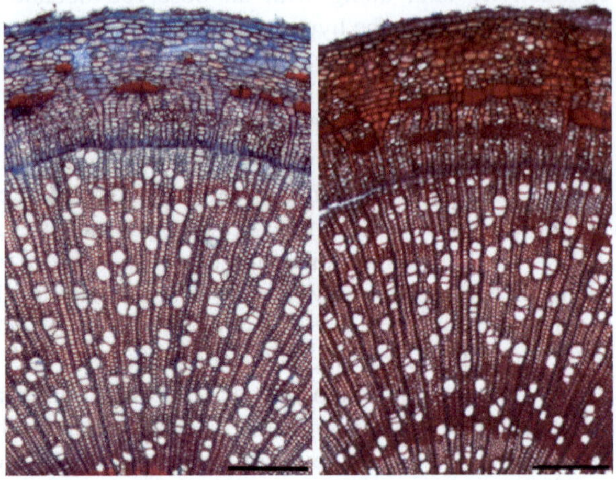

**Fig. 2** Salinity-induced changes in xylem anatomy of *P.* × *canescens* (for growth conditions see Janz et al. 2012). *Left*: stem cross section of control tree. Right: tree grown for two weeks under gradually increasing salt regimes (up to 100 mM NaCl) shows reduced vessel size but increased vessel number. Sections were stained with safranin and astra blue; bars represent 250 μm

numbers per area increased, so that the overall water conductivity remained more or less unaffected (Fig. 3; Janz et al. 2012). As a further method to prevent against vessel collapse under osmotic stress, trees reinforce the cell wall strength of conducting cells (Hacke and Sperry 2001). This phenomenon is also found in poplar vessel cell walls, which show a significant increase in strength when exposed to salinity (Junghans et al. 2006). The extent of these changes in vessel structure may, however, be subject to intraspecific variations. Within the genus poplar, for example, salt-sensitive species react significantly more quickly and in a more pronounced way to salt exposure than a salt-tolerant species like *P. euphratica*, which exhibit only minor anatomical changes even under severe salinity (Chen and Polle 2010; Janz et al. 2012). Interestingly, a similar effect occurs in salt-adapted mangrove species but not in trees of the same species in non-mangrove environments (Janssonius 1950). Anatomical changes in angiosperms upon salt exposure seem predominately to affect the vessel system, whereas the formation of fibres and ray cells do not seem to be altered under these stress conditions (Janz et al. 2012; Escalante-Perez et al. 2009).

In gymnosperm trees, air seeding in the water-conducting xylem elements depends on the mechanical properties of the surrounding tracheid walls. Hence, in order to increase cavitation resistance, cell walls have to thicken, which again leads to an increased wood density (Hacke et al. 2001). Accordingly, increased wood density is found in xylem formed under salinity stress (Stiller 2009). It is not only increased xylem density that occurs under salinity but also enhanced reduction of tracheid fibre length can be observed in various gymnosperm species in increasingly saline conditions (Khamis and Hammad 2007).

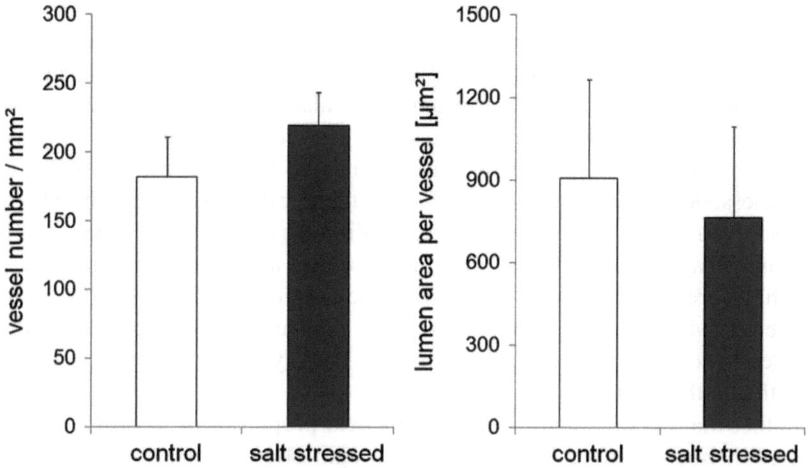


**Fig. 3** Salinity-induced changes in xylem anatomy of *P.* × *canescens* (for growth conditions see Janz et al. 2012). Calculated average vessel lumen area (n>380) significantly decreased under salinity (*right*), whereas the vessel density clearly increased under salt stress (*left*)

# 4 Impact on Wood Growth and Wood Mechanical Properties

Depending on their sensitivity to saline and drought conditions, trees generally reveal a distinct reduction in radial growth. In hardwood tree species, ring width chronologies gained from core analysis of mature oak trees grown under different climatic conditions display a reduction in radial growth under xeric conditions (Gea-Izquierdo et al. 2012). This decline in year ring widths is also accounted for by the shortened growth period under drought stress, when cessation of xylem cell formation has been detected up to five weeks earlier than in irrigated trees (Eilmann et al. 2011; Dünisch and Bauch 1994). Concomitantly, a shift in earlywood/latewood transition has also been detected, starting two to four weeks earlier under drought stress than in irrigated trees (Eilmann et al. 2011; Dünisch and Bauch 1994; Whitmore and Zahner 1967). Even despite a climate change-related extension of the growth period in the Northern Hemisphere, mainly reflected in an earlier onset of spring events (Menzel and Fabian 1999; Menzel et al. 2006), but also in prolonged warm autumns, annual xylem increments still do not necessarily increase. This is most likely based on accompanying drought stress and the consequent effect on xylem differentiation during summer (Eilmann et al. 2011; Dunn et al. 2007) but may also be due to an earlier cessation of carbon uptake in autumn in water-deficient conditions (Piao et al. 2008). A distinct reduction in wood biomass production has also been detected in ring width analysis of different poplar clones grown under water deficiency (Cocozza et al. 2011).

A major aspect in xylem cell wall formation is the angle cellulose microfibrils form in the vertical direction in the secondary cell wall layer ($S_2$). This microfibril angle (MFA) largely determines key mechanical properties, such as modulus of elasticity (MOE) and shrinkage behaviour (Cave and Walker 1994; Harris and Meylan 1965). Apart from the general variation of the MFAs within a growth period, which tend towards higher angles during earlywood formation and lower angles during latewood formation, water stress also strongly influences the MFA. Drought exposure has been shown to result in a reduced MFA in earlywood xylem cells, which show only a 10° to 14° angle compared to 12° to 19° in control trees. Nonetheless, developing xylem cells retained their ability to increase the MFA as a response to irrigation after drought stress (Wimmer et al. 2002). It has also been shown in softwood tracheids that a lower microfibril angle results in greater tensile strength, whereas a greater angle improves the elasticity of the wood (Watson and Dadswell 1964; Mark and Gillis 1973). The changes in MFA orientation due to water deficiency will, therefore, also influence the key mechanical properties of the xylem that is then built.

Another major value defining the mechanical properties of wood is its density. Changes in tracheid fibre lumina under drought stress, as described in Sect. 3, ultimately alter wood density. Larger lumen diameter along with thinner cell walls leads to a reduction in wood density, which might not only influence the physiological processes in the standing tree but also persist in the woody body when it is utilised, thus affecting the biomechanical properties and technological utilisation of the resulting timber (Irvine and Grace 1997; Holtta et al. 2002). Intra-annual fluctuations in xylem density lead to the occurrence of false growth rings, and these have frequently been reported for both angiosperm as well as gymnosperm tree species as a result of environmental stress situations, especially drought stress, but also salinity (Glerum 1970; Hoffer and Tardif 2009; Battipaglia et al. 2010; Chen and Polle 2010; Marchand and Filion 2012; Palakit et al. 2012; Liphschitz and Waisel 1970). The formation of false growth rings, their occurrence across stands, and also their location within single year rings thereby reveal characteristic variations subject to species, substrate type, and climatic conditions such as precipitation and temperature (Rigling et al. 2001, 2002).

## 5 Impact on the Hydraulic System

One of the major traits in the response of woody plants to drought is their vulnerability to xylem embolism. Comparison of six deciduous angiosperm species with six evergreen angiosperm species growing on the same site in a dry karst forest has revealed an almost three times higher amount of vessels in the sapwood of the evergreen species compared to the deciduous species (Fu et al. 2012). Moderate salt stress results in hydraulic adaptation through increased vessel frequency along with a decrease in vessel lumina (Janz et al. 2012). According to the Hagen Poiseuille law (Tyree and Zimmermann 2002), xylem volume flow increases by the power of four with increasing water-conducting cell lumen areas. Hence, increasing tracheid lumen areas, as found in pine under drought conditions (see Sect. 3), may be able to compensate for the overall reduction in the conductive area due to the decreased radial increment. This increase in water transport efficiency, however, has the concurrent effect of impairing its safety, since liability to cavitations is also increased with the enlargement of lumen diameter. This susceptibility is further enhanced by a reduction of tracheid cell wall thickness, as this frequently coincides with water deficiency and is supposed to be caused by the reduced carbohydrate availability under drought stress (Esteban et al. 2012; Garcia Esteban et al. 2010; Eilmann et al. 2011). Intertracheal pits are another major influence on conductivity (Sperry et al. 2006; Choat et al. 2008), and positive correlations between lumen size and bordered pit diameter show the strong linkage between those two factors (Hacke et al. 2004). Accordingly, a drought-induced increase in tracheid lumina comes along with an increase in bordered pit diameter (Esteban et al. 2012). In angiosperm trees changes in water conducting vessel formation appear to be an adaptive response to drought stress as well as to salinity, since smaller vessels are less susceptible to xylem embolism provoked by stress (Sperry and Saliendra 1994; Hacke and Sauter 1996; Logullo et al. 1995). The formation of smaller vessels not only helps to prevent xylem embolism but also contributes to controlling the regulation of water flow under drought conditions (Lovisolo and Schubert 1998). The increase in vessel numbers under water-limited growth conditions compensates for vessel size reduction, just as it does under saline conditions, leaving the xylem: vessel-area ratio largely unaffected (February et al. 1995; Searson et al. 2004; Lovisolo and Schubert 1998; Arend and Fromm 2007; Janz et al. 2012).

## 6 Impact on Wood Chemistry

The molecular wood composition of poplar species after being exposed to different concentrations of salinity stress has been investigated in the developing xylem using Fourier transform infrared spectroscopy-attenuated total reflection (FTIR-ATR), revealing stress-induced alterations in major wood compounds such as cellulose, hemicelluloses, lignin, and proteins (Janz et al. 2012). The molecular changes in cell wall formation under salinity are particularly characterised by changes in the lignin: hemicellulose ratio, as well as significant changes in the lignin: cellulose ratio. Since these results apply to salt-sensitive poplar species just as much as to salt-tolerant poplar species, it can be assumed that the chemical composition of the wood is not influenced by osmotic changes and also occurs separately from anatomical changes. It has also been found that drought likewise influences molecular cell wall composition in poplar. Lignin concentrations were reduced in water-deficient conditions, suggesting a suppression of lignification in response to water limitations. Even though there was a varying extent of response in different poplar species, the tendency could be confirmed throughout the clones (Cocozza et al. 2011).

As described earlier, resin occurrence increases under drought stress in conifer wood. It is not only the resin frequency that alters, however, but also the composition of the oleoresins. The main components in oleoresins of spruce and pine species are diterpenes (resin acids), together with minor amounts of volatile monoterpenes and sesquiterpenes (Croteau and Johnson 1985). In water-deficient conditions, young *Pinus sylvestris* and *Picea abies* significantly increase their total amount of monoterpenes and resin acids in the woody stem, whereas *Pinus taeda* was found to reduce its concentration in resin acids (Hodges and Lorio 1975; Johnson et al. 1997; Turtola et al. 2003). Apart from changes in the overall concentration of resin acids, specific resin acids have also been shown to alter their concentrations. For example, there is an increase in abietane-type resin acids that are known to be highly toxic to white-rot fungi, hence indicating provision for imminent secondary fungal attack (Micales et al. 1994; Eberhardt et al. 1994; Turtola et al. 2003).

## 7 Changes in Hormone and Protein Physiology

Phytohormones are extracellular signal transmitters operating in very low concentrations by evoking specific physiological responses. Just like in any plant, in trees, they can cover a broad spectrum of activity and strongly affect tree growth (Savidge 2001; Groover and Robischon 2006; Munne-Bosch 2007). In gymnosperms, as in angiosperms, a morphogenetic gradient across the cambial zone of the phytohormone auxin governs cell division and cell expansion of the future xylem elements (Uggla et al. 1996; Tuominen et al. 1997; Uggla et al. 1998; Nilsson et al. 2008). In vascular plants, auxin can be present in its conjugated storage form, indole-3-acetic acid (IAA). Both salt stress and water deficiency lead to osmotic stress in plants, which is known to influence auxin transport (Kaldewey et al. 1974; Sheldrake 1979). Consequently, in poplars modified with an auxin-responsive reporter gene construct, the involvement of auxin-mediated salt stress responses in wood formation can be detected (Teichmann et al. 2008).

Abscisic acid (ABA), another phytohormone, is produced in roots as a response to stress factors such as a decrease in soil water potential or an increase in salinity (Finkelstein et al. 2002); a process which has also been documented for trees (Chang et al. 2006; Chen et al. 2001, 2002). In poplar species, it has been shown that salt-sensitive and salt-tolerant species showed different response velocities upon salt stress induction, with the salt-tolerant *P. euphratica* responding faster and maintaining high ABA levels under long-term salinity (Chen et al. 2001, 2003b). There is evidence that salinity-induced ABA regulation might also influence cellular ion balance, since with potassium homeostasis in grey poplar a *KIN2* expression (*PtKIN2*) was strongly induced in the shoot tissue after two weeks' salinity, with concurrently highly elevated ABA concentrations in the roots (Escalante-Perez et al. 2009). Since a high cytosolic $K^+/Na^+$ ratio appears to be critical to plant salt tolerance (Shabala and Cuin 2008), salt-tolerant poplar trees generally try to keep their $K^+$ concentrations up as long as possible when exposed to salt stress (Chen et al. 2001, 2003a), whereas salt-sensitive species show a clear reduction in relative $K^+$ concentration first in root tissue, followed by shoot tissue

(Escalante-Perez et al. 2009). As might be expected from sodium treatment, a rise in the $Na^+$ content in the shoot can also be observed in tandem with $K^+$ depletion, suggesting a $Na^+/K^+$ antagonism in order to balance the osmotic changes induced by salinity. The resulting reduction of $K^+$ in the cells of the xylem differentiating zone might, therefore, also be involved in the changes in wood anatomy (Langer et al. 2002), so that salt stress-induced alterations in wood formation might be regarded as an indirect effect, involving $Na^+$-mediated $K^+$ depletion. Apart from $Na^+$, $Cl^-$ was also found to increase modestly in poplar shoot tissue grown under salt stress, whereas no changes could be observed with phosphate and sulphate (Escalante-Perez et al. 2009).

It seems conclusive, therefore, that alterations in the osmotic potential also evoke changes in gene expression of transport proteins that may be involved in ion uptake and distribution. Indeed, in shoots, gene expression is up-regulated, for example, several $K^+$-release channels showed remarkable up-regulation along with salinity-derived $K^+$ reduction, and also, a putative vacuolar malate transporter was found induced under salt stress along with osmolarity changes and $Cl^-$ uptake (Escalante-Perez et al. 2009). Transcriptome analysis of developing xylem tissue of poplar species with varying degrees of salt sensitivity has shown that, after exposure to salinity, a salt-sensitive species (*P. canescens*), with 382 differentially expressed genes, reacted much more intensely to this stimulation, than a salt-tolerant species (*P. euphratica*) that revealed merely 39 differentially expressed genes compared to control variations. In the salt-sensitive species, genes that are known for their involvement in oxidation-reduction processes were found to be up-regulated, and such antioxidant activities, as well as secondary metabolism, clearly point to defence-related up-regulation. On the other hand, genes involved in cell wall organisation were often found to be down-regulated in both salt-sensitive and salt-tolerant species when exposed to salt stress (Janz et al. 2012). Most interestingly, however, both salt-tolerant and salt-sensitive poplars exhibited in their developing xylem zone a decreased transcript abundance of a range of genes which are known to be activated during tension wood formation. The wood formed under salt stress, therefore, can also be referred to as a kind of 'pressure wood' (Janz et al. 2012).

Even though genetic analysis of salt-stressed trees has already revealed that numerous genetic traits are affected, tailored molecular breeding of putative economically important salinity-resistant species is unlikely to be achieved yet. This is partly due to the large gene pool of tree species, which presumably contain further traits for salt tolerance that are yet to be explored, and partly also due to the existing findings pointing to salt tolerance being a multigenetic trait (Chen and Polle 2010), which makes clearly defined genetic engineering difficult.

## 8 Conclusions

Depending on their intensity, the abiotic stress factors of drought and salinity can have significant influences on wood formation. As past experience in this field of research has shown, this impact not only is dependent on the tree species but also varies across provenances within a species. These findings will have to be particularly been taken into consideration in order to retain stable stands with desirable vitality, growth rates, and wood production. Additional fundamental research on a lab scale is inevitably required in order further to elucidate the physiological and molecular processes in the wood-forming tissue under conditions of drought and salinity, since our knowledge here is still in its infancy. When it comes to the effect on the mechanical and technological demands that wood as a raw material has to face in its utilisation, future investigations will also have to include research on field trials of established stands, especially when concerning economically significant species.

**Acknowledgements** This work was supported by funding from the Center for a Sustainable University, Universität Hamburg, Germany, providing a fellowship in the Postdoctoral Research Group 'Sustainable Future'.

## References

Abe H, Nakai T (1999) Effect of the water status within a tree on tracheid morphogenesis in *Cryptomeria japonica* D-Don. Trees Struct Funct 14:124–129

Abe H, Nakai T, Utsumi Y, Kagawa A (2003) Temporal water deficit and wood formation in *Cryptomeria japonica*. Tree Physiol 23:859–863

Arend M, Fromm J (2007) Seasonal change in the drought response of wood cell development in poplar. Tree Physiol 27:985–992

Baas P, Schweingruber FH (1987) Ecological trends in the wood anatomy of trees, shrubs and climbers from Europe. IAWA Bull 8:245–274

Baas P, Werker E, Fahn A (1983) Some ecological trends in vessel characters. IAWA Bull 4: 141–159

Bacelar EA, Moutinho-Pereira JM, Goncalves BC, Ferreira HF, Correia CA (2007) Changes in growth, gas exchange, xylem hydraulic properties and water use efficiency of three olive cultivars under contrasting water availability regimes. Environ Exp Bot 60:183–192

Battipaglia G, De Micco V, Brand WA, Linke P, Aronne G, Saurer M, Cherubini P (2010) Variations of vessel diameter and delta 13C in false rings of *Arbutus unedo* L. reflect different environmental conditions. New Phytol 188:1099–1112

Carlquist S, Hoekman DA (1985) Ecological wood anatomy of the woody Southern Californian flora. IAWA Bull 6:319–347

Cave ID, Walker JCF (1994) Stiffness of wood in fast-grown plantation softwoods – the influence of microfibril angle. For Prod J 44:43–48

Chang Y, Chen SL, Yin WL, Wang RG, Liu YF, Shi Y, Shen YY, Li Y, Jiang J, Liu Y (2006) Growth, gas exchange, abscisic acid, and calmodulin response to salt stress in three poplars. J Integr Plant Biol 48:286–293

Chen S, Polle A (2010) Salinity tolerance of *Populus*. Plant Biol 12:317–333

Chen SL, Li JK, Wang SS, Hüttermann A, Altman A (2001) Salt, nutrient uptake and transport, and ABA of *Populus euphratica*; a hybrid in response to increasing soil NaCl. Trees Struct Funct 15:186–194

Chen SL, Li JK, Wang TH, Wang SS, Polle A, Hüttermann A (2002) Osmotic stress and ion-specific effects on xylem abscisic acid and the relevance to salinity tolerance in poplar. J Plant Growth Regul 21:224–233

Chen SL, Li JK, Wang SS, Fritz E, Hüttermann A, Altman A (2003a) Effects of NaCl on shoot growth, transpiration, ion compartmentation, and transport in regenerated plants of *Populus euphratica* and *Populus tomentosa*. Can J For Res (Revue Canadienne De Recherche Forestiere) 33:967–975

Chen SL, Li JK, Wang TH, Wang SS, Polle A, Hüttermann A (2003b) Gas exchange, xylem ions and abscisic acid response to $Na^{+}$-salts and $Cl^{-}$-salts in *Populus euphratica*. Acta Bot Sin 45: 561–566

Choat B, Cobb AR, Jansen S (2008) Structure and function of bordered pits: new discoveries and impacts on whole-plant hydraulic function. New Phytol 177:608–625

Clifton NC (1969) Resin pockets in Canterbury Radiata pine G. N Z J For 14:38–49

Cocozza C, Giovannelli A, Traversi ML, Castro G, Cherubini P, Tognetti R (2011) Do tree-ring traits reflect different water deficit responses in young poplar clones (*Populus* × *canadensis* Monch 'I-214' and *P. deltoides* 'Dvina')? Trees Struct Funct 25:975–985

Corcuera L, Camarero JJ, Gil-Pelegrin E (2004) Effects of a severe drought on *Quercus ilex* radial growth and xylem anatomy. Trees Struct Funct 18:83–92

Cown DJ (1973) Resin pockets their occurrence and formation in New-Zealand forests. N Z J For 18:233–251

Croteau R, Johnson MA (1985) Biosynthesis of terpenoid wood extractives. In: Higuchi T (ed) Biosynthesis and biodegradation of wood components. Academic, Orlando, FL, Xvi+679p

de Luis M, Novak K, Raventos J, Gricar J, Prislan P, Cufar K (2011) Cambial activity, wood formation and sapling survival of *Pinus halepensis* exposed to different irrigation regimes. For Ecol Manage 262:1630–1638

Donaldson LA (2002) Abnormal lignin distribution in wood from severely drought stressed *Pinus radiata* trees. IAWA J 23:161–178

Dünisch O, Bauch J (1994) Influence of soil substrate and drought on wood formation of spruce (*Picea abies* L. karst) under controlled conditions. Holzforschung 48:447–457

Dunn AL, Barford CC, Wofsy SC, Goulden ML, Daube BC (2007) A long-term record of carbon exchange in a boreal black spruce forest: means, responses to interannual variability, and decadal trends. Glob Chang Biol 13:577–590

Eberhardt TL, Han JS, Micales JA, Young RA (1994) Decay resistance in conifer seed cones – role of resin acids as inhibitors of decomposition by white-rot fungi. Holzforschung 48:278–284

Eilmann B, Weber P, Rigling A, Eckstein D (2006) Growth reactions of *Pinus sylvestris* L. and *Quercus pubescens* Willd. to drought years at a xeric site in Valais, Switzerland. Dendrochronologia 23:121–132

Eilmann B, Zweifel R, Buchmann N, Fonti P, Rigling A (2009) Drought-induced adaptation of the xylem in Scots pine and pubescent oak. Tree Physiol 29:1011–1020

Eilmann B, Zweifel R, Buchmann N, Pannatier EG, Rigling A (2011) Drought alters timing, quantity, and quality of wood formation in Scots pine. J Exp Bot 62:2763–2771

Escalante-Perez M, Lautner S, Nehls U, Selle A, Teuber M, Schnitzler JP, Teichmann T, Fayyaz P, Hartung W, Polle A, Fromm J, Hedrich R, Ache P (2009) Salt stress affects xylem differentiation of grey poplar (*Populus* × *canescens*). Planta 229:299–309

Esteban LG, Martin JA, de Palacios P, Fernandez FG (2012) Influence of region of provenance and climate factors on wood anatomical traits of *Pinus nigra* Arn. subsp salzmannii. Eur J For Res 131:633–645

FAO, Food and Agriculture Organization of the United Nations (2007) Dominant type of problem lands. In: Digital Media Series. http://www.fao.org/fileadmin/templates/nr/images/resources/images/Maps/geonetwork/probland.png. Accessed Oct 2012

February EC, Stock WD, Bond WJ, Leroux DJ (1995) Relationships between water availability and selected vessel characteristics in *Eucalyptus grandis* and 2 hybrids. IAWA J 16:269–276

Fereres E, Acevedo E, Henderson DW, Hsiao TC (1978) Seasonal-changes in water potential and turgor maintenance in sorghum and maize under water stress. Physiol Plant 44:261–267

Finkelstein RR, Gampala SSL, Rock CD (2002) Abscisic acid signaling in seeds and seedlings. Plant Cell 14:S15–S45

Frey-Wissling A (1942) Uber die Entstehung von Harztaschen. Schweiz Zeitschr Forstw 93: 101–106

Fu PL, Jiang YJ, Wang AY, Brodribb TJ, Zhang JL, Zhu SD, Cao KF (2012) Stem hydraulic traits and leaf water-stress tolerance are co-ordinated with the leaf phenology of angiosperm trees in an Asian tropical dry karst forest. Ann Bot 110:189–199

Garcia Esteban L, Antonio Martin J, de Palacios P, Garcia Fernandez F, Lopez R (2010) Adaptive anatomy of *Pinus halepensis* trees from different Mediterranean environments in Spain. Trees Struct Funct 24:19–30

Garcia-Gonzalez I, Eckstein D (2003) Climatic signal of earlywood vessels of oak on a maritime site. Tree Physiol 23:497–504

Gea-Izquierdo G, Fonti P, Cherubini P, Martin-Benito D, Chaar H, Canellas I (2012) Xylem hydraulic adjustment and growth response of *Quercus canariensis* Willd. to climatic variability. Tree Physiol 32:401–413

Glerum C (1970) Drought ring formation in conifers. For Sci 16:246–252

Groover A, Robischon M (2006) Developmental mechanisms regulating secondary growth in woody plants. Curr Opin Plant Biol 9:55–58

Hacke U, Sauter JJ (1996) Drought-induced xylem dysfunction in petioles, branches, and roots of *Populus balsamifera* L. and *Alnus glutinosa* (L.) Gaertn. Plant Physiol 111:413–417

Hacke UG, Sperry JS (2001) Functional and ecological xylem anatomy. Perspect Plant Ecol Evol Syst 4:97–115

Hacke UG, Sperry JS, Pockman WT, Davis SD, McCulloch KA (2001) Trends in wood density and structure are linked to prevention of xylem implosion by negative pressure. Oecologia 126: 457–461

Hacke UG, Sperry JS, Pittermann J (2004) Analysis of circular bordered pit function – II. Gymnosperm tracheids with torus-margo pit membranes. Am J Bot 91:386–400

Harris JM, Meylan BA (1965) Influence of microfibril angle on longitudinal and tangential shrinkage in *Pinus radiata*. Holzforschung 19:144–152

Hodges JD, Lorio PL (1975) Moisture stress and composition of xylem oleoresin in loblolly-pine. For Sci 21:283–290

Hoffer M, Tardif JC (2009) False rings in jack pine and black spruce trees from eastern Manitoba as indicators of dry summers. Can J For Res (Revue Canadienne De Recherche Forestiere) 39:1722–1736

Holtta T, Vesala T, Peramaki M, Nikinmaa E (2002) Relationships between embolism, stem water tension, and diameter changes. J Theor Biol 215:23–38

Irvine J, Grace J (1997) Continuous measurements of water tensions in the xylem of trees based on the elastic properties of wood. Planta 202:455–461

Janssonius HH (1950) The vessels in the wood of Javan mangrove trees. Blumea 6:465–469

Janz D, Lautner S, Wildhagen H, Behnke K, Schnitzler JP, Rennenberg H, Fromm J, Polle A (2012) Salt stress induces the formation of a novel type of 'pressure wood' in two *Populus* species. New Phytol 194:129–141

Johnson RH, Young BL, Alstad DN (1997) Responses of ponderosa pine growth and volatile terpene concentrations to manipulation of soil water and sunlight availability. Can J For Res 27:1794–1804

Junghans U, Polle A, Duechting P, Weiler E, Kuhlman B, Gruber F, Teichmann T (2006) Adaptation to high salinity in poplar involves changes in xylem anatomy and auxin physiology. Plant Cell Environ 29:1519–1531

Kaldewey H, Ginkel U, Wawczyniak G (1974) Auxin transport and water stress in pea (*Pisum sativum* L.). Berichte Der Deutschen Botanischen Gesellschaft 87:563–576
Khamis MH, Hammad HH (2007) Effect of irrigation by saline ground water on the growth of some conifer seedlings: mortality, growth, biomass and physical wood properties, vol 58. Bulletin of Faculty of Agriculture, Cairo University, pp 36–45
Langer K, Ache P, Geiger D, Stinzing A, Arend M, Wind C, Regan S, Fromm J, Hedrich R (2002) Poplar potassium transporters capable of controlling K(+) homeostasis and K(+)-dependent xylogenesis. Plant J 32:997–1009
Liphschitz N, Waisel Y (1970) The effect of water stresses on radial growth of *Populus euphratica* D. La-Yaaran 20:53–61
Logullo MA, Salleo S, Piaceri EC, Rosso R (1995) Relations between vulnerability to xylem embolism and xylem conduit dimensions in young trees of *Quercus cerris*. Plant Cell Environ 18:661–669
Lovisolo C, Schubert A (1998) Effects of water stress on vessel size and xylem hydraulic conductivity in *Vitis vinifera* L. J Exp Bot 49:693–700
Maherali H, DeLucia EH (2000) Xylem conductivity and vulnerability to cavitation of ponderosa pine growing in contrasting climates. Tree Physiol 20:859–867
Marchand N, Filion L (2012) False rings in the white pine (*Pinus strobus*) of the Outaouais Hills, Quebec (Canada), as indicators of water stress. Can J For Res (Revue Canadienne De Recherche Forestiere) 42:12–22
Mark RE, Gillis PP (1973) Relationship between fiber modulus and S2 angle. Tappi 56:164–167
Menzel A, Fabian P (1999) Growing season extended in Europe. Nature 397:659
Menzel A, Sparks TH, Estrella N, Koch E, Aasa A, Ahas R, Alm-Kuebler K, Bissolli P, Braslavska OG, Briede A, Chmielewski FM, Crepinsek Z, Curnel Y, Dahl A, Defila C, Donnelly A, Filella Y, Jatcza K, Mage F, Mestre A, Nordli O, Penuelas J, Pirinen P, Remisova V, Scheifinger H, Striz M, Susnik A, Van Vliet AJH, Wielgolaski FE, Zach S, Zust A (2006) European phenological response to climate change matches the warming pattern. Glob Chang Biol 12: 1969–1976
Meyer RF, Boyer JS (1972) Sensitivity of cell-division and cell elongation to low water potentials in soybean hypocotyls. Planta 108:77
Micales JA, Han JS, Davis JL, Young RA (1994) Chemical composition and fungitoxic activities of pine cone extractives. Biodeterioration research 4. Mycotoxins, wood decay, plant stress, biocorrosion, and general biodeterioration. Plenum, New York, pp 317–332
Munne-Bosch S (2007) Aging in perennials. Crit Rev Plant Sci 26:123–138
Munns R (2002) Comparative physiology of salt and water stress. Plant Cell Environ 25:239–250
Munns R, Tester M (2008) Mechanisms of salinity tolerance. Annu Rev Plant Biol 59:651–681
Nicholls JWP, Waring HD (1977) Effect of environmental-factors on wood characteristics – .4. Irrigation and partial droughting of *Pinus radiata*. Silvae Genetica 26:107–111
Nilsson J, Karlberg A, Antti H, Lopez-Vernaza M, Mellerowicz E, Perrot-Rechenmann C, Sandberg G, Bhalerao RP (2008) Dissecting the molecular basis of the regulation of wood formation by auxin in hybrid aspen. Plant Cell 20:843–855
Palakit K, Siripattanadilok S, Duangsathaporn K (2012) False ring occurrences and their identification in teak (*Tectona grandis*) in North-Eastern Thailand. J Trop For Sci 24:387–398
Piao S, Ciais P, Friedlingstein P, Peylin P, Reichstein M, Luyssaert S, Margolis H, Fang J, Barr A, Chen A, Grelle A, Hollinger DY, Laurila T, Lindroth A, Richardson AD, Vesala T (2008) Net carbon dioxide losses of northern ecosystems in response to autumn warming. Nature 451: 49–52
Rigling A, Waldner PO, Forster T, Brasker OU, Pouttu A (2001) Ecological interpretation of tree-ring width and intraannual density fluctuations in *Pinus sylvestris* on dry sites in the central Alps and Siberia. Can J For Res (Revue Canadienne De Recherche Forestiere) 31:18–31
Rigling A, Braker O, Schneiter G, Schweingruber F (2002) Intra-annual tree-ring parameters indicating differences in drought stress of *Pinus sylvestris* forests within the Erico-Pinion in the Valais (Switzerland). Plant Ecol 163:105–121

Rigling A, Bruhlhart H, Braker OU, Forster T, Schweingruber FH (2003) Effects of irrigation on diameter growth and vertical resin duct production in *Pinus sylvestris* L. on dry sites in the central Alps, Switzerland. For Ecol Manage 175:285–296

Savidge RA (2001) Intrinsic regulation of cambial growth. J Plant Growth Regul 20:52–77

Searson MJ, Thomas DS, Montagu KD, Conroy JP (2004) Wood density and anatomy of water-limited eucalypts. Tree Physiol 24:1295–1302

Seifert T, Breibeck J, Seifert S, Biber P (2010) Resin pocket occurrence in Norway spruce depending on tree and climate variables. For Ecol Manage 260:302–312

Shabala S, Cuin TA (2008) Potassium transport and plant salt tolerance. Physiol Plant 133: 651–669

Sheldrake AR (1979) Effects of osmotic-stress on polar auxin transport in *Avena* mesocotyl sections. Planta 145:113–117

Sheriff DW, Whitehead D (1984) Photosynthesis and wood structure in *Pinus radiata* D-Don during dehydration and immediately after rewatering. Plant Cell Environ 7:53–62

Sperry JS, Saliendra NZ (1994) Intra-plant and inter-plant variation in xylem cavitation in *Betula occidentalis*. Plant Cell Environ 17:1233–1241

Sperry JS, Hacke UG, Pittermann J (2006) Size and function in conifer tracheids and angiosperm vessels. Am J Bot 93:1490–1500

Sterck FJ, Zweifel R, Sass-Klaassen U, Chowdhury Q (2008) Persisting soil drought reduces leaf specific conductivity in Scots pine (*Pinus sylvestris*) and pubescent oak (*Quercus pubescens*). Tree Physiol 28:529–536

Stiller V (2009) Soil salinity and drought alter wood density and vulnerability to xylem cavitation of baldcypress (*Taxodium distichum* (L.) Rich.) seedlings. Environ Exp Bot 67:164–171

Teichmann T, Bolu-Arianto WH, Olbrich A, Langenfeld-Heyser R, Goebel C, Grzeganek P, Feussner I, Haensch R, Polle A (2008) GH3: GUS reflects cell-specific developmental patterns and stress-induced changes in wood anatomy in the poplar stem. Tree Physiol 28:1305–1315

Tuominen H, Puech L, Fink S, Sundberg B (1997) A radial concentration gradient of indole-3-acetic acid is related to secondary xylem development in hybrid aspen. Plant Physiol 115: 577–585

Turtola S, Manninen AM, Rikala R, Kainulainen P (2003) Drought stress alters the concentration of wood terpenoids in Scots pine and Norway spruce seedlings. J Chem Ecol 29:1981–1995

Tyree MT, Zimmermann MH (2002) Xylem structure and the ascent of sap., 2nd edn, Springer series in wood science. Springer, Berlin

Uggla C, Moritz T, Sandberg G, Sundberg B (1996) Auxin as a positional signal in pattern formation in plants. Proc Natl Acad Sci USA 93:9282–9286

Uggla C, Mellerowicz EJ, Sundberg B (1998) Indole-3-acetic acid controls cambial growth in Scots pine by positional signaling. Plant Physiol 117:113–121

Villar-Salvador P, Castro-Diez P, Perez-Rontome C, Montserrat-Marti G (1997) Stem xylem features in three Quercus (Fagaceae) species along a climatic gradient in NE Spain. Trees Struct Funct 12:90–96

Watson AJ, Dadswell HE (1964) Influence of fibre morphology on paper properties. 3. Length: diameter (L/D) ratio. 4. Micellar spiral angle, vol 17. APPITA, Melbourne, pp 146–156

Whitmore FW, Zahner R (1967) Evidence for a direct effect of water stress on tracheid cell wall metabolism in pine. For Sci 13:397

Wimmer R, Downes GM, Evans R (2002) Temporal variation of microfibril angle in *Eucalyptus nitens* grown in different irrigation regimes. Tree Physiol 22:449–457

# Biology, Chemistry and Structure of Tension Wood

**Judith Felten and Björn Sundberg**

**Abstract** Trees maintain and adjust their stature by developing reaction wood in stems and branches. The physical properties of reaction wood result in a higher strain than in normal wood. Because reaction wood is only formed at one side of the stem, this unilateral strain creates a force and hence a movement of the stem or branches towards a more favorable position. The spectacular modification of cambial growth, cell shape, cell-wall chemistry, and ultrastructure observed in reaction wood has attracted generations of scientists to study its features and molecular regulation. In the early literature, the physiology of reaction wood induction was much studied, especially the relative importance of positional and mechanical sensing for its induction. Even today this is still a matter of debate and confusion, as discussed in the first part of this chapter. In angiosperm trees, reaction wood is denoted tension wood (TW), and in many tree species TW fibers develop an inner cellulose-rich gelatinous layer (G-fibers). Much research has been devoted to understand the chemistry and ultrastructure of the gelatinous layer and its function in creating tension stress in the wood. Less attention has been paid to TW without G-fibers, although it has similar physical properties and function as TW with G-fibers. The chemistry and structural variation of TW, and their importance for TW function, are discussed in the second part of this chapter. Not much is known about the molecular control of TW formation. However, some information has been gained about the role of plant hormones as signaling components in TW induction. The last part of the chapter summarizes this knowledge.

J. Felten • B. Sundberg (✉)
Department of Forest Genetics and Plant Physiology, Umeå Plant Science Center, Swedish University of Agricultural Sciences, 90183 Umeå, Sweden
e-mail: bjorn.sundberg@slu.se

J. Fromm (ed.), *Cellular Aspects of Wood Formation*, Plant Cell Monographs 20,
DOI 10.1007/978-3-642-36491-4_8, 

# 1 Physiology of Sensing and Signaling Mechanisms for Reaction Wood Induction

Primary shoots of trees and other plants grow in response to inherent patterning mechanisms (e.g., apical dominance) and environmental factors (e.g., light and gravity) to establish a branching architecture that is optimal for growth and reproduction. However, large and long-lived organisms like trees often have a requirement to adjust the position of their stems and branches, for example, when crown architecture or the entire tree has been perturbed by abiotic or biotic factors (e.g., wind, snow, grazing, self-weight). Such growth correction of stems is accomplished by the unilateral formation of reaction wood, which exerts a strain that induces a directional movement of the stem towards a more favorable position (Scurfield 1973). With few exceptions (Kojima et al. 2012), reaction wood in angiosperm trees is under tension and denoted tension wood (TW), and, in gymnosperms, it is under compression and denoted compression wood. Development of reaction wood is commonly associated with increased cambial growth, a modified wood anatomy and xylem cell morphology, and a change in the ultrastructure and chemistry of the tracheid/fiber secondary cell wall (Timell 1986; Jourez et al. 2001; Mellerowicz and Sundberg 2008; Donaldson and Singh 2013). In this chapter, we discuss some aspects pertaining to the biology of reaction wood induction and signaling. Although TW and compression wood are different both from a structural and mechanical sense, the basic mechanisms for sensing the stimuli and forming the reaction wood seem to be similar in angiosperms and gymnosperms and are therefore discussed in a broader sense. We then focus on the anatomy, chemistry, and ultrastructure specifically for TW and finally summarize knowledge on the role of plant hormones for its formation.

It has long been discussed in the literature whether reaction wood is induced in response to positional or mechanical sensing. A main reason for this is that experimental studies on reaction wood induction often involve displacement of stems that normally cause both a positional change and mechanical stress in the organ. Jaccard (1938) performed pioneering studies to distinguish between positional or mechanical sensing in the reaction wood response. By bending stems into loops (Jaccard's loop), he demonstrated that reaction wood formed unilaterally in the upper and lower part of the loop where cambial tissues were under tension and compression, respectively, and not along the whole loop although all tissues in the loop were under mechanical stress. He concluded, in accordance with studies from Hartmann (1932), that reaction wood forms at places where the resulting strain will make the stem move towards its original position, irrespectively of tension or compression forces in cambial tissues. These findings by Jaccard, later supported in other studies (reviewed in Timell 1986), strongly suggest that positional sensing is a primary mechanism for the induction of reaction wood. Thus, the fact that TW is formed at the upper side of leaning stems (and vice versa for compression wood) is not because this side is under tension stress, but because this will induce the desired movement of the stem towards the vertical.

A similar positional memory was also demonstrated for gravitropic responses during primary shoot growth of branches, where displaced shoots restore their angle towards the "gravitropic set-point angle" (Digby and Firn 1995). The gravitropic set-point angle was defined as "the angle with respect to gravity at which an organ shows no gravity induced differential growth in order to correct its orientation." This observation suggests a common sensing and signaling mechanism in gravitational responses in both primary shoots and secondary stems. In primary organs, this is translated into asymmetric elongation growth, and in secondary stems, it will induce unilateral reaction wood formation in a way that will result in the movement towards a favorable position.

The relative importance of gravitational and mechanical stimuli for reaction wood induction has been evaluated in many experiments. Experiments in which mechanical stimuli were applied to stems in the absence of any gravity factor, by rotating plants on a clinostat, showed that the gravity factor is important for reaction wood induction. Nevertheless other experiments have pointed towards an importance for mechanical sensing (Timell 1986; Kwon et al. 2001). It is also observed that trees exposed to wind sway often respond by reaction wood formation, which supports the idea that the mechanical stimulus is an important factor (Timell 1986; Telewski 2006). Even in vertically growing trees, self-weight of the crown during growth will induce both mechanical forces and displacement of the stem from vertical, and the tree needs to develop reaction wood in order to maintain itself upright (Almeras and Fournier 2009). This phenomenon is particularly important in slender stems of, for example, young trees. Arcs of TW frequently form in greenhouse grown *Populus* sp., which are often used as a model species in molecular research on trees. Such wood heterogeneity may cause problems when stems are used for experimental studies on wood biology and wood chemistry (Bjurhager et al. 2010).

In addition to gravitational and mechanical stimuli, reaction wood is also formed in response to inherent growth patterning. For example, removal of the leader shoot of trees induces reaction wood in the underlying whorl of branches and an upward movement towards vertical in order to replace the leader shoot (Hartig 1901; Hartmann 1932; Wilson and Archer 1981, 1983). This response evidently takes place without any change in position or any mechanical stimulus. Taken together, reaction wood develops in response to inherent patterning mechanisms, to a positional change, and probably also to mechanical stimuli. The relative importance of the latter two is not resolved, and they may very well act in concert.

The positional change or mechanical stimulus that results in the induction of reaction wood formation in a displaced stem will be sensed along the whole stem. But the development of reaction wood is directed to one side, and even to a specific position of the stem as observed in Jaccard's loops, in order to induce a movement towards the favorable position. This suggests that there is a mechanism for spatial separation of sensing the inductive stimuli and forming reaction wood. Moreover, the inductive signal is active across cells at all stages of xylem development and not only in cells at early developmental stages in cambial tissues. This was deduced from experiments where conifer stems were displaced and reverted back to the original position. It was observed that developing tracheids that were in the stage of expansion and secondary wall formation at the time of displacement were responding by partly exhibiting compression wood characteristics (Kennedy and Farrar 1965; Yoshizawa et al. 1984). It does not seem unlikely that this observation holds true also for TW formation in angiosperm trees, although similar experiments to show this are missing. Little, if anything, is known about the molecular mechanisms underlying sensing and signaling of the reaction wood response. However, all reaction wood characteristics are not always observed together (i.e., altered wood anatomy/chemistry can be observed in the absence of increased growth). Consequently, inductive signal(s) can independently act on different stages of xylem cell development, probably by interacting with the complex genetic and molecular machinery regulating cell division, cell expansion, ultrastructure, and biosynthesis of secondary cell-wall formation. This will lead to the remarkable wood plasticity presented in reaction wood.

## 2 The Chemical and Structural Variation of Tension Wood Fibers

The TW response often (but not always) results in an increased eccentric cambial growth and a decreased frequency of vessel development (see Fig. 1). This will increase the xylem area of TW fibers, which is the load-bearing cell type in the wood and responsible for exerting the tensile strain in the tissue. However, variation in TW anatomy, chemistry, and fiber wall ultrastructure is diverse between species (Onaka 1949; Côté et al. 1969; Clair et al. 2006b; Ruelle et al. 2006, 2007). In particular, the secondary cell wall of TW fibers has attracted large interest because it is highly modified in order to exert a longitudinal shrinkage that leads to the tensile stress in the wood.

### *2.1 Tension Wood with a Gelatinous Layer*

The most spectacular TW fibers observed in many species are those that develop a cellulose-rich gelatinous layer (G-layer) on the lumen side of the fiber (so-called G-fibers) (Norberg and Meier 1966; Côté et al. 1969; Clair et al. 2006b; Mellerowicz and Gorshkova 2012) (Fig. 1). Depending on the tree species, G-fibers can arrange in very different patterns, e.g., be restricted to earlywood as seen in *Fagus grandifolia*, be absent from fibers in the vicinity of vessels like in *Ulmus americana*, or be present either in clusters or as sparse isolated G-fibers like in certain tropical species (Côté et al. 1969; Clair et al. 2006b). Furthermore the G-layer can vary in its thickness and can replace either wholly or partially the S2 or the S3 cell-wall layer, or be added in addition to those layers (Wardrop and Dadswell 1955). The gelatinous-looking structure of the G-layer inspired researchers early on to investigate its ultrastructure and composition. Based on microscopy and optical birefringence

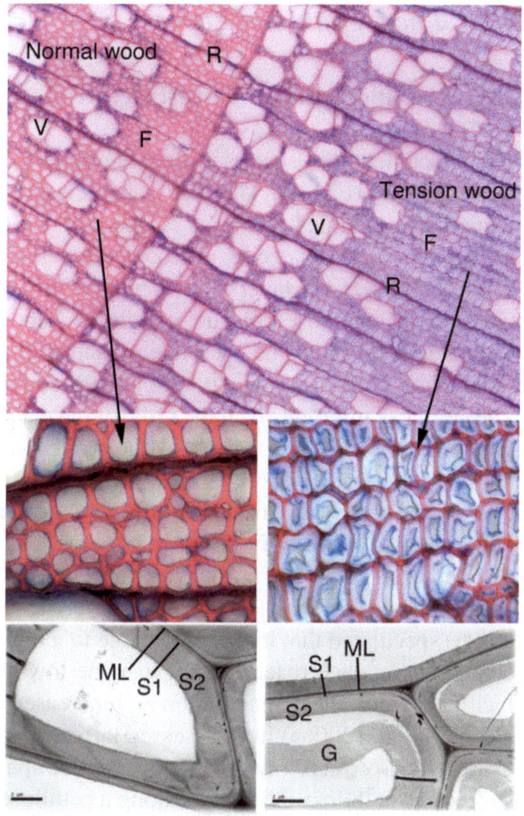


**Fig. 1** Normal wood and tension wood across two annual rings of *Populus tremula*. Sections are stained with safranin/alcian blue for light microscopy. A decrease in vessel frequency and size is visible in TW. Magnification of NW and TW areas shows the presence of G-layers (*light blue*) in TW, whereas in NW, only secondary cell-wall layers (*pink*) are visible. TEM analyses highlight the cell-wall structure in NW and TW. The thick, detached G-layer in TW is visible. Scale bars are 2 μm. Vessel (V), fiber (F), ray (R), middle lamella (ML).

analyses, Norberg and Meier (1966) suggested the existence of numerous water-filled pores in G-layers. A high water content of G-layers was later observed by Raman spectroscopy (Gierlinger and Schwanninger 2006). The mesoporosity (mesopores are pores smaller than 50 nm) of TW was analyzed using nitrogen adsorption by Clair et al. (2008) in *Castanea sativa* and by Chang et al. (2009) in six different tropical tree species. Chang et al. found a large variation in sizes, numbers, networks, and shapes of the pores between G-layers of different species. A common conclusion from both authors was that porosity in TW is normally higher than in NW within the same species. They associated this high porosity mostly with the G-layer itself. Thus, it seems evident that the high porosity of G-layers gives room for water incorporation, which causes the gelatinous appearance. The high water content provides a capacity of the G-layer to swell or shrink transversally. How important these hygroscopic properties are for the longitudinal shrinkage and tension stress in TW is, however, not known (Clair et al. 2005a; Mellerowicz and Gorshkova 2012).

## 2.2 *The Gelatinous Layer*

### 2.2.1 Cellulose and Its Ultrastructure

The attachment of G-layers to the outer layers of the secondary wall is weak (Clair et al. 2005b) (Fig. 1). G-layers can therefore be isolated from transverse wood sections for chemical analysis. Norberg and Meier (1966) analyzed G-layers from *Populus tremula* isolated by ultrasonication in ethanol. Based on paper chromatography of hydrolyzed G-layers, they detected almost only glucan and concluded that the G-layer was composed of pure cellulose. Although this view is modified by recent studies that reported the presence of about 25 % matrix polymers in the G-layer (Nishikubo et al. 2007; Kaku et al. 2009), cellulose is by far the major component. The cellulose microfibrils in the G-layer have a parallel or almost parallel orientation (low microfibril angle, MFA) to the axis of the fiber in all species investigated (Norberg and Meier 1966; Côté et al. 1969; Clair et al. 2011; Lautner et al. 2012). The cellulose crystals in microfibrils are larger in the G-layer than in the S2-layer, with diameter values of 6.5 nm in G-layers and 3.1 nm in S2-layers of the G-fiber (Muller et al. 2006). Translated into cellulose chain number of microfibrils, this corresponds to a fourfold increase in G-layers compared to S2-layers. Muller et al. (2006) speculated that this could be due to a higher number of cellulose biosynthesis complexes in each rosette or because the low hemicellulose content of the G-layer facilitates microfibril aggregation and increased crystallization.

Clair et al. (2006a) have assessed lattice spacing of cellulose microfibrils in TW using synchrotron X-ray diffraction. Lattice spacing describes the distance between successive cellulose monomers along a cellulose microfibril. They found that lattice spacing increased with the mechanical stress measured in TW and was related to deformation of the microfibrils. In a later study, lattice spacing was recorded in developing wood at successive stages of cell-wall formation in TW and NW fibers using synchrotron radiation microdiffraction (Clair et al. 2011). In this study, the authors concluded that lattice spacing was increased in the G-layer already during its biosynthesis. This suggests that the tensile stress measured on a macroscopic level in TW can be traced down to the ultrastructure of the cellulose in the G-layer. It also provides evidence for the idea that the tensile stress is caused by longitudinal shrinkage of the cellulose crystals, established when TW is formed (Okuyama et al. 1994; Clair et al. 2011).

### 2.2.2 Noncellulosic Polymers

Lignin

Even though cellulose is a major compound of the G-layer and its longitudinal shrinkage is thought to be a major factor for stress generation in TW, the G-layer has a more complex composition that involves other polymers. Different studies have reported the presence of lignin, xyloglucans, pectins, and arabinogalactan proteins (AGPs) in G-layers. The presence of lignin in the G-layer has been much debated despite its very low abundance (Pilate et al. 2004). Several studies have used UV microscopy and stainings to indicate the presence of lignin in the G-layer in different woody species such as *Tristania conferta, Grevillea robusta, Hakea laurina, Fraxinus mandshurica, Robinia pseudoacacia,* and *Populus euramericana* (Scurfield and Wardrop 1963; Araki et al. 1982; Prodhan et al. 1995; Yoshida et al. 2002a). Immunolocalization with antibodies against synthesized lignin epitopes also indicated the presence of particularly S-type lignin at the lumen side of the G-layer in *Populus deltoides* × *P. trichocarpa* (Joseleau et al. 2004). The observation of lignin towards the lumen side of the G-layer was also supported by Raman spectroscopy imaging in *Populus nigra* × *P. deltoides* (Gierlinger and Schwanninger 2006). The Raman spectra contained several bands characteristic for lignin, and the authors claimed that this demonstrates the presence of a lignin polymer rather than mono- or oligolignols. Lignin traces in G-layers of other species, such as *Acer* ssp., *Fagus sylvatica,* or *Quercus robur,* were detected by Raman spectroscopy in concentric rings or spots within the G-layer (Lehringer et al. 2008). Prodhan et al. (1995) reported stronger lignin signals towards the S-layer side of G-layers in *Fraxinus mandshurica* using $KMnO_4$ staining and TEM. In addition to these chemical indications of lignin in the G-layer, a protein analysis of the G-layer revealed that some of the most abundant proteins were related to metabolism of lignin and/or phenolic compounds (Kaku et al. 2009). Although lignin, or lignin-like compounds, can be detected in the G-layer, it has always been reported in low abundances and sometimes not detected at all. It is possible that studies that fail to detect lignin used methods with insufficient sensitivity. But it is also possible that the lignin polymer is not a general feature of G-layers. Conclusive analytical evidence for the presence of a lignin polymer in the G-layer is still missing, and it cannot be excluded that stainings and UV spectroscopy detect phenolic compounds rather than lignin. Also, the trace amount of lignin-like compounds that may be present in the G-layer suggests little significance of lignin for TW function.

### Hemicelluloses

A role for hemicellulose in stress generation in TW fibers has been proposed from recent studies (Nishikubo et al. 2007; Mellerowicz and Gorshkova 2012). Norberg and Meier (1966) found in addition to cellulose a low amount (1.5 %) of xylose in their chemical analysis of isolated G-layers but did not investigate this any further. The presence of significant amounts (5–18 %) of xylose in G-layers of *Populus* ssp. was demonstrated later by different authors (Furuya et al. 1970; Nishikubo et al. 2007; Kaku et al. 2009). Xylose is a building block of the hemicelluloses xylan and xyloglucan. Results of immunolocalization (Nishikubo et al. 2007; Sandquist et al. 2010) and linkage analysis of extracted polymers have proposed that xyloglucan is present in G-layers (Nishikubo et al. 2007; Kaku et al. 2009). Further support of xyloglucan in *Populus* G-layers is the presence of xyloglucan:xyloglucosyl transferase (XET) enzyme and activity in the G-layer shown by proteomics (Kaku et al. 2009), immunolocalization, and incorporation of a labeled xyloglucan oligosaccharide into fresh wood sections, respectively (Nishikubo et al. 2007). The linkage analysis of polymers extracted from the G-layer did not detect xylan, and likewise immunolocalization with xylan antibodies (LM10, AX1 (arabinoxylan)) indicated that xylan was absent from G-layers, in contrast to S2-layers where it is highly abundant (Bowling and Vaughn 2008; Decou et al. 2009; Kim et al. 2012). Opposed to these findings, Kim et al. (2012) claimed the presence of xylan in the G-layer based on positive signals from the LM11 antibody. The presence of small amounts of mannan, another hemicellulose, was detected in isolated G-layers (Nishikubo et al. 2007), and glucomannan was also indicated using the LM21, LM22, and BGM C6 antibodies (Kim et al. 2012).

Tension stress is thought to be induced by the longitudinal shrinkage of the G-layer. The tension needs to be transferred from the G- to the S2-layer and to the entire fiber, which is a potential problem as these layers are weakly linked. It was hypothesized that xyloglucan is located at the border between the G- and S2-layer to facilitate this stress transfer. Broken bonds due to tension stresses could continuously be repaired by XET activity (Nishikubo et al. 2007; Mellerowicz and Gorshkova 2012). This hypothesis was strengthened by Sandquist et al. (2010) who used TEM-/SEM-combined immunolocalization with CCRC-M1 antibodies (binding to xyloglucan and rhamnogalacturonan) and reported labeling at the boundary between S2- and G-layer in *Populus* where the transfer of tension stress would occur. However, using another antibody against (fucosylated) xyloglucan (CBM FXG-14b), they observed signals only at the lumen side of the G-layer. In studies of *Liquidambar styraciflua* and *Celtis occidentalis,* Bowling and Vaughn (2008) detected labeling with CCRC-M1 in a thin layer at the lumen side of the G-layer but not in the G-layer or at the G-S interface. Taken together, most data indicate that xyloglucan is the major hemicellulose in the G-layer. A role for xyloglucan in transferring tension stress from G- to S-layer has been suggested in *Populus*, but whether this is a general function in other species remains to be experimentally investigated.

### Pectins and Arabinogalactan Proteins

Early studies showed the presence of higher amounts of galactose in TW compared to NW in different species (Gustafsson et al. 1952; Meier 1962). Further studies identified galactans with complex structures that were not characterized in detail but nevertheless suggested an increased amount of pectins in TW (Meier 1962; Kuo and Timell 1969). Pectins are matrix polymers usually found in primary walls. The cell-wall localization of the detected pectinous structures in these early studies could not be determined since whole wood rather than isolated G-layers was analyzed. Pectins can be grouped into three classes based on the subunits of the polymer: homogalacturonan, rhamnogalacturonan I, and rhamnogalacturonan II.

Specific antibodies against these pectin types were used to examine the G-layer for the presence of pectins. No homogalacturonan was detected by immunolocalization (JIM5 antibodies) in G-layers (Arend 2008; Bowling and Vaughn 2008). However, Bowling and Vaughn detected strong labeling of G-layers in *L. styraciflua* or *C. occidentalis* with CCRC-M10 and CCRC-M22 antibodies that bind rhamnogalacturonan I (RG I). RG I can have different side chains that can be identified by specific antibodies. Arend (2008) detected labeling of *Populus* G-layers with LM5 antibodies that bind to epitopes on 1,4-(β)-galactan sidechains of RG I. Interestingly the signal was weak within the G-layer but strongest at the interface of G-and S2-layer, like for xyloglucan as discussed in the previous section. Again there appears to be variation between species since Bowling and Vaughn (2008) did not observe any signal with LM5 antibodies in G-layers in *L. styraciflua*.

Several *fasciclin-like AGPs* (*FLAs*) are among the most upregulated genes during TW formation in *Populus* based on transcript analysis (Lafarguette et al. 2004; Andersson-Gunnerås et al. 2006). Moreover, both protein analysis and sugar linkage analysis of polymers extracted from isolated G-layers have identified the presence of FLAs (Nishikubo et al. 2007; Kaku et al. 2009). Immunolocalization with JIM14 antibodies suggested the localization of FLAs to a thin layer that lines the G-layer on its lumen side in *P. tremula* × *P. alba* (Lafarguette et al. 2004). Also Bowling and Vaughn (2008) detected a signal with this antibody in G-layers of *L. styraciflua* and *C. occidentalis*. They observed the signal in a patchy pattern across the G-layer. In *Populus*, *FLA*s belong to an expanded gene family with members that have high and specific expression in developing secondary xylem (Lafarguette et al. 2004; Andersson-Gunnerås et al. 2006). The large set of wood specific *Populus FLA*s has only two close homologs in Arabidopsis (*AtFLA11* and *12*) that are also specifically expressed during secondary wall formation. A double At*fla11/12* mutant was demonstrated to have decreased tensile stiffness and strength of the stem, together with modified cell-wall composition (particularly decreased cellulose content) (MacMillan et al. 2010). From this observation, it was suggested that AtFLA11/12 were important for both cellulose biosynthesis and as a structural component of the cell-wall matrix. The role of FLAs in TW is not known, but their high expression suggests that they may be an important matrix component and have an influence on cellulose structure and high MFA in the G-layer.

In summary, it can be concluded that the early view by Norberg and Meier (1966) that the G-layer was almost entirely composed of cellulose has been revised by the use of different analytical tools. Xyloglucans, pectins, AGPs, and lignin-like compounds have all been detected by chemical analysis in isolated G-layers or by microanalytical tools such as immunolocalization or Raman spectroscopy. Their detection and localization may be dependent on the analytical tools used but may also vary between species. Monoclonal antibodies against specific sugar epitopes are a highly powerful method, but results must be carefully interpreted due to possible cross-reactions with nontarget epitopes and different epitope accessibility in the distinct layers. It should also be noted that it is difficult to estimate the abundance of minor cell-wall polymers in the G-layer because isolation of G-layers by sonication in ethanol will unavoidably dissolve some of the extractives (such as phenolic compounds, pectins, and AGPs) and immunolocalization and spectroscopic studies give at best only a crude estimate of quantities.

## 2.3 The S2-Layer of Gelatinous Fibers

Many studies have investigated the chemistry and ultrastructure of the G-layer, but less is known about the S2-layer in G-fibers and whether this is modified in comparison to the S2-layer of fibers in NW. The overall chemical composition of TW shows an increase in the concentration of cellulose and a corresponding decrease in both the major and minor matrix polymers (Timell 1969; Hedenström et al. 2009). This is primarily a result from the additional cellulose-rich G-layer, and the extent of cellulose enrichment in the TW will depend on the proportion of G- to S-layer in the G-fibers. In the early literature, it was correctly stated that the concentration of matrix polymers is not necessarily decreased in the S2-layer per se of the G-fiber (Timell 1969). This author further claimed that the concentration of lignin and xylan is higher in the S2-layer of G-fibers compared to the NW fibers, but this statement was based on unsupported assumptions of similar cellulose content in the two types of wood. Bentum et al. (1969) further visualized lignin in the S2-layer of G-fibers and NW fibers after removal of polysaccharides with hydrofluoric acid. This showed abundant lignin in the S2-layer of both fiber types, but quantitative comparisons are not possible from the published information. It is however not trivial to compare the S2-layer of G-fibers with that of normal fibers because the S2-layer per se is not homogenous. Thus, a fully formed S2-layer in NW may be different from a partly formed S2-layer in G-fibers, even if the outer (and comparable) layer of the S2 wall is similar in the two types of wood. Therefore similar sub-layers of the S2 should be compared between the two fiber types to come to a valid conclusion. Using immunogold labelling combined with electron microscopy, Kim et al. (2012) found no difference between S2 of TW fibers and outer S2 of NW fibers using the anti-xylan antibodies LM10 and LM11, and also no clear difference could be observed for the anti-glucomannan antibodies LM21, LM22, and BGM C6 when comparing the corresponding S2-sublayers. Another approach to compare corresponding S2-sublayers in G-fibers and NW fibers is to record successive stages of fiber cell-wall formation during S2 formation in developing wood. This was done in a study by Clair et al. (2011) who recorded MFA in *Populus deltoides* × *P. trichocarpa*, using synchrotron radiation microdiffraction. They found a MFA of about 25° to the longitudinal fiber axis in both NW and TW when comparing S2 walls at the same (early) developmental stage. In TW, the MFA decreased dramatically towards 0° when the G-layer started to form. But they also concluded a decrease in the MFA to about 10° of the inner layer of S2 of NW fibers. In contrary, Goswami et al. (2008) measured a MFA of about 36° in the S2-layer of matured TW fibers from *Populus nigra* × *P. deltoides* after enzymatic removal of the G-layer. However, as pointed out by Clair et al. (2011), it cannot be excluded that this measure of average MFA was also influenced by the high MFA in the S1 layer (estimated to 45° in Clair et al. 2011). Taken together, there are no consistent and conclusive data of quantitative or qualitative differences between comparable S2-layers in TW G-fibers and NW, although such differences cannot be excluded.

## *2.4 TW Without Gelatinous Fibers*

Even though the presence of G-fibers often is used as an indicator of TW, a significant number of tree species form TW with high tensile strain but without typical G-fibers (Onaka 1949; Okuyama et al. 1994; Bailleres et al. 1995; Yoshida et al. 2002b; Clair et al. 2006b; Ruelle et al. 2006; Sultana et al. 2010). As a further example of the variation of TW fibers between species, several species within the *Flacourtiaceae* family form a multilayered secondary cell-wall layer in the TW fibers (Clair et al. 2006b; Ruelle et al. 2006). This TW also exerts high tensile growth strain. Yet, despite the lack of a G-layer, the TW fibers in these species are in many cases different from the corresponding opposite wood fibers. The emerging picture is that they have a lower MFA, an increased cellulose to lignin ratio, and, for certain species, an increased S to G ratio (Sugiyama et al. 1993; Okuyama et al. 1994; Bailleres et al. 1995; Yoshizawa et al. 2000; Yoshida et al. 2002b; Ruelle et al. 2007). In TW from *Eucalyptus globulus*, Aguayo et al. (2010) found that the decrease in lignin was followed by an increase in xylose, whereas the cellulose concentration was not affected. The high mesoporosity of the G-layer is not shared by the TW fibers lacking a G-layer (Chang et al. 2009). This suggests that the swelling properties of the G-layer are not critical for creating tension stress. Despite the high diversity in TW anatomy and ultrastructure of TW fibers among tree species, all types of TW are able to exert the required tensile stress to induce a movement in the displaced stem. No correlation between TW structure (with and without G-fibers) and the occurrence of the tensile stress that develops in TW was found when comparing several species (Clair et al. 2006b). In species with G-layers, it is clear that this layer is critical for the tensile stress in TW resulting from the longitudinal shrinkage of the cellulose crystals and always combined with a low MFA (Clair and Thibaut 2001; Fang et al. 2008; Goswami et al. 2008; Clair et al. 2011). The low MFA and increased cellulose in the S2-layer of fibers in TW without G-layer strongly suggest that these are the general attributes of TW, and that tensile stresses are generated by similar mechanisms as in G-fibers. Thus, there is no obvious advantage of a G-layer as such (McLean et al. 2012). The G-layer could rather be a complication since the tensile stress in the G-layer caused by cellulose shrinkage must be transmitted to the S2-layer. The exact mechanism behind tensile stress generation in the cellulose microfibrils of TW fibers, mostly investigated in species with G-fiber, is still a matter of debate (Goswami et al. 2008; Clair et al. 2011; Mellerowicz and Gorshkova 2012).

# 3 Hormone Signaling and TW Development

Most experimental approaches on inductive signaling molecules involved in TW formation have been dealing with plant hormones, typically auxin, gibberellins (GAs), and ethylene. It seems plausible that these in turn interact with the transcriptional machinery regulating wood development and further downstream with the regulatory machinery controlling cell division, expansion, and cell-wall biosynthesis (Nieminen et al. 2012; Zhong and Ye 2013). The large number of gene transcripts involved in all these processes are reflected in several studies comparing transcript abundance in TW- and NW-forming tissues (Dejardin et al. 2004; Paux et al. 2005; Andersson-Gunnerås et al. 2006; Hobson et al. 2010; Jin et al. 2011). Here we will discuss experiments evaluating the role of plant hormones in TW formation in trees.

The classical approach to study plant hormone function is by applying the hormone or inhibitors of its biosynthesis, response, or transport and observing their effects on molecular events and various aspects of growth and development. But application studies may also cause aberrant compartmentalization and/or unnormal tissue concentrations of the hormone, and such experiments must therefore be interpreted with caution. However, more conclusive interpretations can be made from application experiments when the resulting hormone balance is measured in the tissue of interest. Correlations between the endogenous hormone balance and the physiological event(s) studied can further provide information about hormone function. More solid evidence about hormone function, however, requires mutants either blocked in the biosynthetic pathway or in the perception of the hormone. This approach is problematic for hormones such as auxin and GAs due to their requirement in many aspects of plant development. Unless inducible systems and tissue-/cell-type-specific promoters are used, such mutants will unavoidably result in plants with more or less aberrant growth, where primary and secondary effects of hormone action are difficult to distinguish.

## *3.1 Auxin*

Auxin affects most aspects of wood development when applied to cambial tissues (reviewed in Sundberg et al. (2000)). Indole-3-acetic acid (IAA) that is apically supplied to decapitated stems enters the polar auxin transport pathway (Sundberg and Uggla 1998) and stimulates cambial cell division and xylogenesis basipetally from the application site along the internode. In such experiments a dose response correlation between the resulting internal IAA levels and the cambial growth was established both in conifers (Scots pine) and angiosperm trees (hybrid aspen) (Sundberg and Little 1990; Björklund et al. 2007). The importance of auxin for most aspects of wood formation is also reflected in the large part of the transcriptome that is affected when auxin is applied to auxin-depleted internodes of hybrid aspen (Björklund et al. 2007; Nilsson et al. 2008). Endogenous IAA is exhibiting a steep concentration gradient across wood-forming tissues with a peak level in the cambial zone both in conifers and angiosperm trees. The auxin concentration gradient was suggested to provide positional signaling required for coordinated wood development (Uggla et al. 1996; Tuominen et al. 1997). In Scots pine a correlation was found between cambial growth and the amount of IAA across the cambial tissues, rather than with its concentration in the cambial zone. In line with the idea that auxin provides positional information, it was suggested that it regulates the radial number of dividing cambial cells, which is important for the rate of wood cell production (Uggla et al. 1998). The multiple roles of auxin in wood development are also suggested from the expression of putative auxin-signaling genes across the different wood-forming tissues (Moyle et al. 2002). The expression pattern of auxin-signaling genes varied for the different genes across the developing wood, hence they did not correlate with the auxin concentration gradient. A similar non-overlap between endogenous auxin and the auxin-responsive *GH3* promoter was shown in *Populus* trees expressing a *GH3::GUS* construct (Teichmann et al. 2008). Transgenic hybrid aspen trees expressing a mutated version of the IAA signaling gene *PttIAA3* exhibited a decrease in auxin responsiveness and also in cambial growth and xylem cell size (Nilsson et al. 2008). It is however unclear how the decreased auxin responsiveness in these trees affected overall plant growth and development and whether the effect on cambial growth was a result of dwarfing.

The potential of applied auxin to affect all aspects of wood development suggests that it could be involved in signaling in the TW response. In angiosperm trees, application of auxin and auxin transport inhibitors has been observed to induce TW fibers with characteristic G-layers, suggesting a role for auxin in the TW response (Cronshaw and Morey 1968; Morey and Cronshaw 1968). However, the results from these and similar experiments are inconsistent and contradictory and therefore difficult to interpret [summarized in Little and Savidge (1987) and Hellgren et al. (2004)].

When the endogenous IAA distribution across wood-forming tissues was compared in TW and normal wood of *Populus tremula*, no clear difference was observed, despite the higher cambial growth rate in the TW-forming trees (Hellgren et al. 2004). However, at the opposite, lower side of the leaning trees IAA levels were decreased, as was cambial growth. Despite any obvious change in IAA balance across wood-forming tissues during TW formation, the transcript abundance of several putative auxin transporters and signaling genes is affected in association to the TW response in *Populus* (Moyle et al. 2002; Paux et al. 2005; Andersson-Gunnerås et al. 2006). This suggests that auxin-signaling pathways somehow are involved in TW formation. Possibly the responsiveness to auxin is modified in leaning stems as suggested from experiments where IAA was applied to horizontal shoots and stimulated cambial growth at the upper but not at the lower side (Wareing et al. 1964). However, experimental indications supporting a role for auxin per se in TW signaling are meager.

## 3.2 Gibberellins

GAs stimulate both cambial cell division and fiber elongation when applied to woody tissues or when GA levels or perception is increased in genetically modified *Populus* trees (Little and Savidge 1987; Ridoutt et al. 1996; Eriksson et al. 2000; Dünisch et al. 2006; Mauriat and Moritz 2009; Gou et al. 2011). GA stimulation of cambial growth is strongly synergistic with auxin, and at a certain auxin-to-GA balance fewer and smaller vessels are formed similar to what is observed in TW (Digby and Wareing 1966; Little and Savidge 1987; Björklund et al. 2007). The synergy with auxin can at least partly be explained by the stimulating effect of GAs on polar auxin transport (Björklund et al. 2007). When GAs are applied on their own to auxin-depleted decapitated stems, they still stimulate cell division. But the produced cells form dedifferentiated structures of parenchyma-like cells, and consequently differentiation into typical fibers or vessels was not observed (Little and Savidge 1987; Björklund et al. 2007). Transgenic *Populus* trees deficient in GAs due to increased activity of GA2 oxidase (which catabolizes active GAs) are dwarfs and, as a consequent of dwarfism, also have decreased cambial growth (Busov et al. 2003; Mauriat and Moritz 2009; Gou et al. 2011).

Endogenous bioactive GAs measured across developing wood tissues in *Populus* exhibit a peak concentration localized to the zone of cell expansion, with low levels in the cambial meristem (Israelsson et al. 2005). This observation questions whether the main function of endogenous GAs is to stimulate cambial cell division. It can not be excluded that application experiments and ectopic transgene expression of GA metabolism/perception genes could result in abnormal GA distribution/perception, and the resulting phenotype may therefore not reflect endogenous GA function. Information about the balance of endogenous GAs or IAA-to-GAs ratios in wood-forming tissues of angiosperm trees with different growth rates is not yet available.

Several application experiments have been reported that point towards a role for GAs in TW formation. Weeping Japanese cherry (*Prunus spachiana*) is a variety with slender weeping branches due to poor cambial growth, and no TW is observed in these branches. When GAs were applied to weeping branches, they stimulated cambial growth and induced upright movement of the branch as a consequence of TW formation at the upper side (Nakamura et al. 1994; Baba et al. 1995; Yoshida et al. 1999). The mutation causing the weeping phenotype is not known. But these experiments suggest that the weeping phenotype is due to a lack of GA biosynthesis and that GAs are required for TW formation. When bioactive GAs were applied to tilted stems of *Fraxinus mandshuricas* and *Acacia mangium*, they stimulated the upward bending of the stem compared to control trees by increasing cambial growth and the amount of TW formed (Jiang et al. 1998; Nugroho et al. 2012). When GA biosynthesis inhibitors (paclobutrazole and/or uniconazole-P) were applied to the same experimental systems, the upward bending was negated and the formation of TW much reduced. However, the limited amount of TW that did form exhibited typical G-fibers. Together these studies indicate that the induced cambial growth associated to TW formation is dependent on GAs, but they do not provide evidence that GAs are involved in G-fiber differentiation. However, when GAs were applied to wood-forming tissues of upright trees from *Quercus, Kalopanax, Fraxinus,* and *Populus* ssp., they stimulated cambial growth and induced fibers with an inner cellulose-rich layer similar to G-fibers, despite the absence of any gravitational stimulus (Funada et al. 2008). Thus, applied GAs also have the potential to induce the developmental program of G-fibers. Altogether, these application studies strongly indicate a role for GAs in TW formation. Applied GAs stimulate cambial growth, enhance fiber to vessel ratio, and can trigger the induction of G-fibers. They also stimulate the TW response in tilted stems, and applied GA biosynthesis inhibitors inhibit the TW wood response. Supporting evidence from measurements of endogenous GA-to-IAA balance during TW formation or from GA insensitive mutants is still missing.

## 3.3 Ethylene

Applied ethylene (or its precursor ACC) stimulates cambial growth, modifies xylem cell shape, and, in *Populus*, induces a xylem anatomy with fewer and smaller vessels as observed in TW (Little and Savidge 1987; Du and Yamamoto 2007; Love et al. 2009). However, applied ethylene has not yet been reported to induce G-fibers in angiosperm trees. Ethylene is the only plant hormone that has been shown to increase in association to reaction wood formation, both in conifers and angiosperm trees (Du and Yamamoto 2007). In *Populus* ACC oxidase, ACO (the last enzyme in ethylene biosynthesis) is heavily induced in TW-forming tissues and probably responsible for the unilateral induction of ethylene biosynthesis in leaning stems (Andersson-Gunnerås et al. 2003).

Ethylene is sensed by ER-membrane bound receptors (Alonso and Stepanova 2004). A dominant negative mutation in the Arabidopsis ETR1 receptor (*etr1-1*) gives rise to ethylene insensitivity (Bleecker et al. 1998). Studies of ethylene insensitive Arabidopsis mutants showed that ethylene is not required for normal vegetative growth or xylogenesis (Tholen et al. 2004). The physiological function of environmentally induced ethylene is therefore ideally studied in ethylene insensitive transgenic plants. Ethylene insensitive hybrid aspen trees were produced by heterologous expression of *etr1-1* and used in combination with experiments based on application of the gas 1-Methylcyclopropene (1-MCP) that blocks ethylene perception on wild-type trees to dissect the physiological role of ethylene in the TW response (Love et al. 2009). Ethylene insensitivity did not inhibit the formation of fibers with typical G-layers in response to leaning. But it was demonstrated that the stimulation of cambial growth and stem eccentricity was at least partly a response to endogenous ethylene. This is the first conclusive demonstration of a role for a plant hormone in the TW response. The function of endogenous ethylene in other aspects of the TW response needs further investigation.

Downstream ethylene signaling pathways have been unraveled in Arabidopsis, and the transcription factors ethylene response factors (ERFs) have been shown to be involved in this signaling (Solano et al. 1998). Applied ethylene (or its precursor ACC) was found to modify the transcript abundance (fold change >5 times) of 72 out of 169 *ERF*s in *Populus* stems (Vahala J, Felten J, Love J, Gorzsás A, Gerber L, Lamminmäki A, Kangasjärvi J and Sundberg B, unpublished). Some of these *ERF*s were also induced in TW-forming tissues in leaning stems and had the potential to modify different aspects of wood formation when overexpressed in transgenic hybrid aspen, suggesting that they may be part of the transcriptional network regulating TW formation.

## 3.4 Reflections on the Role of Plant Hormones in the TW Response

Both IAA and GAs are required for normal cambial growth and xylem development. The question is whether a drastic change in the level of one or both of them is part of the TW signaling. There is little evidence from application studies or measurements of endogenous levels that a change in IAA concentration is part of the TW induction per se. But the decrease in IAA level on the lower side of tilted stems, and the different growth response of upper and lower side when IAA is applied to horizontal stems, suggest that auxin transport and/or signaling is directly or indirectly part of the TW signaling (Hellgren et al. 2004; Wareing et al. 1964). On the same theme, it can be pointed out that redistribution of auxin transport is of great importance for gravitropic responses of primary shoots (Friml et al. 2002). The GA application experiments with weeping Japanese cherry and tilted seedlings of *Acacia* and *Fraxinus* support the idea that an increase in GAs is part of TW signaling (Nakamura et al. 1994; Baba et al. 1995; Yoshida et al. 1999; Jiang et al. 1998; Nugroho et al. 2012). Moreover, applied GAs mimic the induction of G-fibers in several tree species (Funada et al. 2008). The resulting concentration of GAs and IAA in the developing wood tissues is unknown in these experiments, which complicates their interpretation. However, in the presence of auxin, applied GAs caused a large stimulation on cambial growth in *Populus* (resembling the TW response), and resulting internal concentrations were found to be physiological relevant (Björklund et al. 2007). Information on endogenous GA-to-IAA balance during TW formation would be helpful to evaluate their roles in TW signaling. Ethylene levels do increase during TW formation and are important for the increased cambial growth associated to TW formation (Love et al. 2009). Applied ethylene also induce typical TW anatomy with fewer and smaller vessels, but conclusive evidence for other roles of ethylene in TW formation needs further studies. Despite the indications that both GAs and ethylene are part of TW signaling, neither ethylene insensitivity nor GA inhibitors could inhibit the formation of G-layers. It cannot be excluded that an increase in both hormones takes place during TW formation and that their function is partly redundant. Moreover, most likely TW signaling involves cross-talk between plant hormones as well as additional, yet not identified, components. The plant hormones are, however, most likely rather upstream primary responses to TW sensing.

**Acknowledgments** We kindly thank Kjell Olofsson and Dr. Melissa Roach for providing graphical material for Fig. 1 and Drs Urs Fischer and Totte Niittylä for reviewing this chapter. We also thank FORMAS, Swedish Research Council, VINNOVA, and Bio4Energy (the Swedish Programme for renewable energy) for funding.

## References

Aguayo MG, Quintupill L, Castillo R, Baeza J, Freer J, Mendonca RT (2010) Determination of differences in anatomical and chemical characteristics of tension and opposite wood of 8-year old *Eucalyptus globulus*. Maderas-Ciencia Y Tecnologia 12:241–251

Almeras T, Fournier M (2009) Biomechanical design and long-term stability of trees: morphological and wood traits involved in the balance between weight increase and the gravitropic reaction. J Theor Biol 256:370–381

Alonso JM, Stepanova AN (2004) The ethylene signaling pathway. Science 306:1513–1515

Andersson-Gunnerås S, Hellgren JM, Björklund S, Regan S, Moritz T, Sundberg B (2003) Asymmetric expression of a poplar ACC oxidase controls ethylene production during gravitational induction of tension wood. Plant J 34:339–349

Andersson-Gunnerås S, Mellerowicz EJ, Love J, Segerman B, Ohmiya Y, Coutinho PM et al (2006) Biosynthesis of cellulose-enriched tension wood in Populus: global analysis of transcripts and metabolites identifies biochemical and developmental regulators in secondary wall biosynthesis. Plant J 45:144–165

Araki N, Fujita M, Saiki H, Harada H (1982) Transition of the fiber wall structure from normal wood to tension wood in *Robinia pseudoacacia* L. and *Populus euramericana* Guinier. Mokuzai Gakhaishi 28:267–273

Arend M (2008) Immunolocalization of (1,4)-beta-galactan in tension wood fibers of poplar. Tree Physiol 28:1263–1267

Baba K, Adachi K, Take T, Yokoyama T, Itoh T, Nakamura T (1995) Induction of tension wood in Ga(3)-treated branches of the weeping type of Japanese-cherry, *Prunus spachiana*. Plant Cell Physiol 36:983–988

Bailleres H, Chanson B, Fournier M, Tollier MT, Monties B (1995) Wood structure, chemical-composition and growth strains in Eucalyptus clones. Annales Des Sciences Forestieres 52: 157–172

Bentum ALK, Cote WA, Day AC, Timell TE (1969) Distribution of lignin in normal and tension wood. Wood Sci Technol 3:218–231

Björklund S, Antti H, Uddestrand I, Moritz T, Sundberg B (2007) Cross-talk between gibberellin and auxin in development of Populus wood: gibberellin stimulates polar auxin transport and has a common transcriptome with auxin. Plant J 52:499–511

Bjurhager I, Olsson AM, Zhang B, Gerber L, Kumar M, Berglund LA et al (2010) Ultrastructure and mechanical properties of populus wood with reduced lignin content caused by transgenic down-regulation of cinnamate 4-hydroxylase. Biomacromolecules 11:2359–2365

Bleecker AB, Esch JJ, Hall AE, Rodriguez FI, Binder BM (1998) The ethylene-receptor family from Arabidopsis: structure and function. Philos Trans R Soc Lond B Biol Sci 353:1405–1412

Bowling AJ, Vaughn KC (2008) Immunocytochemical characterization of tension wood: gelatinous fibers contain more than just cellulose. Am J Bot 95:655–663

Busov VB, Meilan R, Pearce DW, Ma CP, Rood SB, Strauss SH (2003) Activation tagging of a dominant gibberellin catabolism gene (GA 2-oxidase) from poplar that regulates tree stature. Plant Physiol 132:1283–1291

Chang SS, Clair B, Ruelle J, Beauchene J, Di Renzo F, Quignard F et al (2009) Mesoporosity as a new parameter for understanding tension stress generation in trees. J Exp Bot 60:3023–3030

Clair B, Thibaut B (2001) Shrinkage of the gelatinous layer of poplar and beech tension wood. IAWA J 22:121–131

Clair B, Gril J, Baba K, Thibaut B, Sugiyama J (2005a) Precautions for the structural analysis of the gelatinous layer in tension wood. IAWA J 26:189–195

Clair B, Thibaut B, Sugiyama J (2005b) On the detachment of the gelatinous layer in tension wood fiber. J Wood Sci 51:218–221

Clair B, Almeras T, Yamamoto H, Okuyama T, Sugiyama J (2006a) Mechanical behavior of cellulose microfibrils in tension wood, in relation with maturation stress generation. Biophys J 91:1128–1135

Clair B, Ruelle J, Beauchene J, Prevost MF, Fournier M (2006b) Tension wood and opposite wood in 21 tropical rain forest species 1. Occurrence and efficiency of the G-layer. IAWA J 27: 329–338

Clair B, Gril J, Di Renzo F, Yamamoto H, Quignard F (2008) Characterization of a gel in the cell wall to elucidate the paradoxical shrinkage of tension wood. Biomacromolecules 9:494–498

Clair B, Almeras T, Pilate G, Jullien D, Sugiyama J, Riekel C (2011) Maturation stress generation in poplar tension wood studied by synchrotron radiation microdiffraction. Plant Physiol 155: 562–570

Côté WA, Day AC, Timell TE (1969) A contribution to ultrastructure of tension wood fibers. Wood Sci Technol 3:257–271

Cronshaw J, Morey PR (1968) Effect of plant growth substances on development of tension wood in horizontally inclined stems of *Acer rubrum* seedlings. Protoplasma 65:379–391

Decou R, Lhernould S, Laurans F, Sulpice E, Leple JC, Dejardin A et al (2009) Cloning and expression analysis of a wood-associated xylosidase gene (PtaBXL1) in poplar tension wood. Phytochemistry 70:163–172

Dejardin A, Leple JC, Lesage-Descauses MC, Costa G, Pilate G (2004) Expressed sequence tags from poplar wood tissues – a comparative analysis from multiple libraries. Plant Biol 6:55–64

Digby J, Firn RD (1995) The gravitropic set-point angle (GSA): the identification of an important developmentally controlled variable governing plant architecture. Plant Cell Environ 18:1434–1440

Digby J, Wareing PF (1966) Effect of applied growth hormones on cambial division and differentiation of cambial derivatives. Ann Bot 30:539–548

Donaldson L, Singh A (2013) Formation and structure of compression wood. In: Fromm J (ed) Cellular aspects of wood formation, vol 20. Springer, Heidelberg, pp 225–256

Du S, Yamamoto F (2007) An overview of the biology of reaction wood formation. J Integr Plant Biol 49:131–143

Dünisch O, Fladung M, Nakaba S, Watanabe Y, Funada R (2006) Influence of overexpression of a gibberellin 20-oxidase gene on the kinetics of xylem cell development in hybrid poplar (*Populus tremula* L. and *P. tremuloides* Michx.). Holzforschung 60:608–617

Eriksson ME, Israelsson M, Olsson O, Moritz T (2000) Increased gibberellin biosynthesis in transgenic trees promotes growth, biomass production and xylem fiber length. Nat Biotechnol 18:784–788

Fang CH, Clair B, Gril J, Liu SQ (2008) Growth stresses are highly controlled by the amount of G-layer in poplar tension wood. IAWA J 29:237–246

Friml J, Wisniewska J, Benkova E, Mendgen K, Palme K (2002) Lateral relocation of auxin efflux regulator PIN3 mediates tropism in Arabidopsis. Nature 415:806–809

Funada R, Miura T, Shimizu Y, Kinase T, Nakaba S, Kubo T et al (2008) Gibberellin-induced formation of tension wood in angiosperm trees. Planta 227:1409–1414
Furuya N, Takahashi S, Miyazaki M (1970) The chemical composition of gelatinous layer from the tension wood of *Populus euroamericana*. Mokuzai Gakkaishi 16:26–30
Gierlinger N, Schwanninger M (2006) Chemical imaging of poplar wood cell walls by confocal Raman microscopy. Plant Physiol 140:1246–1254
Goswami L, Dunlop JWC, Jungnikl K, Eder M, Gierlinger N, Coutand C et al (2008) Stress generation in the tension wood of poplar is based on the lateral swelling power of the G-layer. Plant J 56:531–538
Gou JQ, Ma C, Kadmiel M, Gai Y, Strauss S, Jiang XN et al (2011) Tissue-specific expression of Populus C-19 GA 2-oxidases differentially regulate above- and below-ground biomass growth through control of bioactive GA concentrations. New Phytol 192:626–639
Gustafsson C, Ollinmaa PJ, Saarnio J (1952) The carbohydrates in birchwood. Acta Chem Scand 6:1299–1300
Hartig R (1901) Holzuntersuchungen. Altes und Neues. Julius Springer Verlag, 99
Hartmann F (1932) Ursachen und Gesetzmässigkeit exzentrischen Dickenwachstums bei Nadel- und Laubbäumen. Forstwissenschaftliches Centralblatt 54:497
Hedenström M, Wiklund-Lindstrom S, Oman T, Lu FC, Gerber L, Schatz P et al (2009) Identification of lignin and polysaccharide modifications in Populus wood by chemometric analysis of 2D NMR spectra from dissolved cell walls. Mol Plant 2:933–942
Hellgren JM, Olofsson K, Sundberg B (2004) Patterns of auxin distribution during gravitational induction of reaction wood in poplar and pine. Plant Physiol 135:212–220
Hobson N, Roach MJ, Deyholos MK (2010) Gene expression in tension wood and bast fibres. Russ J Plant Physiol 57:321–327
Israelsson M, Sundberg B, Moritz T (2005) Tissue-specific localization of gibberellins and expression of gibberellin-biosynthetic and signaling genes in wood-forming tissues in aspen. Plant J 44:494–504
Jaccard P (1938) Exzentrisches Dickenwachstum und anatomische-histologische Differenzierung des Holzes. Berichte der Schweizerischen Botanischen Gesellschaft 48:491–537
Jiang S, Furukawa I, Honma T, Mori M, Nakamura T, Yamamoto F (1998) Effects of applied gibberellins and uniconazole-P on gravitropism and xylem formation in horizontally positioned Fraxinus mandshurica seedlings. J Wood Sci 44:385–391
Jin H, Do J, Moon D, Noh EW, Kim W, Kwon M (2011) EST analysis of functional genes associated with cell wall biosynthesis and modification in the secondary xylem of the yellow poplar (*Liriodendron tulipifera*) stem during early stage of tension wood formation. Planta 234: 959–977
Joseleau JP, Imai T, Kuroda K, Ruel K (2004) Detection in situ and characterization of lignin in the G-layer of tension wood fibres of *Populus deltoides*. Planta 219:338–345
Jourez B, Riboux A, Leclercq A (2001) Anatomical characteristics of tension wood and opposite wood in young inclined stems of poplar (Populus euramericana cv 'Ghoy'). IAWA J 22: 133–157
Kaku T, Serada S, Baba K, Tanaka F, Hayashi T (2009) Proteomic analysis of the G-layer in poplar tension wood. J Wood Sci 55:250–257
Kennedy RW, Farrar JL (1965) Tracheid development in tilted seedlings. In: Coté, WA Jr (ed) Cellular ultrastructure of woody plants. Proceedings, Advanced Science Seminar, New York 1964. Pages: xii + 603 pp. Syracuse University Press; Syracuse; USA
Kim JS, Sandquist D, Sundberg B, Daniel G (2012) Spatial and temporal variability of xylan distribution in differentiating secondary xylem of hybrid aspen. Planta 235:1315–1330
Kojima M, Becker VK, Altaner CM (2012) An unusual form of reaction wood in Koromiko [Hebe salicifolia G. Forst. (Pennell)], a southern hemisphere angiosperm. Planta 235:289–297
Kuo CM, Timell TE (1969) Isolation and Characterization of a Galactan from Tension Wood of American Beech (Fagus Grandifolia Ehrl). Svensk Papperstidning-Nordisk Cellulosa 72:703

Kwon M, Bedgar DL, Piastuch W, Davin LB, Lewis NG (2001) Induced compression wood formation in Douglas fir (*Pseudotsuga menziesii*) in microgravity. Phytochemistry 57:847–857

Lafarguette F, Leple JC, Dejardin A, Laurans F, Costa G, Lesage-Descauses MC et al (2004) Poplar genes encoding fasciclin-like arabinogalactan proteins are highly expressed in tension wood. New Phytol 164:107–121

Lautner S, Zollfrank C, Fromm J (2012) Microfibril angle distribution of poplar tension wood. IAWA J 33:431–439

Lehringer C, Gierlinger N, Koch G (2008) Topochemical investigation on tension wood fibres of *Acer* spp., *Fagus sylvatica* L. and *Quercus robur* L. Holzforschung 62:255–263

Little CHA, Savidge RA (1987) The role of plant-growth regulators in forest tree cambial growth. Plant Growth Regul 6:137–169

Love J, Bjorklund S, Vahala J, Hertzberg M, Kangasjarvi J, Sundberg B (2009) Ethylene is an endogenous stimulator of cell division in the cambial meristem of Populus. Proc Natl Acad Sci USA 106:5984–5989

MacMillan CP, Mansfield SD, Stachurski ZH, Evans R, Southerton SG (2010) Fasciclin-like arabinogalactan proteins: specialization for stem biomechanics and cell wall architecture in Arabidopsis and Eucalyptus. Plant J 62:689–703

Mauriat M, Moritz T (2009) Analyses of GA20ox- and GID1-over-expressing aspen suggest that gibberellins play two distinct roles in wood formation. Plant J 58:989–1003

McLean JP, Arnould O, Beauchene J, Clair B (2012) The effect of the G-layer on the viscoelastic properties of tropical hardwoods. Ann For Sci 69:399–408

Meier H (1962) Studies on a galactan from tension wood of beech (*Fagus-Sylvatica* L.). Acta Chem Scand 16:2275–2283

Mellerowicz EJ, Gorshkova TA (2012) Tensional stress generation in gelatinous fibres: a review and possible mechanism based on cell-wall structure and composition. J Exp Bot 63:551–565

Mellerowicz EJ, Sundberg B (2008) Wood cell walls: biosynthesis, developmental dynamics and their implications for wood properties. Curr Opin Plant Biol 11:293–300

Morey PR, Cronshaw J (1968) Induction of tension wood by 2,4-dinitrophenol and auxins. Protoplasma 65:393–405

Moyle R, Schrader J, Stenberg A, Olsson O, Saxena S, Sandberg G et al (2002) Environmental and auxin regulation of wood formation involves members of the Aux/IAA gene family in hybrid Aspen. Plant J 31:675–685

Muller M, Burghammer M, Sugiyama J (2006) Direct investigation of the structural properties of tension wood cellulose microfibrils using microbeam X-ray fibre diffraction. Holzforschung 60:474–479

Nakamura T, Saotome M, Ishiguro Y, Itoh R, Higurashi S, Hosono M et al (1994) The effects of Ga(3) on weeping of growing shoots of the Japanese-cherry, Prunus-spachiana. Plant Cell Physiol 35:523–527

Nieminen K, Robischon M, Immanen J, Helariutta Y (2012) Towards optimizing wood development in bioenergy trees. New Phytol 194:46–53

Nilsson J, Karlberg A, Antti H, Lopez-Vernaza M, Mellerowicz E, Perrot-Rechenmann C et al (2008) Dissecting the molecular basis of the regulation of wood formation by auxin in hybrid aspen. Plant Cell 20:843–855

Nishikubo N, Awano T, Banasiak A, Bourquin V, Ibatullin F, Funada R et al (2007) Xyloglucan endo-transglycosylase (XET) functions in gelatinous layers of tension wood fibers in poplar – a glimpse into the mechanism of the balancing act of trees. Plant Cell Physiol 48:843–855

Norberg PH, Meier H (1966) Physical and chemical properties of gelatinous layer in tension wood fibres of aspen (*Populus Tremula* L.). Holzforschung 20:174–178

Nugroho WD, Yamagishi Y, Nakaba S, Fukuhara S, Begum S, Marsoem SN et al (2012) Gibberellin is required for the formation of tension wood and stem gravitropism in *Acacia mangium* seedlings. Ann Bot 110:887–895

Okuyama T, Yamamoto H, Yoshida M, Hattori Y, Archer RR (1994) Growth stresses in tension wood – role of microfibrils and lignification. Annales Des Sciences Forestieres 51:291–300

Onaka F (1949) Studies on compression and tension wood. Wood Research, Bulletin of the Wood Research Institute, Kyoto University, Japan, vol 24, pp 1–88

Paux E, Carocha V, Marques C, de Sousa AM, Borralho N, Sivadon P et al (2005) Transcript profiling of *Eucalyptus xylem* genes during tension wood formation. New Phytol 167:89–100

Pilate G, Chabbert B, Cathala B, Yoshinaga A, Leple JC, Laurans F et al (2004) Lignification and tension wood. C R Biol 327:889–901

Prodhan AKMA, Ohtani J, Funada R, Abe H, Fukazawa K (1995) Ultrastructural investigation of tension wood fiber in *Fraxinus-Mandshurica* Rupr. Var. *japonica* Maxim. Ann Bot 75: 311–317

Ridoutt BG, Pharis RP, Sands R (1996) Fibre length and gibberellins A(1) and A(20) are decreased in *Eucalyptus globulus* by acylcyclohexanedione injected into the stem. Physiol Plant 96: 559–566

Ruelle J, Clair B, Beauchene J, Prevost MF, Fournier M (2006) Tension wood and opposite wood in 21 tropical rain forest species 2. Comparison of some anatomical and ultrastructural criteria. IAWA J 27:341–376

Ruelle J, Yoshida M, Clair B, Thibaut B (2007) Peculiar tension wood structure in *Laetia procera* (Poepp.) Eichl. (Flacourtiaceae). Trees Struct Funct 21:345–355

Sandquist D, Filonova L, von Schantz L, Ohlin M, Daniel G (2010) Microdistribution of xyloglucan in differentiating poplar cells. Bioresources 5:796–807

Scurfield G (1973) Reaction wood – its structure and function. Science 179:647–655

Scurfield G, Wardrop AB (1963) The nature of reaction wood. VII. Lignification in reaction wood. Aust J Bot 11:107–116

Solano R, Stepanova A, Chao QM, Ecker JR (1998) Nuclear events in ethylene signaling: a transcriptional cascade mediated by ETHYLENE-INSENSITIVE3 and ETHYLENE-RESPONSE-FACTOR1. Genes Dev 12:3703–3714

Sugiyama K, Okuyama T, Yamamoto H, Yoshida M (1993) Generation process of growth stresses in cell-walls – relation between longitudinal released strain and chemical-composition. Wood Sci Technol 27:257–262

Sultana RS, Ishiguri F, Yokota S, Iizuka K, Hiraiwa T, Yoshizawa N (2010) Wood anatomy of nine Japanese hardwood species forming reaction wood without gelatinous fibers. IAWA J 31: 191–202

Sundberg B, Little CHA (1990) Tracheid production in response to changes in the internal level of indole-3-acetic-acid in 1-year-old shoots of Scots pine. Plant Physiol 94:1721–1727

Sundberg B, Uggla C (1998) Origin and dynamics of indoleacetic acid under polar transport in *Pinus sylvestris*. Physiol Plant 104:22–29

Sundberg B, Uggla C, Tuominen H, (2000) Cambial growth and auxin gradients. In Savidge RA, Barnett, JR, Napier R (ed): Cell and molecular biology of wood formation, p169. Experimental Biology Reviews, Bios Scientific Publishers LTD, Oxon, England

Teichmann T, Bolu-Arianto WH, Olbrich A, Langenfeld-Heyser R, Gobel C, Grzeganek P et al (2008) GH3: GUS reflects cell-specific developmental patterns and stress-induced changes in wood anatomy in the poplar stem. Tree Physiol 28:1305–1315

Telewski FW (2006) A unified hypothesis of mechanoperception in plants. Am J Bot 93: 1466–1476

Tholen D, Voesenek LACJ, Poorter H (2004) Ethylene insensitivity does not increase leaf area or relative growth rate in arabidopsis, *Nicotiana tabacum*, and *Petunia* × *hybrida*. Plant Physiol 134:1803–1812

Timell TE (1969) Chemical composition of tension wood. Svensk Papperstidning-Nordisk Cellulosa 72:173–181

Timell TE (1986) Compression wood in gymnosperms, vol 2. Springer, Berlin

Tuominen H, Puech L, Fink S, Sundberg B (1997) A radial concentration gradient of indole-3-acetic acid is related to secondary xylem development in hybrid aspen. Plant Physiol 115: 577–585

Uggla C, Moritz T, Sandberg G, Sundberg B (1996) Auxin as a positional signal in pattern formation in plants. Proc Natl Acad Sci USA 93:9282–9286

Uggla C, Mellerowicz EJ, Sundberg B (1998) Indole-3-acetic acid controls cambial growth in Scots pine by positional signaling. Plant Physiol 117:113–121

Wardrop AB, Dadswell HE (1955) The nature of reaction wood. IV. Variations in cell wall organization in tension wood fibers. Aust J Bot 3:177–189

Wareing PF, Hanney CEA, Digby J (1964) The role of endogenous hormones in cambial activity and xylem differentiation. In: Zimmermann MH (ed) The formation of wood in forest trees. Academic, New York, pp 323–344

Wilson BF, Archer RR (1981) Apical control of branch movements in white-pine – biological aspects. Plant Physiol 68:1285–1288

Wilson BF, Archer RR (1983) Apical control of branch movements and tension wood in black cherry and white ash trees. Can J For Res (Revue Canadienne De Recherche Forestiere) 13:594–600

Yoshida M, Nakamura T, Yamamoto H, Okuyama T (1999) Negative gravitropism and growth stress in GA(3)-treated branches of *Prunus spachiana* Kitamura f. *spachiana* cv. *Plenarosea.* J Wood Sci 45:368–372

Yoshida M, Ohta H, Okuyama T (2002a) Tensile growth stress and lignin distribution in the cell walls of black locust (*Robinia pseudoacacia*). J Wood Sci 48:99–105

Yoshida M, Ohta H, Yamamoto H, Okuyama T (2002b) Tensile growth stress and lignin distribution in the cell walls of yellow poplar, Liriodendron tulipifera Linn. Trees Struct Funct 16: 457–464

Yoshizawa N, Koike S, Idei T (1984) Structural changes of tracheid wall accompanied by compression wood formation in *Taxus cuspidata* and *Torreya nucifera.* Bull Utsunomiya Univ For 20:59–76

Yoshizawa N, Inami A, Miyake S, Ishiguri F, Yokota S (2000) Anatomy and lignin distribution of reaction wood in two Magnolia species. Wood Sci Technol 34:183–196

Zhong R, Ye ZH (2013) Transcriptional regulation of wood formation in tree species. In: Fromm J (ed) Cellular aspects of wood formation, vol 20. Springer, Heidelberg, pp 141–158

# Formation and Structure of Compression Wood

**L.A. Donaldson and A.P. Singh**

**Abstract** Compression wood is a hard, dark-coloured wood typically found on the lower side of leaning stems and branches in conifers, *Taxus* and *Ginkgo*. This reaction wood is the result of the geotropic response of the tree, usually resulting from stem lean or the effect of stem flexing caused by wind. Compression wood is characterised by anatomical and compositional features that vary in a continuum between normal wood and severe compression wood. The main characteristics of compression wood are altered cell wall structure especially increased microfibril orientation, presence of helical cavities and intercellular spaces and increased lignification associated with significant amounts of $(1 \rightarrow 4)$-β-galactan in the secondary wall. This chapter briefly reviews the formation, structure and composition of compression wood with an emphasis on recent advances.

## Abbreviations

| | |
|---|---|
| DNA | Deoxyribonucleic acid |
| EST | Expressed sequence tag |
| IAA | Indole acetic acid |
| MFA | Cellulose microfibril angle |
| NAA | 1-Naphthaleneacetic acid |
| NPA | 1-N-naphthylphthalamic acid |
| SDS-PAGE | Sodium dodecyl sulphate polyacrylamide gel electrophoresis |
| TEM | Transmission electron microscopy |
| XET | Xyloglucan endotransglycosylase |

L.A. Donaldson (✉) • A.P. Singh
Scion, Private Bag 3020, Rotorua 3010, New Zealand
e-mail: lloyd.donaldson@scionresearch.com; adya.singh@scionresearch.com

J. Fromm (ed.), *Cellular Aspects of Wood Formation*, Plant Cell Monographs 20, DOI 10.1007/978-3-642-36491-4_9, 

# 1 Compression Wood Occurrence

Compression wood is a hard, dark-coloured wood typically found on the lower side of leaning stems and branches (Fig. 1). This reaction wood is the result of the geotropic response of the tree, usually resulting from stem lean or the effect of stem flexing caused by wind. It is generally associated with eccentric growth thus forming wide, crescent shaped regions of dark-coloured wood which are responsible for generating the growth stress to correct stem lean (Timell 1986). Compression wood can also occur in apparently straight stems, as a result of either corrected lean or stimulated growth, and is often associated with branch whorls (Cown 1974). In certain circumstances, compression wood can also occur on the upper side of leaning stems as a response to stress caused by upward flexure of the stem (Starbuck and Roberts 1983; Yoshizawa et al. 1986a).

Compression wood occurs in the secondary xylem of all conifers (Yoshizawa et al. 1986b; Yoshizawa and Idei 1987), as well as *Taxus* and *Ginkgo* (Timell 1978a, b), but is absent from other gymnosperms such as Cycadales and Gnetales (Westing 1965; Tomlinson 2001; Fisher and Marler 2006; Altaner et al. 2010). *Ginkgo*, *Taxus* and Araucariaceae have less well-developed compression wood features with helical cavities being absent or indistinct (Westing 1965; Timell 1978a, b). *Taxus* retains its characteristic helical thickenings in compression wood, whereas in Douglas fir (*Pseudotsuga menziesii* [Mirb.] Franco), helical thickenings which occur in normal wood are absent in compression wood. Compression wood therefore seems to have undergone some recent evolutionary development (Timell 1983). Some features of compression wood, including increased lignification and absence of an S3 layer, are present in reaction xylem of the primitive angiosperm *Buxus microphylla* Nakai (Yoshizawa et al. 1993a, b).

Severe compression wood has been of much interest to wood technologists and processors because of its negative impact on wood processing in solid wood, composite products and pulp and paper manufacture (Warensjö 2003; Ban et al. 2004; Akbulut and Nadir 2006; Tarmian et al. 2008). This is the result of both the anatomical and compositional features of severe compression wood, which are very different from normal wood. For example, greater longitudinal shrinkage of compression wood compared to normal wood is considered to be related to increased cellulose microfibril angle (MFA) in the S2 layer of compression wood tracheids (Harris & Meylan 1965; Meylan 1972; Harris 1977; Megraw et al. 1998; Jiang et al. 2002; Cown et al. 2004) and to its increased lignin and galactan content (Floyd 2005; Brennan et al. 2012).

The literature on compression wood is extensive and the reader is referred to the 3-volume monograph by Timell (1986) for a detailed review of historical aspects of compression wood biology and chemistry. This chapter briefly reviews the formation, structure and composition of compression wood with an emphasis on recent advances.

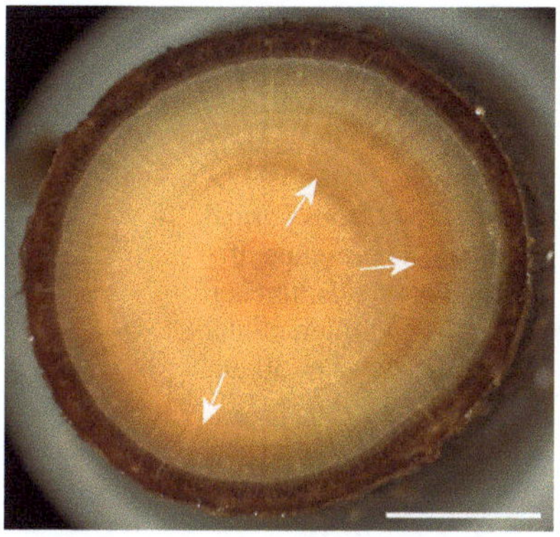

**Fig. 1** A 2-year-old radiata pine stem showing arcs of compression wood (*arrows*). Scalebar = 5 mm

## 2 Compression Wood Structure

The majority of descriptions of compression wood in the literature relate to severe compression wood, which develops on the underside of leaning stems and branches (Timell 1986). Compression wood typically has wide annual rings and increased density compared to normal wood, as a result of tracheids with small lumens and thick, highly lignified cell walls that differ in ultrastructure from those of normal wood (Timell 1986). Compression wood tracheids have characteristic differences in structure compared to normal wood tracheids (Figs. 2, 3 and 4). These anatomical or ultrastructural features include:

- Thicker cell walls with a rounded shape
- Intercellular spaces in the middle lamella at the cell corners
- A thicker S1 layer
- An S2 layer divided into an outer S2L region and an inner S2 region containing helical cavities that may extend from the inner S2 into the outer S2L and often show a branching structure
- High microfibril angle in the S2 region
- Extended pit apertures on bordered and crossfield pits
- Absence of an S3 layer (Wardrop and Dadswell 1950)

Helical cavities may not develop in the compression wood of some gymnosperms such as *Ginkgo* and *Taxus* (Timell 1978a, b). In species that normally have resin canals, large areas of severe compression wood that fill an entire growth ring will often lack resin canals, but in small areas of transient compression wood or in mild compression wood resin canals occur as normal. Compression wood tracheids are shorter than normal or opposite wood tracheids (Shelbourne and Ritchie 1968; Siripatanadilok and Leney 1985; Yoshizawa et al. 1985, 1987; Yoshizawa and Idei 1987) and may have deformed tips, indicating that intrusive growth may be restricted in compression wood (Yoshizawa et al. 1987).

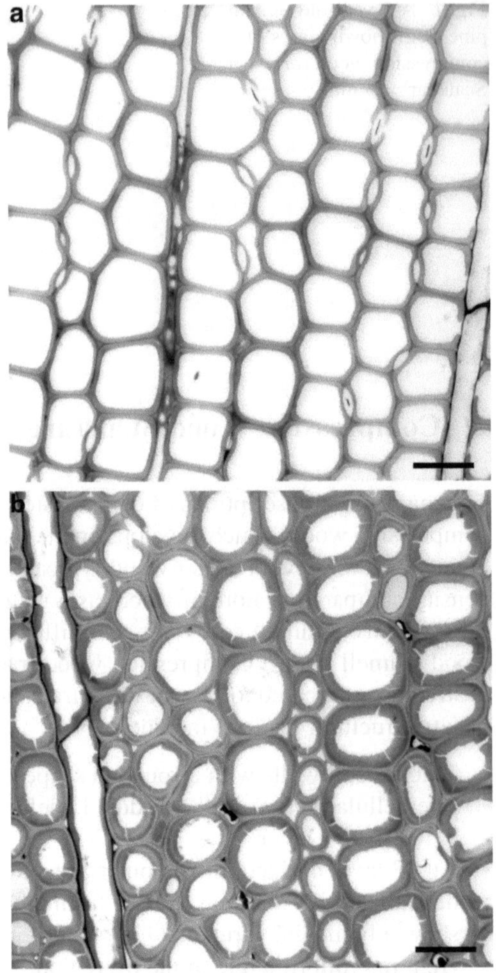


**Fig. 2** Transverse sections of normal wood (**a**) and compression wood (**b**) stained with toluidine blue. Compression wood is distinguished from normal wood by the presence of rounded tracheids, thicker cell walls and intercellular spaces. Scalebar = 30 μm

There are also some differences between compression wood and normal wood in inter-tracheid pitting. Pits in compression wood are smaller and fewer (Cockrell 1974; Lee and Eom 1988; Mayr et al. 2005; Tarmian et al. 2011), and the pit aperture is narrower featuring extensions oriented parallel to the cellulose microfibril orientation (Fig. 4c) (Cockrell 1974; Mayr et al. 2005). The narrower lumens, smaller pit apertures and fewer pits may result in reduced water conductibility in compression wood (Spicer and Gartner 1998; Mayr and Cochard 2003; Mayr et al. 2005).

The above description is for typical severe compression wood tracheids, but any of these features may be of variable occurrence in particular examples of compression wood irrespective of the severity of the compression wood. Intercellular

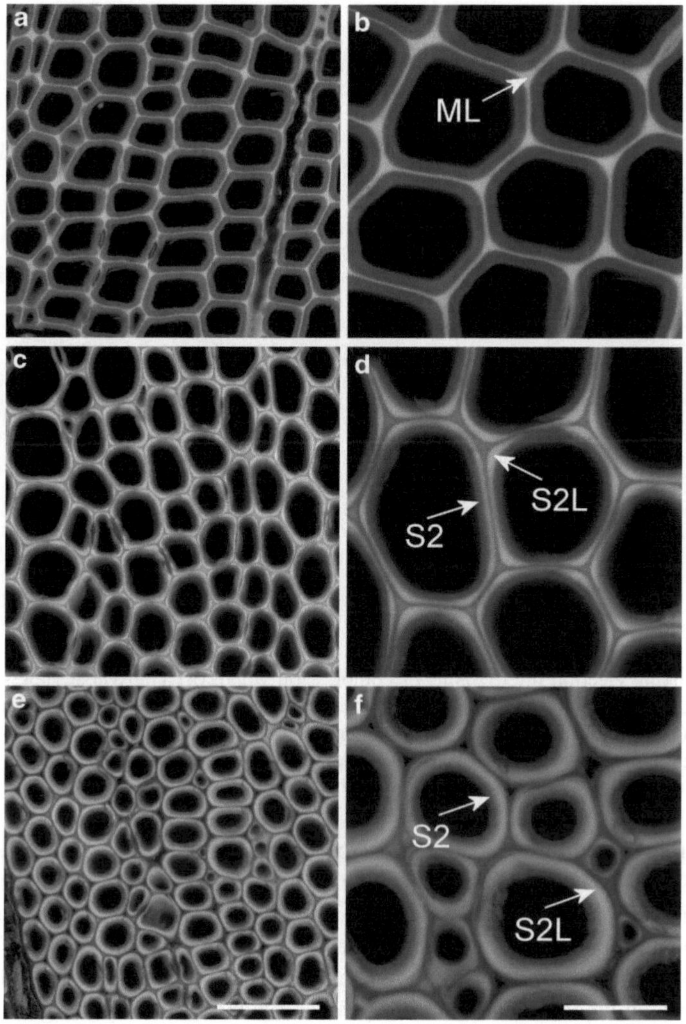


**Fig. 3** Confocal fluorescence images of radiata pine in transverse view showing lignin autofluorescence in normal wood (**a**, **b**), mild (**c**, **d**) and severe (**e**, **f**) compression wood. Both mild and severe compression wood are characterised by reduced lignification of the middle lamella (ML) and by the presence of a highly lignified S2L region in the outer secondary wall. (**a**, **c**, **e**) Scalebar = 75 μm. (**b**, **d**, **f**) Scalebar = 25 μm

spaces, for example, may not always be present. Some features however vary specifically in relation to the severity of the compression wood (e.g. see this chapter Sect. 5). Compression wood therefore has a variable structure depending on species, growing conditions, tree age and the cause of the compression wood (stem lean, wind, environmental stress).

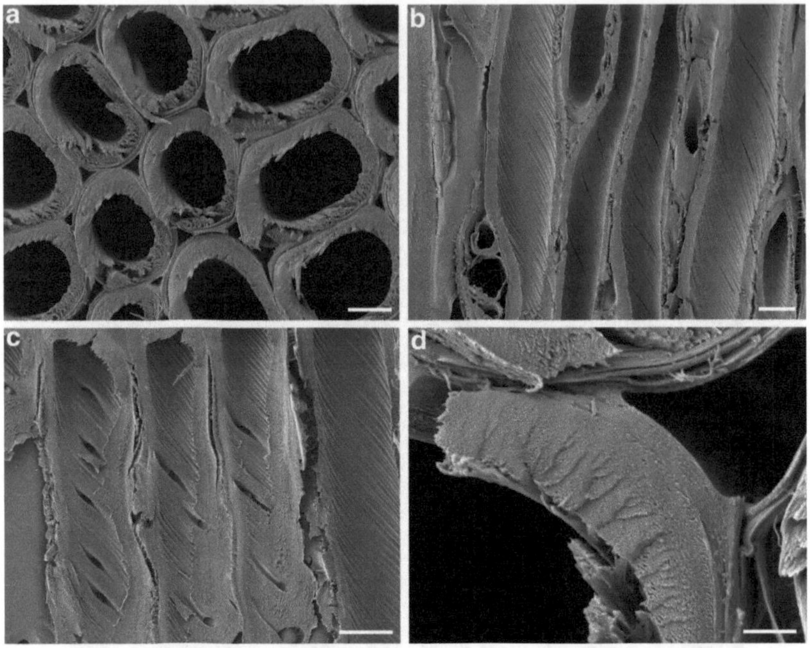


**Fig. 4** Scanning electron micrographs of compression wood. (**a**) Transverse view showing the rounded shape of tracheids and intercellular spaces. Scalebar = 10 μm. (**b**) Tangential longitudinal view showing helical cavities in the secondary cell wall of tracheids. Scalebar = 20 μm. (**c**) Radial longitudinal view showing extended bordered pit apertures. Scalebar = 20 μm. (**d**) Transverse view of helical cavities showing the branched structure. Scalebar = 2 μm

Compared to normal wood, compression wood contains higher amounts of lignin and (1 → 4)-β-galactan and proportionately lower amounts of cellulose, mannan and xylan. Compression wood cell wall composition varies in relation to severity, with greater amounts of lignin and galactan present in more severe forms and reduced amounts in mild forms (Nanayakkara et al. 2009). In the mildest forms of compression wood, the amount of lignin and galactan is only marginally greater than in normal wood. Thus, cell wall composition has been proposed as a useful quantitative chemical indicator for assessing the severity of compression wood (Nanayakkara et al. 2009).

Many early studies on compression wood focused on lignin distribution, using UV absorbance microscopy to characterise compression wood cell walls (Parham and Côté 1971; Fukazawa 1974). More recent studies have benefitted from the use of fluorescence microscopy and transmission electron microscopy (TEM) based on lignin autofluorescence and potassium permanganate staining, respectively (Singh 1997; Donaldson et al. 1999; Singh and Donaldson 1999).

In recent years, the focus of work has shifted towards understanding the spatial distribution of cell wall components other than lignin in differentiating and mature compression wood cell walls. Galactan is of particular interest because in

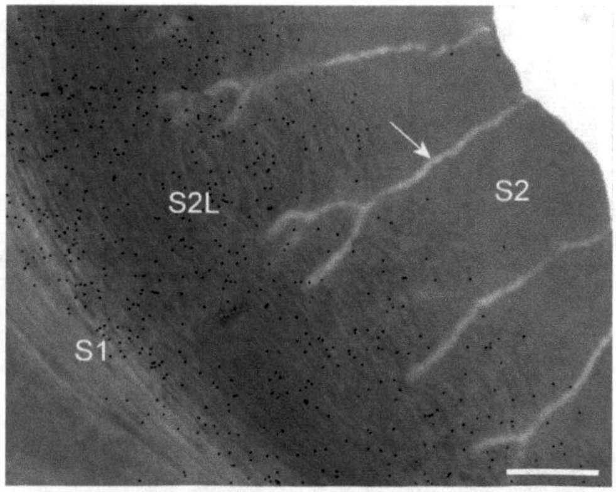


**Fig. 5** Transmission electron micrograph of radiata pine compression wood in transverse view labelled with LM5 antibody showing the distribution of galactan epitope by gold labelling (*black dots*) and stained with potassium permanganate to show lignin. The galactan epitope is localised mainly to the S2L region where it is associated with increased lignification. The *arrow* shows a helical cavity. Scalebar = 1 μm

compression wood, it is present in unusually high amounts, up to 14 % w/w in severe compression wood (Nanayakkara et al. 2009). Immunolocalisation studies have shown that galactan is preferentially located in the outer regions of the S2 wall that are also more highly lignified (Figs. 5 and 6), leading to suggestions that galactan may play a role in lignification (Schmitt et al. 2006; Altaner et al. 2007a; Mast et al. 2009; Kim et al. 2010; Donaldson and Knox 2012). Galactan can also serve as a useful chemical indicator of compression wood (Altaner et al. 2009; Nanayakkara et al. 2009).

# 3 Lignification

Compression wood shows marked changes in the distribution of lignin across the cell wall with reduced lignification of the middle lamella and increased lignification of the outer secondary cell wall, a region known as the S2L layer (Lange 1950; Wardrop and Davies 1964; Côté et al. 1966, 1968; Parham and Côté 1971; Wood and Goring 1971; Fukazawa 1974; Donaldson et al. 1999; Singh and Donaldson 1999; Wi et al. 2000). In mild compression wood, the middle lamella is less lignified despite the absence of intercellular spaces, while the S2L region is most obvious near the corners of cells (Fig. 3) (Donaldson et al. 1999, 2004). In severe compression wood, intercellular spaces reduce the contribution of middle lamella lignin to overall lignin content which is nevertheless increased by the greater lignification of the S2L layer. In this wood type, the S2L region extends all the way around the tracheid and is of more or less uniform thickness, especially when tracheids have a rounded shape (Lange 1950; Wardrop and Davies 1964; Côté et al.

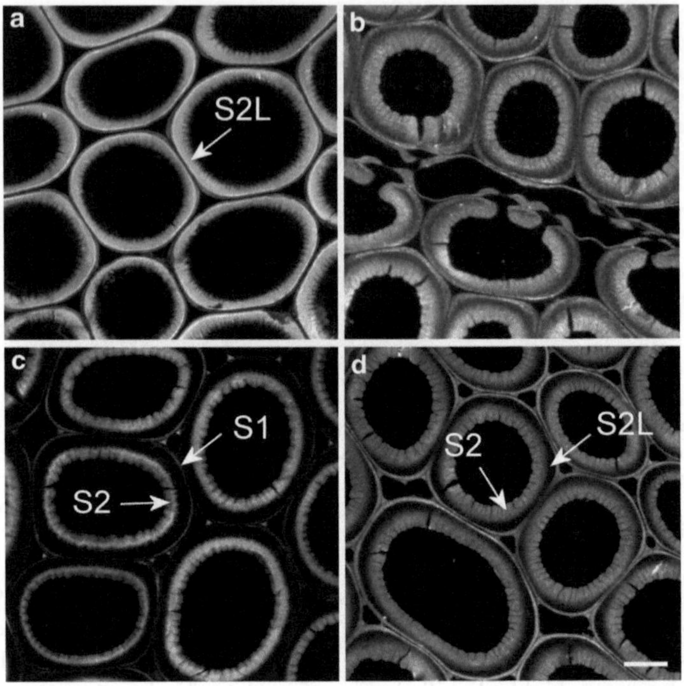


**Fig. 6** Transverse sections of radiata pine compression wood fluorescently immunolabeled with (**a**) LM 5 (galactan); (**b**) LM22 (mannan); (**c**) LM10 (low or unsubstituted xylan); and (**d**) LM 11 (xylan). Scalebar = 10 μm

1966, 1968; Wood and Goring 1971; Fukazawa 1974; Wi et al. 2000). Compression wood contains greater amounts of *p*-hydroxyphenyl lignin units synthesised from *p*-coumaryl alcohol, and studies using fluorescence spectroscopy and microautoradiography have shown that this lignin type is localised to both the S2L region and middle lamella (Fukushima and Terashima 1991; Donaldson et al. 2010). Time-of-flight secondary ion mass spectrometry also indicates that *p*-hydroxyphenyl lignin units are localised in the outer secondary wall (Tokareva et al. 2007). In normal wood, *p*-hydroxyphenyl lignin units are localised exclusively in the middle lamella (Fujita and Harada 1979; Whiting and Goring 1982; Westermark 1985; Saito and Fukushima 2005). In Scots pine, juvenile wood and compression wood show different distributions of dibenzodioxocin units in lignin which occur in the S3 layer in juvenile wood and in the S1 layer in compression wood as localised by immunocytochemistry (Kukkola et al. 2008).

The S2L region is characteristic of both mild and severe compression wood and can thus be used to microscopically distinguish these wood types from normal or opposite wood (Donaldson et al. 1999, 2004). This feature occurs not only in the conifers but also in *Ginkgo* and *Taxus* (Timell 1978a, b; Yumoto et al. 1983; Yoshizawa et al. 1992; Wi et al. 2000) and is thought to contribute the longitudinal maturation strain which acts to correct stem lean through lignification (Boyd 1972,

1973b; Okuyama 1993; Sugiyama et al. 1993; Okuyama et al. 1998). Helical checks develop during secondary wall formation (Fujita et al. 1973, 1978a) and facilitate the swelling of the outer secondary wall caused by lignification which is enhanced by the steep orientation of the cellulose microfibrils. The inner S2 region of compression wood tracheids shows radial lignin striations associated with helical cavities which appear to form between these more highly lignified regions (Singh et al. 1998). The inner part of the secondary wall in compression wood tracheids may be slightly more lignified compared to the secondary wall of normal wood. Donaldson et al. (1999) found increased lignification of both inner and outer S2 regions in radiata pine (*Pinus radiata* D. Don) using interference microscopy. In two different samples of mild compression wood, lignin concentrations in different regions of the cell wall were 26 %, 46 % and 57 % v/v, respectively, for the S2, S2L and middle lamella regions and 20 %, 29 % and 46 % for the same regions in a second sample. The inner S2 region in mild compression wood was therefore more lignified than normal wood in at least one of these samples, the value for normal wood being 21 % (Donaldson 1985).

# 4 Polysaccharides

## *4.1 Cellulose Microfibrils*

Compression wood typically has a high microfibril angle in the S2 layer compared to normal or opposite wood (Wardrop and Dadswell 1950; Gorišek and Torelli 1999; Donaldson et al. 2004; Yeh et al. 2005, 2006a; Donaldson 2008). However, in juvenile wood where MFA is already high, the MFA of compression wood may often be similar to normal or opposite wood (Donaldson et al. 2004). Even in some examples of mature wood, MFA may be the same in opposite and compression wood within single growth rings (Nečesaný 1955; Harris 1977; Donaldson and Burdon 1995; Donaldson et al. 2004). In mild compression wood of radiata pine, MFA is about 5° higher on average than opposite wood, while in severe compression wood it is 8° higher (Donaldson et al. 2004). In loblolly pine (*P. taeda* L.), Yeh et al. (2006a) found that compression wood MFA was higher than juvenile wood MFA and that juvenile compression wood MFA was higher than corresponding mature compression wood MFA.

MFA in growth rings formed after the compression wood zone may be either higher than or similar to opposite wood (Wardrop and Dadswell 1950; Donaldson et al. 2004; Kibblewhite et al. 2007). In growth rings containing bands of transient compression wood, within-ring MFA patterns may differ from those in normal or opposite wood rings with local increases in MFA associated with the transient compression wood (Hiller 1964a, b; Park et al. 1979; Donaldson et al. 2004; Kibblewhite et al. 2007).

Cellulose crystallinity is lower in compression wood than in normal wood (Lee 1961; Scurfield 1973; Timell 1986). In severe compression wood, the S3 layer is absent, but the S1 layer is thicker and may have a higher and less variable microfibril angle than in normal wood (Brändström 2004). In compression wood, the S1 layer is 25–30 % of the width of the secondary wall compared to 10 % in normal wood (Timell 1986). Mild compression wood tracheids may still have an S3 layer (Singh et al. 2003). It is not known if microfibril orientation varies across the S2 layer or between the inner less lignified and outer more lignified regions. However, the ultrastructural organisation of the wall does vary between these two regions. For example, the diameter and arrangement of microfibril clusters (macrofibrils) change between inner and outer S2 regions, being smaller in the inner S2 and larger in the outer S2L region (Donaldson 2007).

## 4.2 Galactan

Compression wood is characterised by significant amounts of (1 → 4)-β-galactan which has been immunolocalised to the outer secondary wall using the LM5 antibody (Figs. 5 and 6) (Jones et al. 1997), including both the S2L region and to a lesser extent the S1 layer (Schmitt et al. 2006; Altaner et al. 2007a; Mast et al. 2009; Kim et al. 2010; Donaldson and Knox 2012). This polysaccharide is therefore associated with the increased lignification of the outer secondary wall in both mild and severe compression wood tracheids (Fig. 5). In normal wood, (1 → 4)-β-galactan is absent from the secondary wall but can be detected in small amounts in the primary wall/middle lamella (Donaldson and Knox 2012).

## 4.3 Other Polysaccharides

Compression wood is characterised by reduced amounts of mannans and xylans (Nanayakkara et al. 2009). Using immunolocalisation, these polysaccharides have been shown to be reduced in the S2L region of severe compression wood tracheids (Fig. 6) (Kim et al. 2011; Donaldson and Knox 2012). The reduction in mannan/xylan is therefore spatially associated with increased galactan and lignification (Donaldson and Knox 2012). In radiata pine normal and compression wood, lignin, galactose, mannose and glucose contents are strongly and linearly correlated with each other, while xylose and arabinose show non-linear correlations (Kibblewhite et al. 2007, 2010). (1 → 5)-α-arabinan is associated with intercellular spaces in compression wood of sitka spruce (*Picea sitchensis* Bong. Carrière) as demonstrated by immunolocalisation (Altaner et al. 2007b).

Laricinan is a (1 → 3)-β-glucan first isolated in compression wood of Tamarack (*Larix laricina* [Du Roi] K.Koch) where it represents 2–3 % w/w of the wood composition, occurring in both tracheids and ray cells (Hoffmann and Timell 1970,

1972). Laricinan has been localised to the helical cavities of compression wood tracheids in a number of softwood species using aniline blue staining which is specific for (1 → 3)-β-glucans (Brodski 1972; Waterkeyn et al. 1982; Włoch and Hejnowicz 1983). This has been confirmed more recently by Altaner et al. (2007b) using immunolocalisation. Laricinan was originally thought to be involved in strain generation in compression wood (Brodski 1972; Włoch 1975), but this was later discounted by Boyd (1978) in favour of a lignification-based strain generation mechanism (Boyd 1972, 1973b).

# 5 Compression Wood Severity

While typical compression wood tracheids with a rounded shape, helical checking and intercellular spaces form the classical severe compression wood found in leaning stems and in branches, there is increasing recognition of a continuum of compression wood types between normal wood and severe compression wood (Côté et al. 1967; Yumoto et al. 1983; Donaldson et al. 1999). This mild compression wood is often associated with fast growth and may not be visibly distinguishable from normal wood (Donaldson et al. 1999, 2004; Nanayakkara et al. 2009). Mild compression wood can also occur in the transition zone between normal wood and more severe compression wood (Yumoto et al. 1983). Microscopically the tracheids of mild compression wood have an S2L region of varying extent but may lack the other cellular characteristics of compression wood. Mild compression wood tracheids may have a distinct S3 layer which is absent in severe compression wood (Singh et al. 2003). The presence of an S3 layer in mild compression wood cells displaying normal or near-normal morphology has been confirmed by transmission electron microscopy (Singh et al. 2003). However, the appearance and thickness of the S3 layer can vary greatly, and even adjoining cells display a range in form, from being extremely thin and rudimentary to normal in appearance and thickness (Singh et al. 2003).

Early studies often mention compression wood severity in a different context from the more modern meaning. The term mild compression wood often referred to samples containing only small amounts of compression wood rather than wood with properties intermediate between severe compression wood and opposite or normal wood (Pillow 1941; Pillow and Luxford 1937; Perem 1958; Low 1964; Burdon 1975; Seth and Jain 1978; Timell 1982).

In perhaps the first mention of mild compression wood, Pillow et al. (1936) recognised mild compression wood as assessed by microscopic examination. Using a light transmission technique combined with anatomical and physical measurements, Shelbourne and Ritchie (1968) were able to distinguish different grades of compression wood. In short-leaf pine (*Pinus echinata* Mill.), mild compression wood has a more abrupt latewood transition than severe compression wood and is thus similar to normal or opposite wood (Harris 1976).

Although it is often difficult to detect mild compression wood in transverse discs from stems and branches on the basis of eccentricity of the secondary xylem in relation to pith position, and colour differences from normal wood, its presence can be readily detected by examining lignin autofluorescence in transverse sections using fluorescence microscopy with little or no sample preparation (Fig. 3) (Donaldson et al. 1999, 2004). This compares favourably to UV microscopy or TEM where resin embedding and sectioning are a requirement making the later techniques less suitable for rapid screening. Although mild compression wood can be detected by conventional staining and light microscopy, differences are much less obvious, requiring very careful examination of sections. Greater lignification of the S2L region is the first and also the most reliable diagnostic feature for detecting mild compression wood (Yumoto et al. 1983; Donaldson et al. 1999; Singh and Donaldson 1999), with cell corner regions being consistently more highly lignified.

Yumoto et al. (1983) have developed a detailed system for grading compression wood based on anatomical features. In this classification, the primary feature of compression wood is the increased lignification of the outer secondary wall which these authors observed by means of UV absorbance microscopy. The presence of well-developed helical cavities was also considered useful as a measure of severity. In milder forms of compression wood, increased lignification is restricted to the corners of cells, while in more severe forms, increased lignification extends around the circumference of the cell. Other anatomical features such as intercellular spaces, helical cavities and rounded shape are more characteristic of severe compression wood (Yumoto et al. 1983).

In a study of one leaning and one non-leaning tree of radiata pine, both ramets from the same clone, Donaldson et al. (2004) found both mild and severe compression wood in both trees although there was more severe compression wood in the leaning tree. Reduced lignification of the middle lamella was the earliest sign of compression wood as observed by fluorescence microscopy. Both mild and severe compression wood types showed an increase in microfibril angle with an average increase of $4^{o}$ and $8^{o}$, respectively, compared to opposite wood. The difference in MFA between normal and compression wood was less in juvenile wood, where MFA was already high, as compared to mature wood. Basic density was 22 % greater in severe compression wood compared to opposite wood, mostly related to increased wall thickness. Mild compression wood did not show any difference in cell dimensions or basic density from opposite wood.

Recently some effort has been applied to finding a method to quantitatively measure compression wood severity in order to facilitate studies of the relationship between severity and wood properties/quality. Some attempts have been made to differentiate compression wood by colour analysis, but this approach often confuses compression wood with latewood and may not reliably identify mild compression wood (Andersson & Walter 1995). In this study the authors mention that colour analysis was able to detect mild compression wood, but there is no indication of how mild compression wood was defined.

Hyperspectral image analysis has been used to detect and classify compression wood in Norway spruce using reflected light at visible and near-infrared wavelengths by comparison with reference spectra (Duncker and Spiecker 2008, 2009). This technique allowed classification of normal, mild and severe compression wood as well as measurement of cross-sectional area and relationship with growth increment.

Scanning fourier transform infrared microspectroscopy has been used to measure compression wood severity using increment core samples that have been converted into sawdust in a way that preserves spatial location (Altaner et al. 2009). This compression wood indicator correlated well with other indicators including microfibril angle and immunolabeling for detection of (1 → 4)-β-galactan.

Near-infrared spectroscopy has been used to measure the percentage of compression wood on increment cores based on the compositional differences between normal and compression wood, especially lignin and galactan contents (Chen et al. 2007).

Image analysis of tracheid cross sections has been used to distinguish compression wood from normal wood. Using a method based on the fast fourier transform, Moëll and Fujita (2004) were able to distinguish severe compression wood but not mild compression wood from normal or opposite wood.

Chemical differences have also been used to characterise compression wood severity (Newman et al. 2005; Chen et al. 2007; Nanayakkara et al. 2009). Using samples of compression wood graded by colour, Newman et al. (2005) found a good correlation between colour and galactan content. Chemical microanalysis was used to measure the amount of compression wood in juvenile wood samples in order to adjust property measurements for the amount of compression wood. In a similar study (Nanayakkara et al. 2009), fluorescence microscopy was used to identify the presence and extent of the S2L region in tracheids from wood samples of differing types. Both lignin content, especially *p*-hydroxyphenyl β-ethers and uncondensed *p*-hydroxyphenyl C-9 units, and galactose content increased with compression wood severity while mannose and glucose content decreased. The galactose and *p*-hydroxyphenyl lignin contents were linearly correlated with lignin content, but the best predictor of compression wood severity was *p*-hydroxyphenyl unit content.

Fluorescence spectroscopy of lignin has shown characteristic differences between normal wood and compression wood using spectral confocal microscopy which are probably related to *p*-hydroxyphenyl unit content. By calculating a ratio of fluorescence peaks at 435 nm and 485 nm, a numerical index of compression wood severity was obtained using anatomical characterisation as a benchmark (Donaldson et al. 2010).

# 6 Similarities to Juvenile Wood

Compression wood in conifers is often associated with juvenile wood, and both wood types have similar properties. Yeh et al. (2005) have compared the properties of juvenile and compression wood in clonal cuttings of loblolly pine grown under normal conditions, controlled bending, and under windy conditions. Morphological and chemical properties were found to be significantly different in juvenile wood and compression wood. Wood formed under controlled wind conditions was found to be mild compression wood.

In a subsequent study, Yeh et al. (2006a) compared the properties of juvenile and mature compression wood in loblolly pine. There was no difference in tracheid length between juvenile and mature compression wood, but tracheids in juvenile compression wood were narrower. Mature compression wood tracheids had much greater curl and kink indices than juvenile compression wood tracheids. Juvenile compression wood tracheids had a higher microfibril angle than mature compression wood tracheids in agreement with other studies (Marton et al. 1972; Megraw 1985; Donaldson 1992; Sahlberg et al. 1997; Kretschmann et al. 1998; Donaldson et al. 2004). Mature compression wood may contain slightly more extractives than juvenile compression wood, but the dark colour of compression wood is due to increased lignin content and thicker cell walls rather than increased extractives (Yeh et al. 2006a). Compression wood from both locations contained increased lignin and galactan content with mature compression wood lignin having a slightly greater total OH content (Yeh et al. 2006a).

Brennan et al. (2012) studied the composition and structure of four juvenile wood types from young trees of two clones of *Pinus radiata* grown in a glasshouse, including normal juvenile wood, juvenile opposite wood, juvenile compression wood and juvenile flexure wood produced by rocking the trees. Among these wood types, microfibril angle was the same, but juvenile compression wood contained elevated lignin and galactan and showed the highest longitudinal swelling from oven dry to moisture saturated state. Longitudinal swelling was related to both lignin content and galactan content. Juvenile compression wood was also distinguished from the other wood types by its *p*-hydroxyphenyl lignin content.

# 7 Compression Wood Formation

## *7.1 Cytoplasmic Organelles*

Several studies have examined the ultrastructure of cytoplasm in the cambium and in differentiating tracheids during compression wood formation (Wardrop and Davies 1964; Fujita et al. 1978b; Timell 1980; Furusawa et al. 1998). These studies have used chemical fixation rather than more modern techniques such as high pressure freezing and freeze substitution (Samuels et al. 2002), so preservation of organelles may have been less than ideal especially in cells undergoing programmed cell death (Fujita et al. 1973).

Fujita et al. (1978b) examined cell organelles in developing compression wood of Sugi (*Cryptomeria japonica* D. Don). They found a weakly developed cytoplasm in cells at the stage of S1 layer formation with an increase in cytoplasmic density during S2 layer formation. Microtubules and golgi vesicles were abundant during this stage of development with a decrease in the number of vesicles at the lignification stage due to fusion with the plasma membrane. Large vesicles were associated with formation of the helical cavities/ridges, the cytoplasm and plasma membrane penetrating the helical cavities.

Timell (1980) made similar observations on dormant cambium of compression wood in Norway spruce (*Picea abies* [L.] H.Karst), showing a clear difference between the dense cytoplasm of inactive cambial cells and the degraded residual cytoplasm of adjacent almost fully formed compression wood tracheids which were probably undergoing programmed cell death.

Furusawa et al. (1998) demonstrated the orientation of cortical microtubules in developing compression wood tracheids of Japanese yew (*Taxus cuspidata* Siebold & Zucc.). Using confocal microscopy and immunocytochemistry, microtubule orientation was found to match adjacent cellulose microfibril orientation in developing compression wood cell walls as expected from many other studies on non-tree species.

A number of questions about the role of cytoplasmic organelles in compression wood cell wall formation remain to be resolved. The location of biosynthesis for polysaccharides such as $(1 \rightarrow 4)$-β-galactan characteristic of compression wood (Mast et al. 2010) and for lignin monomers in compression wood (Morikawa et al. 2010) is still poorly understood. Modern techniques that have been applied to understand normal wood formation processes would also be useful in compression wood studies. For example, studies aimed at tracking monolignols during wood formation suggest that the export of monolignol is likely to be mediated by membrane transporters and not golgi vesicles (Kaneda et al. 2008).

## 7.2 Cell Wall Thickening and Lignification

Secondary walls of compression wood tracheids are characterised by the presence of helical cavities which are oriented parallel to the cellulose microfibrils, and the origin of these cavities has been the subject of several studies. Cavities extend from the lumen to the outer S2L region and often show a characteristic branched structure (Figs. 4d and 5). A schizogenous origin has been suggested as a result of swelling in the outer part of the secondary wall due to increased lignification (Boyd 1973a) or alternatively resulting from contraction of the inner S2 region (Wardrop and Davies 1964). However, observations of intimate cytoplasmic association, large vesicles and indented plasma membrane, following the contours of the helical ridges as they appear in early stages of development (Fujita et al. 1973, 1978b), are suggestive of a more active method of ridge/cavity formation. Fujita et al. (1973) were able to distinguish ridges/cavities at all stages of secondary wall development indicating that they are an inherent part of the secondary wall structure and don't form after wall formation as a result of strain within the secondary wall. The branched structure of helical cavities supports the idea that these structures facilitate longitudinal swelling of compression wood secondary walls rather than being the result of such swelling. The observation of helical cavities in Juniper wood in the absence of modified lignification supports this conclusion (Hänninen et al. 2012).

Fujita et al. (1978a, 1979) and Takabe et al. (1986) examined the process of cell wall thickening and lignification in compression wood of Sugi. They demonstrated that secondary wall formation in compression wood tracheids occurs in three distinct phases: (1) deposition of the S1 layer and lignification of the middle lamella and primary cell wall, (2) thickening of the S2 layer and (3) lignification of the secondary wall. This differs from the development of normal wood tracheids where wall formation and lignification are a more continuous process (Donaldson 2001).

The lignin in compression wood contains increased levels of *p*-hydroxyphenyl units derived from *p*-coumaryl alcohol (Yasuda and Sakakibara 1975; Nanayakkara et al. 2009). Fukushima and Terashima (1991) studied the formation of lignin in compression wood of Japanese black pine (*P. thunbergii* Parl.) using tritiated gluco-*p*-coumaryl alcohol visualised using autoradiography. Lignin deposition in compression wood was characterised by an extended period of deposition of *p*-hydroxyphenyl units in both middle lamella and secondary wall in contrast to the brief period of deposition in the middle lamella region that occurs during normal or opposite wood formation.

## 7.3 *Polysaccharides*

The development of antibodies directed against cell wall polysaccharides has recently allowed immunolocalisation of specific polysaccharides in developing and mature cell walls of both normal and compression wood (Donaldson 2009). Kim et al. (2011) have investigated the distribution of mannan and xylan in differentiating compression wood tracheids of Sugi. The deposition of polysaccharides follows the pattern of wall thickening and lignification described above (Fig. 7). During S1 formation, xylans were deposited in both the middle lamella/primary wall and in the S1 layer. Mannans were also deposited in the S1 layer at the same time. As the S2 layer was formed, reduced amounts of xylan were deposited in the outer S2 region and greater amounts in the inner S2 during helical cavity formation. Mannans were found to be more evenly distributed in the S2 layer. Donaldson and Knox (2012) subsequently found that in radiata pine, mannan is also reduced in the outer S2 region of compression wood tracheids using a different antibody for detection (Fig. 6).

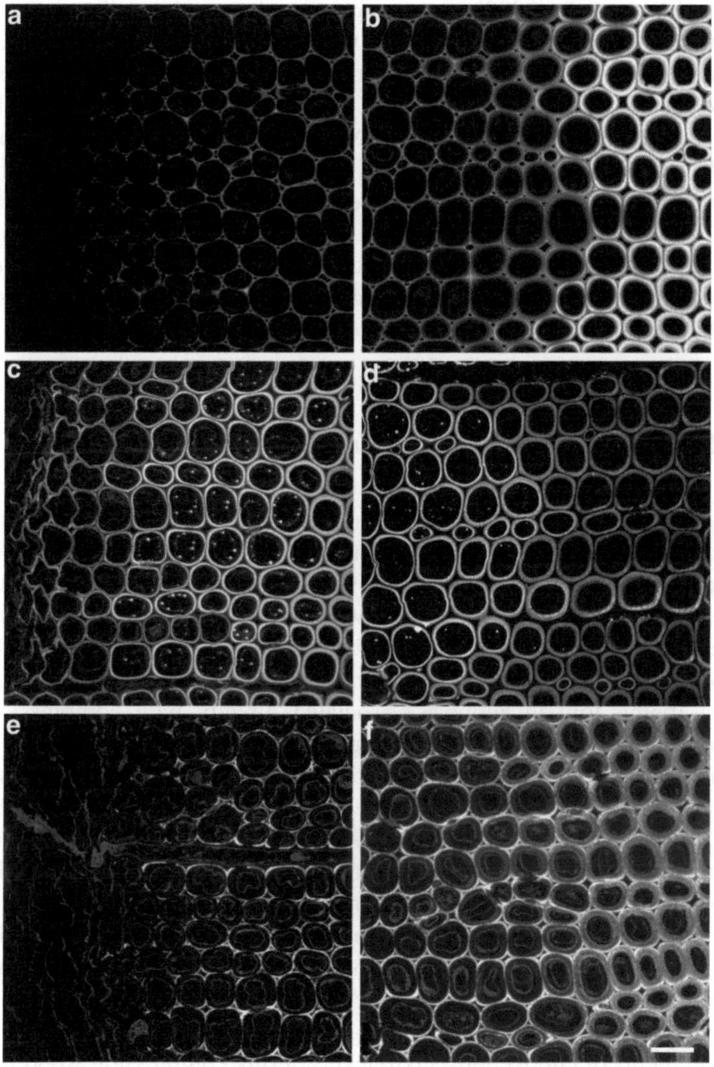


**Fig. 7** Transverse sections of developing xylem in radiata pine showing lignification (**a** and **b**, lignin autofluorescence); deposition of galactan (**c** and **d**, LM5 epitope by immunofluorescence); and deposition of mannan (**e** and **f**, LM22 epitope by immunofluorescence). Scalebar = 30 μm. The cambium is towards the left with maturity increasing towards the right. The galactan epitope is masked by the onset of lignification as shown by the decreasing brightness near the mature xylem to the right in (**d**)

Kim et al. (2010) examined the distribution of (1 → 4)-β-galactan in developing compression wood tracheids. Galactan deposition begins during S1 formation but is greatest in the outer S2 layer. The amount of galactan epitope detected in the S1 layer decreases during lignification of the secondary wall, possibly due to masking by lignin. Mast et al. (2009) also examined galactan deposition in relation to lignification of developing compression wood tracheids in radiata pine. Galactan is deposited in the secondary wall prior to lignification (Fig. 7), and the amount of epitope detected is reduced at the onset of lignification suggesting that lignin is masking the galactan epitope (Fig. 7) (Mast et al. 2009). Galactan seems to be evenly distributed across the unlignified S2 layer prior to lignification, but after lignification, galactan epitope was only detected in the outer S2 region (Mast et al. 2009; Donaldson and Knox 2012). Increased galactan and reduced mannan and xylan are therefore associated with the increased lignification found in the outer S2 layer of compression wood tracheids suggesting that these polysaccharides may be involved in controlling cell wall formation, especially lignification.

## *7.4 Hormonal Aspects of Compression Wood Formation*

A number of studies have shown the involvement of plant hormones including ethylene and auxin in compression wood formation (Du and Yamamoto 2007). In a tilting experiment, Little and Eklund (1999) found increased ethylene production in tilted stems associated with compression wood formation. The auxin inhibitor NAA prevented compression wood formation below the site of application to the stem. Evolution of ethylene from the cambial region was increased when compression wood was being formed and was associated directly with compression wood formation rather than increased tracheid production.

Application of auxin to vertical stems of conifers induces compression wood and this effect may be mediated by increased ethylene (Wershing and Bailey 1942; Balch et al. 1964; Wardrop and Davies 1964; Nečesaný 1958; Starbuck and Phelps 1986; Sundberg and Little 1990; Du et al. 2004; Tsai et al. 2010). However, exogenously applied ethylene does not induce compression wood so its role may be indirect (Telewski et al. 1983; Yamamoto and Kozlowski 1987a, b; Eklund and Little 1996). Auxin transport inhibitors also induce compression wood in the region of stem above their application (Phelps et al. 1974; Yamaguchi et al. 1980, 1983; Sundberg et al. 1994).

Measurements of endogenous auxin levels during compression wood formation have provided variable results. Wilson et al. (1989) found no relationship between endogenous auxin levels in the cambium and compression wood formation, while Funada et al. (1990) and Du et al. (2004) found the contrary result that endogenous auxin levels were elevated on the side of the stem forming compression wood. Sundberg et al. (1994) found that auxin does not increase in the cambium above the application point of the auxin inhibitor NPA and concluded that the NPA receptor is involved in compression wood formation by modulating IAA levels in specific cells. Other studies have also found no indication of changes in auxin level or distribution associated with compression wood formation (Hellgren et al. 2004). Despite these negative results, auxin is still considered to be the main regulating factor in compression wood formation.

Little is known about the perception of gravitational signals in conifers. There are two potential candidates for perception including sedimentation of amyloplasts (Nakamura et al. 2001) and mechanosensitive ion channels on the plasma membrane (Hoson et al. 2005).

## 7.5 Molecular Aspects of Compression Wood Formation

Molecular biology is providing insight into compression wood formation through gene discovery, differential expression of genes in cambium forming compression wood compared to normal wood, and protein and metabolite profiling (Plomion et al. 2001). Compression wood is an almost ideal system for this type of investigation because it can so easily be induced. While large numbers of gene sequences are differentially expressed in compression wood, less than half can be identified in terms of function by comparison with gene libraries (Allona et al. 1998; Yamashita et al. 2008). Not surprisingly many upregulated genes are involved in lignin biosynthesis, cell wall proteins including biosynthetic enzymes, enzymes involved in carbohydrate metabolism, and regulatory proteins (Allona et al. 1998; Whetten et al. 2001; Bedon et al. 2007; Koutaniemi et al. 2007; Mast et al. 2010).

In Norway spruce, the same set of monolignol biosynthetic genes were expressed in compression wood compared to normal wood, but differences were observed in peroxidase expression between the two wood types (Koutaniemi et al. 2007). Using protein extraction and characterisation by SDS-PAGE, McDougall (2000) also found evidence for a novel peroxidase expressed in developing compression wood suggesting separate functions for these enzymes. Proteins extracted from normal and compression wood were equally able to oxidise *p*-coumaryl alcohol, but the compression wood extract had less activity against coniferyl alcohol supporting the theory proposed by Dean et al. (1998) that differential oxidase activity is responsible for deposition of different lignin types in normal and compression wood.

Several members of the R2R3-MYB family of transcription factors are known to act as regulators of lignin metabolism during wood formation, and three such genes are preferentially upregulated in developing compression wood within 76 h of induction (Bedon et al. 2007).

In a detailed proteomic study of normal and compression wood in Maritime pine (*P. pinaster* Ait.) Plomion et al. (2000) characterised 137 proteins of which 19 % were associated with a growth strain effect. Upregulated proteins included an ethylene forming enzyme (1-aminocyclopropane-1-carboxylate oxidase), a transcription factor, 2 lignification genes (caffeic *O*-methyltransferase and caffeoyl CoA-*O*-methyltransferase), members of the *S*-adenosyl-L-methionine-synthase gene family and enzymes involved in nitrogen and carbon assimilation (glutamine synthetase and fructokinase). The upregulation of an ethylene forming enzyme supports the association of ethylene with the gravitropic response in conifers. In a similar study using differential EST expression in loblolly pine, Whetten et al.

(2001) found that of the 69 most abundant transcripts, 33 were more abundant in compression wood than normal wood, while three were more abundant in normal wood compared to compression wood. Those more abundant in compression wood included cDNA's related to monolignol biosynthesis (4-coumarate CoA ligase, caffeoyl CoA-*O*-methyltransferase, glycine hydroxymethyltransferase and *S*-adenosyl methionine synthetase). The three cDNA's more abundant in normal wood included genes for a Skp1-like protein, a xyloglucan endotransglycosylase (XET)-like protein, and a protein similar to pollen allergens. Skp1 does not seem to have an obvious connection with cell wall formation so its function is unclear in this context. Other studies have also identified cDNA's associated with novel arabinogalactan proteins and proline rich proteins also in loblolly pine (Zhang et al. 2000).

Further work is needed to identify the many unknown genes found in differential expression studies using combined genomic, proteomic and metabolomic approaches (Yeh et al. 2006b).

## 7.6 Growth Response to Tilting in Young Trees

Induction of compression wood by tilting of young trees is a gradual process taking at least 10 days after the stimulus is applied as indicated by a progressive increase in the lignification phase of the primary cell wall and middle lamella in differentiating tissue. Mild compression wood is formed first followed by a progressive increase in severity (Yumoto et al. 1982).

In a study comparing five conifer species, Yoshizawa et al. (1986b) found some variation in growth response to tilting, indicating that the response begins near the top of the stem and extends downwards with a corresponding formation of compression wood. With prolonged tilting, compression wood formation declines on the underside of the stem and may sometimes occur on the upper side in slightly inclined stems. Starbuck and Roberts (1983) also reported compression wood on the upper side of inclined stems in Douglas fir (*Pseudotsuga menziesii* [Mirb.] Franco) seedlings.

In a study comparing crooked and straight stems of young radiata pine, trees were tilted to either 15° or 30° and compression wood formation was assessed (Lachenbruch et al. 2010). Crooked stems righted themselves more quickly than straight stems suggesting an overcompensation response. In 1-year-old trees, stem angle showed a threshold of about 10° where more compression wood formed above 10° than below, but older trees showed no relationship between compression wood amount and stem tilt. Likewise, a similar experiment in Sugi found a maximal compression wood response at 30° with no increase after further tilting (Yamashita et al. 2007). Radiata pine seedlings that were continuously rocked rather than tilted formed similar amounts of compression wood to straight trees (Apiolaza et al. 2011a). Maritime pine trees subjected to combinations of both leaning and wind under controlled conditions showed similar rates of apical straightening, but the rate of basal straightening in trees exposed to leaning and wind was 4x greater than in trees exposed to leaning with no wind even though both groups of trees formed similar amounts of compression wood (Berthier and Stokes 2005). This suggests that the severity of compression wood may have been different in the two groups of trees, but this was not assessed.

To compare response to stem tilt in seedlings of maritime pine and loblolly pine, Ba et al. (2010) induced leaning for a 35-day period and observed the growth response. Maritime pine showed a rapid reorientation of stem apices within 24 h, but in loblolly pine, stem reorientation was slower and occurred at the stem base with significantly more compression wood formation. The reorientation process was considered more efficient in loblolly pine even though compression wood was formed immediately after tilting in both species. This difference in response to tilting was related to difference in habitat with maritime pine showing adaptions to arid conditions by reducing its compression wood response and hence maintaining xylem with more efficient water conduction properties. The rapid shoot tip reorientation compensates for the slower stem straightening response and maximises exposure of needles to light during seedling establishment.

## 7.7 *Heritability*

Although compression wood is strongly influenced by environment, there is also a significant heritable component to its occurrence with heritabilities ranging from 0.3 to 0.9 (Einspahr et al. 1964; Shelbourne et al. 1969; Burdon 1975; Cown et al. 1992; Apiolaza et al. 2011b). Stem form and compression wood occurrence show variable relationships among clones, and the correlation between severe compression wood occurrence and stem deviation from vertical or crookedness is often small (Shelbourne et al. 1969; Burdon 1975; Lachenbruch et al. 2010). Compression wood may be correlated with growth rate (Cown 1974; Cown et al. 1992) and microfibril angle (Donaldson and Burdon 1995), both of which are heritable. It seems that studies of compression wood genetics have received little attention in recent literature. More work is needed not just to assess natural compression wood occurrence in clones and families (Burdon and Low 1992) but to do so after deliberately inducing stem lean in order to assess the magnitude of response to geotropic stimulus (Apiolaza et al. 2011a, b). Tools are now available to assess the severity of compression wood response (e.g. see this chapter Sect. 5) which may provide further insight into the potential for moderating the negative effects of compression wood on wood quality without compromising tree growth and form.

## 7.8 Formation in Zero-Gravity Environments

Experiments on compression wood formation have been carried out in the microgravity environment of space with the aim of distinguishing a response to gravity from a response to strain induced by bending. Kwon et al. (2001) subjected 1-year-old plants of Douglas fir and loblolly pine to 17 days growth in the microgravity environment of the Space Shuttle Columbia. Some plants were bent at 45° in an effort to induce compression wood formation compared to control plants growing in the 1 g environment on Earth. Flexed stems from both microgravity and 1 g environments formed compression wood confirming that mechanical stress induces compression wood and that both cambial cells and developing tracheids can respond to this mechanical stress by changing their cell wall structure and composition.

## 7.9 Mechanism of Stress Generation

There are several controversial theories on the mechanism of axial stress generation in compression wood (Boyd 1972; Okuyama 1993; Sugiyama et al. 1993; Yamamoto et al. 1995, Yamamoto 1998; Bamber 2001). An earlier proposal that laricinan located within the helical cavities might be responsible for stress generation (Brodski 1972; Włoch 1975) was discounted by Boyd (1978) due to the ability of the laricinan to expand into the lumen, thus preventing it from generating significant stress. In one theory, swelling due to lignification generates compressive stress (Boyd 1972, 1973b; Okuyama 1993; Sugiyama et al. 1993; Okuyama et al. 1998), while in the alternative theory, cellulose microfibrils are deposited in a compressed state, thus generating compressive stress (Bamber 2001). Bamber argues that the increased roundness of compression wood tracheids is the result of tension rather than expansion and notes that this shape change occurs prior to lignification of the secondary wall. The function of lignification in the latter case is to cement the microfibrils to ensure transmission of the stress and thus increase the compression strength of the cell wall. In both cases, a high microfibril angle ensures that any stress generated is directed axially. Bamber (2001) has proposed that the high microfibril angle in compression wood is a result of the compressive stress during wall formation which would act to limit the reorientation of cellulose microfibrils that normally occurs between the S1 and S2 layers in normal wood. A theory combining both matrix swelling and microfibril contraction has been proposed by Okuyama (1993) and developed further by Yamamoto et al. (1995), Yamamoto (1998) and Guitard et al. (1999) to try and overcome shortcomings of the two separate mechanisms proposed by other investigators.

Recent studies have attempted to resolve the role of structural features in stress generation and to understand the mechanism. Using tensile tests on thin tissue slices of Norway spruce, Reiterer et al. (1999) confirmed that increased microfibril angle optimises the extensibility of wood. Gindl (2002) proposed that the observed increase in lignification of compression wood increases the resistance of the cell walls to compression failure, especially the increased amounts of *p*-hydroxyphenyl units. Burgert et al. (2004) compared the mechanical properties of compression wood from four gymnosperms with variable compression wood features including *Ginkgo, Juniperus, Taxus* and *Picea*. Microfibril angle was found to be a dominant factor in determining extensibility, but this study did not resolve the issue of stress generation. Further studies are needed to measure the effect of lignification, helical cavities and microfibril angle on stress generation, and an obvious way of doing this is to compare compression wood of varying severity. In a comparison of tangential shrinkage in sitka spruce mature wood, juvenile wood and compression wood, Leonardon et al. (2010) found that at constant MFA, compression wood showed lower tangential shrinkage than juvenile wood, thus demonstrating the effect of cell wall composition. Brennan et al. (2012) have demonstrated relationships between both lignification and galactosyl residue content in radiata pine with swelling of cell walls from the oven dry to fully saturated state, thus demonstrating the likely involvement of both matrix components in growth stress generation in compression wood.

The observation that increased lignification is the most consistent feature of compression wood suggests that this is the dominant factor in generating compressive stress as proposed by Boyd (1972). Okuyama et al. (1998) found a correlation between growth stress development and lignification in the secondary wall, especially in the outer secondary wall. This supports the conclusion that lignification contributes to longitudinal growth stress development in compression wood tracheids. Microfibril angle and helical cavities are also likely to be involved as facilitating factors, while the potential roles of galactan and microfibril contraction remain unclear.

## 8 Conclusions

In the 26 years since Timell's monograph on compression wood (Timell 1986), a number of new discoveries have changed our view of compression wood from one based largely on a stereotype severe compression wood to one based on the realisation that compression wood forms a continuum from normal wood to severe forms (Yumoto et al. 1983; Donaldson et al. 1999; Singh et al. 2003; Nanayakkara et al. 2009). Improved detection of the sometimes subtle differences in milder forms has opened the way to relating wood properties to compression wood severity measured on quantitative scales although relatively few studies have taken advantage of this new capability as yet (Donaldson et al. 2004; Nanayakkara et al. 2009; Brennan et al. 2012). Development of antibodies against cell wall polysaccharides and their use in immunocytochemistry of plant cell walls has revealed new details that provide hints on how the assembly of cell walls is controlled (Mast et al. 2009, 2010; Kim et al. 2010; Donaldson and Knox 2012). When combined with improved cytoplasmic fixation techniques such as high pressure freezing or even perhaps live-cell experiments, immunocytochemistry will allow more detailed elucidation of the biosynthetic steps involved in changes from normal cell walls to compression wood cell walls (Samuels et al. 2002; Kaneda et al. 2008). Future developments in molecular biology and genetic manipulation will allow improved understanding of the biological role of compression wood. For example, altering the expression levels of specific polysaccharides, notably galactan, is an obvious way of determining the role of polysaccharides in cell wall assembly and in the altered properties of compression wood (Mast et al. 2010). New more rapid analytical techniques offer scope for relating specific compression wood responses to environment and genotype with potential for improved wood quality.

## References


Akbulut A, Nadir A (2006) Effect of compression wood on surface roughness and surface absorption of medium density fibreboard. Silva Fenn 40:161–167

Allona I, Quinn M, Shoop E, Swope K, Cyr SS, Carlis J, Riedl J, Retzel E, Campbell MM, Sederoff R, Whetten RW (1998) Analysis of xylem formation in pine by cDNA sequencing. Proc Natl Acad Sci USA 95:9693–9698

Altaner C, Hapca AI, Knox JP, Jarvis MC (2007a) Detection of β-1–4-galactan in compression wood of Sitka spruce [*Picea sitchensis* (Bong.) Carrière] by immunofluorescence. Holzforschung 61:311–316

Altaner C, Knox JP, Jarvis MC (2007b) In situ detection of cell wall polysaccharides in Sitka spruce (*Picea sitchensis* [Bong.] Carrière) wood tissue. Bioresources 2:284–295

Altaner CM, Tokareva EN, Wong JCT, Adrian I, Hapca AI, McLean JP, Jarvis MC (2009) Measuring compression wood severity in spruce. Wood Sci Technol 43:279–290

Altaner CM, Jarvis MC, Fisher JB, Marler TE (2010) Molecular xylem cell wall structure of an inclined *Cycas micronesica* stem, a tropical gymnosperm. IAWA J 31:3–11

Andersson C, Walter F (1995) Classification of compression wood using digital image analysis. For Prod J 45:87–92

Apiolaza L, Butterfield B, Chauhan SS, Walker JCF (2011a) Characterization of mechanically perturbed young stems: can it be used for wood quality screening? Ann For Sci 68:407–414

Apiolaza L, Chauhan SS, Walker JCF (2011b) Genetic control of very early compression and opposite wood in *Pinus radiata* and its implications for selection. Tree Genet Genomes 7:563–571

Ba M, Salin F, Fourcaud T, Stokes A (2010) Reorientation strategies in leaning stems of young maritime pine (*Pinus pinaster*) and loblolly pine (*Pinus taeda*). IAWA J 31:465–480

Balch RE, Clark J, Bonga JM (1964) Hormonal action in production of tumours and compression wood by an aphid. Nature 202:721–722

Bamber RK (2001) A general theory for the origin of growth stresses in reaction wood: how trees stay upright. IAWA J 22:205–212

Ban W, Mancoski D, Lucia LA (2004) Evaluation of the pulping response of juvenile and mature black spruce compression wood. Cell Chem Technol 38:79–85

Bedon F, Grima-Pettenati J, Mackay J (2007) Conifer R2R3-MYB transcription factors: sequence analyses and gene expression in wood-forming tissues of white spruce (*Picea glauca*). BMC Plant Biol 7:17. doi:10.1186/1471-2229-7-17

Berthier S, Stokes A (2005) Righting response of artificially inclined maritime pine (*Pinus pinaster*) saplings to wind loading. Tree Physiol 26:73–79

Boyd JD (1972) Tree growth stresses V.: evidence of an origin in differentiation and lignification. Wood Sci Technol 6:251–262
Boyd JD (1973a) Helical fissures in compression wood cells: causative factors and mechanics of development. Wood Sci Technol 7:92–111
Boyd JD (1973b) Compression wood force generation and functional mechanics. N Z J For Sci 3:240–258
Boyd JD (1978) Significance of laricinan in compression wood tracheids. Wood Sci Technol 12:25–35
Brändström J (2004) Microfibril angle of the S1 cell wall layer of Norway spruce compression wood tracheids. IAWA J 25:415–423
Brennan M, McLean JP, Altaner CM, Ralph J, Harris PJ (2012) Cellulose microfibril angles and cell-wall polymers in different wood types of *Pinus radiata*. Cellulose 19:1385–1404
Brodski P (1972) Callose in compression wood tracheids. Acta Soc Bot Pol 41:321–327
Burdon RD (1975) Compression wood in *Pinus radiata* clones on four different sites. N Z J For Sci 5:152–164
Burdon RD, Low CB (1992) Wood properties: variation, heritabilities, and interrelationships with other traits. N Z J For Sci 22:228–245
Burgert I, Frühmann K, Keckes J, Fratzl P, Stanzl-Tschegg S (2004) Structure-function relationships of four compression wood types: micromechanical properties at the tissue and fibre level. Trees Struct Funct 18:480–485
Chen Q-M, Hu Z, Chang H-M, Li B (2007) Micro-analytical methods for determination of compression wood content in loblolly pine. J Wood Chem Technol 27:169–178
Cockrell RA (1974) A comparison of latewood pits, fibril orientation, and shrinkage of normal and compression wood of giant *Sequoia*. Wood Sci Technol 8:197–206
Côté WA, Timell TE, Zabel RA (1966) Studies on compression wood I. Distribution of lignin in compression wood of red spruce (*Picea rubens* Sarg.). Holz Roh-Werkst 24:432–438
Côté WA, Day AC, Kutscha NP, Timell TE (1967) Studies on compression wood V. Nature of the compression wood formed in early springwood of conifers. Holzforschung 21:180–186
Côté WA, Day AC, Timell TE (1968) Studies on compression wood VII. Distribution of lignin in normal and compression wood of tamarack *Larix laricina* (Du Roi, K.Koch). Wood Sci Technol 2:13–37
Cown DJ (1974) Comparison of the effects of two thinning regimes on some wood properties of radiata pine. N Z J For Sci 4:540–551
Cown DJ, Young GD, Burdon RD (1992) Variation in wood characteristics of 20-year-old half-sib families of *Pinus radiata*. N Z J For Sci 22:63–76
Cown DJ, Ball RD, Riddell MJC (2004) Wood density and microfibril angle in 10 *Pinus radiata* clones: distribution and influence on product performance. N Z J For Sci 34:293–315
Dean JFD, LaFayette PR, Rugh C, Tristram AH, Hoopes JT, Eriksson K-EL, Merkle SA (1998) Laccases associated with lignifying vascular tissues. In: Lewis NG, Sarkanen S (eds) Lignin and lignan biosynthesis, vol 697, American Chemical Society Series. American Chemical Society, Washington, DC, pp 96–108
Donaldson LA (1985) Within- and between-tree variation in lignin concentration in the tracheid cell wall of *Pinus radiata*. N Z J For Sci 15:361–369
Donaldson LA (1992) Within- and between-tree variation in microfibril angle in *Pinus radiata*. N Z J For Sci 22:77–86
Donaldson LA (2001) Lignification and lignin topochemistry – an ultrastructural view. Phytochemistry 57:859–873
Donaldson LA (2007) Cellulose microfibril aggregates and their size variation with cell wall type. Wood Sci Technol 41:443–460
Donaldson LA (2008) Microfibril angle: measurement, variation and relationships. IAWA J 29:345–386
Donaldson LA (2009) Immunocytochemistry of wood cell walls: a review. N Z J For Sci 39:161–168

Donaldson LA, Burdon RD (1995) Clonal variation and repeatability of microfibril angle in *Pinus radiata*. N Z J For Sci 25:164–174

Donaldson LA, Knox JP (2012) Localisation of cell wall polysaccharides in normal and compression wood of radiata pine – relationships with lignification and microfibril orientation. Plant Physiol 158:642–653

Donaldson LA, Singh AP, Yoshinaga A, Takabe K (1999) Lignin distribution in mild compression wood of *Pinus radiata* D. Don. Can J Bot 77:41–50

Donaldson LA, Grace JC, Downes G (2004) Within tree variation in anatomical properties of compression wood in radiata pine. IAWA J 25:253–271

Donaldson LA, Radotić K, Kalauzi A, Djikanović D, Jeremić M (2010) Quantification of compression wood severity in tracheids of *Pinus radiata* D Don using confocal fluorescence imaging and spectral deconvolution. J Struct Biol 169:106–115

Du S, Yamamoto F (2007) An overview of the biology of reaction wood formation. J Integr Plant Biol 49:131–143

Du S, Sugano M, Tsushima M, Nakamura T, Yamamoto F (2004) Endogenous indole-3-acetic acid and ethylene evolution in tilted *Metasequoia glyptostroboides* stems in relation to compression-wood formation. J Plant Res 117:171–174

Duncker P, Spiecker H (2008) Cross-sectional compression wood distribution and its relation to eccentric radial growth in *Picea abies* [L.] Karst. Dendrochronologia 26:195–202

Duncker P, Spiecker H (2009) Detection and classification of Norway spruce compression wood in reflected light by means of hyperspectral image analysis. IAWA J 30:59–70

Einspahr DW, Goddard RE, Gardner HS (1964) Slash pine, wood and fiber property heritability study. Silvae Genet 13:103–109

Eklund L, Little CHA (1996) Laterally applied Ethrel causes local increases in radial growth and indole-3-acetic acid concentration in *Abies balsamea* shoots. Tree Physiol 16:509–513

Fisher JB, Marler TE (2006) Eccentric growth but no compression wood in a horizontal stem of *Cycas micronesica* (Cycadales). IAWA J 27:377–382

Floyd S (2005) Effect of hemicellulose on longitudinal shrinkage in wood. In: Entwistle KM, Walker CF (eds) The hemicellulose workshop. University of Canterbury, New Zealand, pp 115–120

Fujita M, Harada H (1979) Autoradiographic investigations of cell wall development. II. Tritiated phenylalanine and ferulic acid assimilation in relation to lignification. Mokuzai Gakkaishi 25:89–94

Fujita M, Saiki H, Harada H (1973) The secondary wall formation of compression wood tracheids. On the helical ridges and cavities. Bull Kyoto Univ For 45:192

Fujita M, Saiki H, Harada H (1978a) The secondary wall formation of compression wood tracheids. II Cell wall thickening and lignification. Mokuzai Gakkaishi 24:158–163

Fujita M, Saiki H, Harada H (1978b) The secondary wall formation of compression wood tracheids. III Cell organelles in relation to cell wall thickening and lignification. Mokuzai Gakkaishi 24:353–361

Fujita M, Saiki H, Sakamoto J, Araki N, Harada H (1979) The secondary wall formation of compression wood tracheid. IV. Cell wall structure of transitional tracheids between normal wood and compression wood. Bull Kyoto Univ For 51:247–56

Fukazawa K (1974) The distribution of lignin in compression- and lateral-wood of *Abies sachalinensis* using ultraviolet microscopy. Res Bull Coll Exp For Hokkaido Univ 31:87–114

Fukushima K, Terashima N (1991) Heterogeneity in formation of lignin. Part XV: formation and structure of lignin in compression wood of *Pinus thunbergii* studied by microautoradiography. Wood Sci Technol 25:371–381

Funada R, Mizukami E, Kubo T, Fushitani M, Sugiyama T (1990) Distribution of indole-3-acetic acid and compression wood formation in the stems of inclined *Cryptomeria japonica*. Holzforschung 44:331–334

Furusawa O, Funada R, Murakami Y, Ohtani J (1998) Arrangement of cortical microtubules in compression wood tracheids of *Taxus cuspidata* visualized by confocal laser microscopy. J Wood Sci 44:230–233

Gindl W (2002) Comparing mechanical properties of normal and compression wood in Norway spruce: the role of lignin in compression parallel to the grain. Holzforschung 56:395–401

Gorišek Z, Torelli N (1999) Microfibril angle in juvenile, adult and compression wood of spruce and silver fir. Phyton 39:129–132

Guitard D, Masse H, Yamamoto H, Okuyama T (1999) Growth stress generation: a new mechanical model of the dimensional change of wood cells during maturation. J Wood Sci 45:384–391

Hänninen T, Tukiainen P, Svedström K, Serimaa R, Saranpää P, Kontturi E, Hughes M, Vuorinen T (2012) Ultrastructural evaluation of compression wood-like properties of common juniper (*Juniperus communis* L.). Holzforschung 66:389–395

Harris RA (1976) Characterization of compression wood severity in *Pinus echinata* Mill. IAWA Bull 1976:47–50

Harris JM (1977) Shrinkage and density of radiata pine compression wood in relation to its anatomy and mode of formation. N Z J For Sci 7:91–106

Harris JM, Meylan BA (1965) The influence of microfibril angle on longitudinal and tangential shrinkage in *Pinus radiata*. Holzforschung 19:144–153

Hellgren JM, Olofsson K, Sundberg B (2004) Patterns of auxin distribution during gravitational induction of reaction wood in poplar and pine. Plant Physiol 135:212–220

Hiller CH (1964a) Correlation of fibril angle with wall thickness of tracheids in summerwood of slash and loblolly pine. Tappi J 47:125–128

Hiller CH (1964b) Estimating size of the fibril angle in latewood tracheids of slash pine. J For 62:249–252

Hoffmann GC, Timell TE (1970) Isolation of a β-1,3-glucan (laricinan) from compression wood of *Larix laricina*. Wood Sci Technol 4:159–162

Hoffmann GC, Timell TE (1972) Polysaccharides in compression wood of Tamarack (*Larix laricina*). Isolation and characterization of laricinan, an acidic glucan. Sven Papperstidn 75:135–142

Hoson T, Saito Y, Soga K, Wakabayashi K (2005) Signal perception, transduction, and response in gravity resistance. Another graviresponse in plants. Adv Space Res 36:1196–1202

Jiang X, Yin Y, Lu J, Hiroyuki U (2002) The influence of S2 microfibril angle on longitudinal and tangential shrinkage in China fir (*Cunninghamia lanceolata*) plantation. Chin For Sci Technol 1:1–7

Jones L, Seymour GB, Knox JP (1997) Localization of pectic galactan in tomato cell walls using a monoclonal antibody specific to (1 → 4)-β-D galactan. Plant Physiol 113:1405–1412

Kaneda M, Rensing KH, Wong JCT, Banno B, Mansfield SD, Samuels AL (2008) Tracking monolignols during wood development in lodgepole pine. Plant Physiol 147:1750–1760

Kibblewhite P, Evans R, Grace J, Riddell M, Suckling I (2007) Growth-layer-level microfibril angle, lignin and carbohydrate variation with radial direction, earlywood and latewood, and compression wood content, for bent and straight ramets of a radiata pine clone. Appita 2007:189–197

Kibblewhite RP, Suckling ID, Evans R, Grace JC, Riddell MJC (2010) Lignin and carbohydrate variation with earlywood, latewood, and compression wood content of bent and straight ramets of a radiata pine clone. Holzforschung 64:101–109

Kim JS, Awano T, Yoshinaga A, Takabe K (2010) Immunolocalization of β-1–4-galactan and its relationship with lignin distribution in developing compression wood of *Cryptomeria japonica*. Planta 232:109–119

Kim JS, Awano T, Yoshinaga A, Takabe K (2011) Occurrence of xylan and mannan polysaccharides and their spatial relationship with other cell wall components in differentiating compression wood tracheids of *Cryptomeria japonica*. Planta 233:721–735

Koutaniemi S, Warinowski T, Kärkönen A, Alatalo E, Fossdal CG, Saranpää P, Laakso T, Fagerstedt KV, Simola LK, Paulin L, Rudd S, Teeri TH (2007) Expression profiling of the

lignin biosynthetic pathway in Norway spruce using EST sequencing and real-time RT-PCR. Plant Mol Biol 65:311–328
Kretschmann DE, Alden HA, Verrill S (1998) Variations of microfibril angle in loblolly pine: comparison of iodine crystallization and X-ray diffraction techniques. In: Butterfield BG (ed) Microfibril angle in wood. University of Canterbury, Christchurch, New Zealand, pp 157–176
Kukkola E, Saranpää P, Fagerstedt K (2008) Juvenile and compression wood cell wall layers differ in lignin structure in Norway spruce and scots pine. IAWA J 29:47–54
Kwon M, Bedgar DL, Piastuch W, Davin LB, Lewis NG (2001) Induced compression wood formation in Douglas fir (*Pseudotsuga menziesii*) in microgravity. Phytochemistry 57:847–857
Lachenbruch B, Droppelmann F, Balocchi C, Peredo M, Perez E (2010) Stem form and compression wood formation in young *Pinus radiata* trees. Can J For Res 40:26–36
Lange PW (1950) The distribution of lignin in the cell wall of normal and reaction wood from spruce and a few hardwoods. Sven Papperstidn 57:525–532
Lee CL (1961) Crystallinity of wood cellulose fibres studied by X-ray methods. For Prod J 11:108–112
Lee PW, Eom YG (1988) Anatomical comparison between compression wood and opposite wood in a branch of Korean pine (*Pinus koraiensis*). IAWA Bull ns 9:275–284
Leonardon M, Altaner CM, Vihermaa L, Jarvis MC (2010) Wood shrinkage: influence of anatomy, cell wall architecture, chemical composition and cambial age. Eur J Wood Prod 68:87–94
Little CHA, Eklund L (1999) Ethylene in relation to compression wood formation in *Abies balsamea* shoots. Trees Struct Funct 13:173–177
Low A (1964) A study of compression wood in Scots pine (*Pinus sylvestris* L.). Forestry 37:179–201
Marton R, Rushton R, Sacco JS, Sumiya K (1972) Dimensions and ultrastructure of growing fibers. Tappi 55:1499–1504
Mast SW, Donaldson LA, Torr K, Phillips L, Flint H, West M, Strabala TJ, Wagner A (2009) Exploring the ultrastructural localization and biosynthesis of β(1,4)-galactan in *Pinus radiata* compression wood. Plant Physiol 150:573–583
Mast S, Peng L, Jordan TW, Flint H, Phillips L, Donaldson L, Strabala TJ, Wagner AW (2010) Proteomic analysis of membrane preparations from developing *Pinus radiata* compression wood. Tree Physiol 30:1456–1468
Mayr S, Cochard H (2003) A new method for vulnerability analysis of small xylem areas reveals that compression wood of Norway spruce has lower hydraulic safety than opposite wood. Plant Cell Environ 26:1365–1371
Mayr S, Bardage S, Brändström J (2005) Hydraulic and anatomical properties of light bands in Norway spruce compression wood. Tree Physiol 26:17–23
McDougall GJ (2000) A comparison of proteins from the developing xylem of compression and non-compression wood of branches of sitka spruce (*Picea sitchensis*) reveals a differentially expressed laccase. J Exp Bot 51:1395–1401
Megraw RA (1985) Wood quality factors in loblolly pine. Tappi, Atlanta, Georgia
Megraw RA, Leaf G, Bremer D (1998) Longitudinal shrinkage and microfibril angle in loblolly pine. In: Butterfield BG (ed) Microfibril angle in wood. IAWA/IUFRO, Christchurch, New Zealand
Meylan BA (1972) The influence of microfibril angle on the longitudinal shrinkage-moisture content relationship. Wood Sci Technol 6:293–301
Moëll MK, Fujita M (2004) Fourier transform methods in image analysis of compression wood at the cellular level. IAWA J 25:311–324
Morikawa Y, Yoshinaga A, Kamitakahara H, Wada M, Takabe K (2010) Cellular distribution of coniferin in differentiating xylem of *Chamaecyparis obtusa* as revealed by Raman microscopy. Holzforschung 64:61–67
Nakamura T, Negishi Y, Funada R, Yamada M (2001) Sedimentation amyloplasts in starch sheath cells of woody stems of Japanese cherry. Adv Space Res 27:957–960

Nanayakkara B, Manley-Harris M, Suckling ID, Donaldson LA (2009) Quantitative chemical indicators to assess the gradation of compression wood. Holzforschung 63:431–439
Nečesaný V (1955) Submicroscopic morphology of the cell walls in the reaction wood of conifers. Biologia 10:647–659
Nečesaný V (1958) Effect of β-indoleacetic acid on the formation of reaction wood. Phyton 11:117–127
Newman RH, Hemmingson JA, Butterfield BG (2005) A micro-analytical approach to the carbohydrate chemistry of compression wood. International symposium on wood, fibre and pulping chemistry, Auckland, New Zealand, 16–19 May 2005, pp 593–596
Okuyama T (1993) Growth stresses in tree. Mokuzai Gakkaishi 39:747–756
Okuyama T, Takeda H, Yamamoto H, Yoshida M (1998) Relation between growth stress and lignin concentration in the cell wall: ultraviolet microscopic spectral analysis. J Wood Sci 44:83–89
Parham RA, Côté WA (1971) Distribution of lignin in normal and compression wood of *Pinus taeda* L. Wood Sci Technol 5:49–62
Park S, Saiki H, Harada H (1979) Structure of branch wood in Akamatsu (*Pinus densiflora* Sieb. et Zucc.). I. Distribution of compression wood, structure of annual ring and tracheid dimensions. Mokuzai Gakkaishi 25:311–317
Perem E (1958) The effect of compression wood on the mechanical properties of white spruce and red pine. For Prod J 8:235–240
Phelps JE, Saniewski M, Smoliński M, Pieniażek J, McGinnes EA (1974) A note on the structure of morphactin induced wood in two coniferous species. Wood Fiber 6:13–17
Pillow MY (1941) A new method for detecting compression wood. J For 39:385–387
Pillow MY, Luxford RF (1937) Structure, occurrence and properties of compression wood. US Dept Agr Tech Bull 546:1–32
Pillow MY, Schafer ER, Pew JC (1936) Occurrence of compression wood in black spruce and its effect on groundwood pulp. USDA Forest Service Report 1–6
Plomion C, Pionneau C, Brach J, Costa P, Baillères H (2000) Compression wood-responsive proteins in developing xylem of maritime pine (*Pinus pinaster* Ait.). Plant Physiol 123:959–969
Plomion C, Leprovost G, Stokes A (2001) Wood formation in trees. Plant Physiol 127:1513–1523
Reiterer A, Lichtenegger H, Tschegg S, Fratzl P (1999) Experimental evidence for a mechanical function of the cellulose microfibril angle in wood cell walls. Philos Mag A 79:2173–2184
Sahlberg U, Salmén L, Oscarsson A (1997) The fibrillar orientation in the S2-layer of wood fibres as determined by X-ray diffraction analysis. Wood Sci Technol 31:77–86
Saito K, Fukushima K (2005) Distribution of lignin interunit bonds in the differentiating xylem of compression and normal woods of *Pinus thunbergii*. J Wood Sci 51:246–251
Samuels AL, Rensing KH, Douglas CJ, Mansfield SD, Dharmawardhana DP, Ellis BE (2002) Cellular machinery of wood production: differentiation of secondary xylem in *Pinus contorta* var. *latifolia*. Planta 216:72–82
Schmitt U, Singh A, Frankenstein C, Möller R (2006) Cell wall modifications in woody stems induced by mechanical stress. N Z J For Sci 36:72–86
Scurfield G (1973) Reaction wood: its structure and function. Science 179:647–655
Seth MK, Jain KK (1978) Percentage of compression wood and specific gravity in blue pine (*Pinus wallichiana* A.B.. Jackson). Wood Sci Technol 12:17–24
Shelbourne CJA, Ritchie KS (1968) Relationships between degree of compression wood development and specific gravity and tracheid characteristics in loblolly pine. Holzforschung 22:185–190
Shelbourne CJA, Zobel BJ, Stonecypher RW (1969) The inheritance of compression wood and its genetic and phenotypic correlations with six other traits in 5-year-old loblolly pine. Silvae Genet 18:1–56
Singh AP (1997) The ultrastructure of the attack of *Pinus radiata* mild compression wood by erosion and tunnelling bacteria. Can J Bot 75:1095–1102

Singh AP, Donaldson LA (1999) Ultrastructure of tracheid cell walls in radiata pine (*Pinus radiata* D. Don) mild compression wood. Can J Bot 77:32–40

Singh AP, Sell J, Schmitt U, Zimmermann T, Dawson B (1998) Radial striation of the S2 layer in mild compression wood tracheids of *Pinus radiata*. Holzforschung 52:563–566

Singh AP, Kim YS, Park BD, Chung GC, Wong AHH (2003) Presence of a distinct S3 layer in mild compression wood tracheids of *Pinus radiata*. Holzforschung 57:243–247

Siripatanadilok S, Leney L (1985) Compression wood in Western Hemlock *Tsuga heterophylla* (RAF.) Sarg. Wood Fiber Sci 17:254–265

Spicer R, Gartner BL (1998) Hydraulic properties of Douglas-fir (*Pseudotsuga menziesii*) branches and branch halves with reference to compression wood. Tree Physiol 18:777–784

Starbuck CJ, Phelps JE (1986) Induction of compression wood in rooted cuttings of *Pseudostuga menziesii* (Mirb.) Franco by indole-3-acetic acid. IAWA Bull 7:13–16

Starbuck CJ, Roberts AN (1983) Compression wood in rooted cuttings of Douglas fir. Physiol Plantarum 57:371–374

Sugiyama K, Okuyama T, Yamamoto H, Yoshida M (1993) Generation process of growth stresses in cell walls: relation between longitudinal released strain and chemical composition. Wood Sci Technol 27:257–262

Sundberg B, Little CHA (1990) Tracheid production in response to changes in the internal level of indole-3-acetic acid in 1-year-old shoots of Scots pine. Plant Physiol 94:1721–1727

Sundberg B, Tuominen H, Little CHA (1994) Effects of the indole-3-acetic acid (IAA) transport inhibitors N-1 -Naphthylphthalamic acid and morphactin on endogenous IAA dynamics in relation to compression wood formation in 1-year-old *Pinus sylvestris* shoots. Plant Physiol 106:469–476

Takabe K, Fujita M, Harada H, Saiki H (1986) Lignification process in *Cryptomeria* (*Cryptomeria japonica* D. Don) tracheid: electron microscopic observation of lignin skeleton of differentiating xylem. Res Bull Coll Exp For 43:783–788

Tarmian A, Remond R, Faezipour M, Karimi A, Perré P (2008) Reaction wood drying kinetics: tension wood in *Fagus sylvatica* and compression wood in *Picea abies*. Wood Sci Technol 43:113–130

Tarmian A, Azadfallah M, Gholamiyan H, Shahverdi M (2011) Inter-tracheid and cross-field pitting in compression wood and opposite wood of Norway spruce (*Picea abies* L.). Notulae Scientia Biologicae 3:145–151

Telewski FW, Wakefield AH, Jaffe MJ (1983) Computer-assisted image analysis of tissues of ethrel-treated *Pinus taeda* seedlings. Plant Physiol 72:177–181

Timell TE (1978a) Helical thickenings and helical cavities in normal and compression wood of *Taxus baccata*. Wood Sci Technol 12:1–15

Timell TE (1978b) Ultrastructure of compression wood in *Ginkgo biloba*. Wood Sci Technol 12:89–103

Timell TE (1980) Organization and ultrastructure of the dormant cambial zone in compression wood of *Picea abies*. Wood Sci Technol 14:161–179

Timell TE (1982) Recent progress in the chemistry and topochemistry of compression wood. Wood Sci Technol 16:83–122

Timell TE (1983) Origin and evolution of compression wood. Holzforschung 37:1–10

Timell TE (1986) Compression wood in gymnosperms. Springer, Berlin

Tokareva EN, Pranovich AV, Fardim P, Daniel G, Holmbom B (2007) Analysis of wood tissues by time-of-flight secondary ion mass spectrometry. Holzforschung 61:647–655

Tomlinson PB (2001) Reaction tissues in *Gnetum gnemon* – a preliminary report. IAWA J 22:401–413

Tsai C-C, Chen S-J, Chien C-T, Kuo-Huang L-L (2010) Induction of compression wood in seedlings of Taiwan incense cedar (*Calocedrus macrolepis* var. *formosana*) during the mid-season growth pause. Bot Stud 51:163–170

Wardrop AB, Dadswell HE (1950) The nature of reaction wood II. The cell wall organisation of compression wood tracheids. Aust J Biol Sci 3:1–13

Wardrop AB, Davies GW (1964) The nature of reaction wood. VIII. The structure and differentiation of compression wood. Aust J Bot 12:24–38

Warensjö M (2003) Compression wood in Scots pine and Norway spruce – distribution in relation to external geometry and the impact on the dimensional stability in sawn wood. Dissertation, Swedish University of Agricultural Sciences

Waterkeyn L, Caeymaex S, Decamps E (1982) La callose des trachéides du bois de compression chez *Pinus silvestris* et *Larix decidua*. Bulletin de la Société Royale de Botanique de Belgique 115:149–155

Wershing HF, Bailey IW (1942) Seedlings as experimental material in the study of "redwood" in conifers. J For 40:411–414

Westermark U (1985) The occurrence of *p*-hydroxyphenylpropane units in the middle-lamella lignin of spruce (*Picea abies*). Wood Sci Technol 19:223–232

Westing AH (1965) Formation and function of compression wood in gymnosperms. Bot Rev 31:381–480

Whetten R, Sun Y-H, Zhang Y, Sederoff R (2001) Functional genomics and cell wall biosynthesis in loblolly pine. Plant Mol Biol 47:275–291

Whiting P, Goring DAI (1982) Chemical characterization of tissue fractions from the middle lamella and secondary wall of black spruce tracheids. Wood Sci Technol 16:261–267

Wi SG, Lim KP, Kim YS (2000) Inhomogenous distribution of lignin in compression wood of *Pinus densiflora* and *Ginkgo biloba*. In: Kim YS (ed) New horizons in wood anatomy. Chonnam National University Press, Kwangju, Korea

Wilson BF, Chien CT, Zaerr JB (1989) Distribution of endogenous indole-3-acetic acid and compression wood formation in reoriented branches of Douglas-fir. Plant Physiol 91:338–344

Włoch W (1975) Longitudinal shrinkage of compression wood in dependence on water content and cell wall structure. Acta Soc Bot Pol 44:217–229

Włoch W, Hejnowicz Z (1983) Location of laricinan in compression wood tracheids. Acta Soc Bot Pol 52:201–203

Wood JR, Goring DAI (1971) The distribution of lignin in stem and branch wood of Douglas fir. Pulp Pap-Canada 72:T95–T102

Yamaguchi K, Itoh T, Shimaji K (1980) Formation of compression wood induced by 1-N-naphthylphthalamic acid (NPA), an IAA transport inhibitor. Wood Sci Technol 14:181–185

Yamaguchi K, Shimaji K, Itoh T (1983) Simultaneous inhibition and induction of compression wood formation by morphactin in artificially inclined stems of Japanese larch (*Larix leptolepis* Gordon). Wood Sci Technol 17:81–89

Yamamoto H (1998) Generation mechanism of growth in wood cell walls: roles of lignin and cellulose microfibril during cell wall maturation. Wood Sci Technol 32:171–182

Yamamoto F, Kozlowski TT (1987a) Effects of flooding, tilting of stems, and ethrel application on growth, stem anatomy and ethylene production of *Pinus densiflora* seedlings. J Exp Bot 38:293–310

Yamamoto F, Kozlowski TT (1987b) Effect of ethrel on growth and stem anatomy of *Pinus halepensis* seedlings. IAWA Bull ns 8:11–19

Yamamoto H, Okuyama T, Yoshida M (1995) Generation process of growth stresses in cell walls. VI. Analysis of growth stress generation by using a cell model having three layers (S1, S2 and I + P). Mokuzai Gakkaishi 41:1–8

Yamashita S, Yoshida M, Takayama S, Okuyama T (2007) Stem-righting mechanism in gymnosperm trees deduced from limitations in compression wood development. Ann Bot London 99:487–493

Yamashita S, Yoshida M, Yamamoto H, Okuyama T (2008) Screening genes that change expression during compression wood formation in *Chamaecyparis obtusa*. Tree Physiol 28:1331–1340

Yasuda S, Sakakibara A (1975) The chemical composition of lignin from compression wood. Mokuzai Gakkaishi 21:363–369

Yeh T-F, Goldfarb B, Chang H-M, Peszlen I, Braun JL, Kadla JF (2005) Comparison of morphological and chemical properties between juvenile wood and compression wood of loblolly pine. Holzforschung 59:669–674

Yeh T-F, Braun JL, Goldfarb B, Chang H-M, Kadla JF (2006a) Morphological and chemical variations between juvenile wood, mature wood, and compression wood of loblolly pine (*Pinus taeda* L.). Holzforschung 60:1–8

Yeh T-F, Morris CR, Goldfarb B, Chang H-M, Kadla JF (2006b) Utilization of polar metabolite profiling in the comparison of juvenile wood and compression wood in loblolly pine (*Pinus taeda*). Tree Physiol 26:1497–1503

Yoshizawa N, Idei T (1987) Some structural and evolutionary aspects of compression wood tracheids. Wood Fiber Sci 19:343–352

Yoshizawa N, Koike S, Idei T (1985) Formation and structure of compression wood tracheids induced by repeated inclination in *Taxus cuspidata*. Mokuzai Gakkaishi 31:325–333

Yoshizawa N, Okamoto Y, Idei T (1986a) Righting movement and xylem development in tilted young conifer trees. Wood Fiber Sci 18:579–589

Yoshizawa N, Tanaka Y, ldei T (1986b) Development of vascular cambium and compression wood formation in the shoot of young spruce (*Picea jezoensis* var. *hondoensis*). IAWA Bull ns 7:21–30

Yoshizawa N, Kiyomiya M, Idei T (1987) Variations in tracheid length and morphological changes in tracheid tips associated with the development of compression wood. Wood Sci Technol 21:1–10

Yoshizawa N, Satoh I, Yokota S, Idei T (1992) Response of differentiating tracheids to stem inclination in young trees of *Taxus cuspidata*. IAWA Bull ns 13:187–194

Yoshizawa N, Satoh M, Yokota S, Idei T (1993a) Formation and structure of reaction wood in *Buxus microphylla* var. insularis Nakai. Wood Sci Technol 27:1–10

Yoshizawa N, Watanabe N, Yokota S, Idei T (1993b) Distribution of guaiacyl and syringyl lignins in normal and compression wood of *Buxus microphylla* var. *insularis* Nakai. IAWA J 14:139–151

Yumoto M, Ishida S, Fukazawa K (1982) Studies on the formation and structure of the compression wood cells induced by artificial inclination in young trees of *Picea glauca*. I. Time course of the compression wood formation following inclination. Res Bull Coll Exp For Hokkaido Univ 39:137–162

Yumoto M, Ishida S, Fukazawa K (1983) Studies on the formation and structure of the compression wood cells induced by artificial inclination in young trees of *Picea glauca* IV. Gradation of the severity of compression wood tracheids. Res Bull Coll Exp For Hokkaido Univ 39:409–454

Zhang Y, Sederoff RR, Allona I (2000) Differential expression of genes encoding cell wall proteins in vascular tissues from vertical and bent loblolly pine trees. Tree Physiol 20:457–466

# Index



J. Fromm (ed.), *Cellular Aspects of Wood Formation*, Plant Cell Monographs 20,
DOI 10.1007/978-3-642-36491-4, © Springer-Verlag Berlin Heidelberg 2013